Frontiers in Statistical Quality Control 6

Hans-Joachim Lenz
Peter-Theodor Wilrich
Editors

Frontiers in Statistical Quality Control 6

With 88 Figures
and 49 Tables

Springer-Verlag Berlin Heidelberg GmbH

Prof. Dr. Hans-Joachim Lenz
Prof. Dr. Peter-Theodor Wilrich
Freie Universität Berlin
Institut für Statistik und Ökonometrie
Garystraße 21
14195 Berlin
Germany

ISBN 978-3-7908-1374-6

Library of Congress Cataloging-in-Publication Data
CIP-Kurztitelaufnahme der Deutschen Bibliothek
Frontiers in Statistical Quality Control 6: with 49 tables / Hans-Joachim Lenz; Peter-Theodor Wilrich ed. – Heidelberg; New York: Physica-Verl., 2001

ISBN 978-3-7908-1374-6 ISBN 978-3-642-57590-7 (eBook)
DOI 10.1007/978-3-642-57590-7

Cover design: Erich Kirchner, Heidelberg

SPIN 10793053 88/2202-5 4 3 2 1 0 – Printed on acid-free paper

Editorial

The VIth International Workshop on Intelligent Statistical Quality Control took place in Würzburg, Germany, and was hosted by Professor E. von Collani, Universität Würzburg.
It was attended by 36 research workers from 13 countries. On behalf of all participants, the editors would like to thank Professor von Collani and his staff for their superb hospitality, the perfect organisation, and the stimulating atmosphere.

The impact of any conference or workshop is mainly shaped by the quality of papers that are read at the meeting. We would like to express our deep gratitude to the following members of the scientific programme committee, who did an excellent job with respect to the recruiting of invited speakers and the refereeing of all the submitted papers:

Prof. Chooichiro Asano, Japan
David Baillie, United Kingdom
Prof. Elart von Collani, Germany
Prof. Hans-J. Lenz, Germany
Prof. Hiroshi Ohta, Japan
Prof. Poul Thyregod, Denmark
Prof. Peter-Th. Wilrich, Germany
Prof. William H. Woodall, United States of America.

The papers collected in this volume fall into four categories: Sampling Inspection, Statistical Process Control, Data Analysis and Process Capability Studies, and Experimental Design.

In the first group there are four papers on **Sampling Inspection**.
Göb presents a rigorous methodology for statistical batch inspection with special emphasis on inspection by variables. ***Baillie and Klaassen*** study a procedure for switching between accept-zero attributes or variables sampling plans that guarantees a specified limit on the average outgoing quality. ***Grzegorzewski*** proposes the design of single sampling plans by attributes with fuzzy boundaries on the producer's and consumer's risk. ***Melgaard and Thyregod*** consider the effect of measurement uncertainty on acceptance sampling by variables.

The second group of papers deals with **Statistical Process Control**, and consists of 11 contributions.
Knoth, Schmid and Schöne extend the well-known Shewhart charts to the first two moments of auto-correlated change-point models. ***Hryniewicz*** presents ISO 3951 variables sampling plans as alternatives to Shewhart control charts. ***Ladany*** studies in detail the reward-optimal set-up of a target-value that depends upon the specification limits and the asymmetry of rewards. ***Nieckula*** analyses Shewhart X-bar charts under a finite set of out-of-control processes like shift, trend, cycles, mixtures etc. ***Sparks*** studies monitoring processes defined on ordinal scales and applies multiple correspondence analysis to visualise results. ***Reynolds and Stoumbos*** evaluate and compare alternatives to Shewhart's p-chart, such as Bernoulli and binomial CUSUM charts. ***Ryan and Faddy*** demonstrate the non-robustness of in-control ARLs if CUSUM procedures are used when the underlying distribution is skewed or heavy-tailed. A way to resolve non-normality problems in process control is suggested by ***Shore***, who uses an inverse normalising transformation with three

functional parameters in designing control charts based on the ARL criterion. ***Stoumbos, Jones, Woodall and Reynolds*** investigate the effect of the size of a reference sample and the size of subgroups on the statistical properties of Liu's (multivariate) control charts, which are based on the data depth measure introduced by Tukey (1975). The paper by ***Mastrangelo, Porter and Baxley*** deals with process monitoring in a highly multivariate environment, presenting a solution based upon a heuristic reduction of the observation space to a set of driving variables to which MEWMA is applied. ***Saniga and Pohlen*** report the main results on quality of service inferred from their empirical investigation on 48 quality or productivity improvement projects.

Part 3 on **Data Analysis and Process Capability Studies** consists of four contributions. ***Nishina*** applies path models, which are a sub-class of graphical models, to causal detection in process analysis. ***Steiner and MacKay*** develop Shewhart control charts when lifetime censoring occurs at fixed levels. ***Vännman*** proposes graphical methods for quality improvement based on a generalised (δ,γ) capability index plot, and a confidence region for (δ,γ), which is implanted into the (δ,γ)-plot. The paper by ***Wallgren*** generalises the C_{pk} - capability index to AR(1)-processes, and uses Monte-Carlo techniques to study the relationship between the effective and nominal confidence levels.

Part 4 deals with **Experimental Design** and consists of three papers. ***Kowalski and Vining*** review the state-of-the-art of the use of split-plot designs in industry, and outline extensions to robust designs and problems of second-order experiments. ***Ojima, Kouda and Okuhara*** propose an overall testing method for the identification of the active factors in saturated fractional factorial experiments. ***Tsui*** votes for a response model approach in the spirit of Miller and Wu (1996), directly modelling the response as a function of control, noise, and signal factors; this is compared with a simple control-factor-based loss model of Taguchi (1986) and with a Miller and Wu type of model.

We would like to close with our cordial thanks to Mrs. J. Röhmel and A. Wnuk, who fought successfully at two fronts: WINWORD and LATEX paper wrapping and integration. Moreover, the smooth collaboration with Physica - Verlag is kindly acknowledged.

The editors are happy and proud to announce that the Workshop on Intelligent Statistical Quality Control will be continued in Ontario, Canada, in the near future.

Berlin, June 2000

Hans-J. Lenz
Peter-Th. Wilrich

Contents

Author Index

Asano, Ch., Prof. Dr., 3-22-12-204 Amanuma, Suginami, Tokyo 167-0032, Japan
e-mail: QYG04340@niftyserve.or.jp

Baillie, David H., 45 Lye Green Road, Chesham, Bucks HP5 3LS, United Kingdom
e-mail: david.baillie@which.net

Baxley, Jr., Robert V.,Solutia, Inc., Pensacola, FL 32560, U.S.A.

Collani, Elart von, Prof. Dr., Universität Würzburg, Volkswirtschaftliches Institut, Sanderring 2, D-97070 Würzburg, Germany
e-mail: collani@mathematik.uni-wuerzburg.de

Faddy, Belinda J., University of Newcastle, Newcastle, NSW, Australia

Göb, Rainer, Prof. Dr., Universität Würzburg, Institut für Angewandte Mathematik u. Statistik, Sanderring 2, D-97070 Würzburg, Germany
e-mail: goeb@mathematik.uni-wuerzburg.de

Grzegorzewski, Przemyslaw, Polish Academy of Sciences, Systems Research Institute, ul. Newelska 6, PL-01447 Warszawa, Poland
e-mail: pgrzeg@ibspan.waw.pl

Hryniewicz, Olgierd, Prof. Dr., Polish Academy of Sciences, Systems Research Institute, ul. Newelska 6, PL-01447 Warszawa, Poland
e-mail:hryniewi@ibspan.waw.pl

Jones, L. Allison, University of Miami, Coral Gables, FL 33124-8237, U.S.A.

Klaassen, Chris A.J., Prof. Dr., Institute for Business and Industrial Statistics, University of Amsterdam, The Netherlands

Knoth, Sven, Departement of Statistics, Europe University Viadrina, D-15207 Frankfurt(Oder), Germany

Kouda, Akira, Faculty of Economics, Higashi Nippon International University, 970-8023, Japan

Kowalski, Scott, University of Florida, Department of Statistics, 103 Griffin-Floyd Hall, P.O. Box 118454, Gainesville, FL 32611-8454, U.S.A.

Koyama, Takeshi, Prof. Dr., Tokushima Bunri University, Faculty of Engineering, Shido-cho, Kagawa, Japan 769-2101
e-mail: koyama@es.bunri-u.ac.jp

Ladany, Shaul P., Prof. Dr., Ben-Gurion University of the Negev, Department of Industrial Engineering and Management, Beer Sheva 84105, P.O. Box 653, Israel
e-mail: ladany@bgumail.bgu.ac.il

Lenz, Hans-J., Prof. Dr., Freie Universität Berlin, Institut f. Statistik u. Ökonometrie, Garystraße 21, D-14195 Berlin, Germany
e-mail: hjlenz@wiwiss.fu-berlin.de

MacKay, R. Jock, University of Waterloo, Department of Statistics and Actuarial Science, Waterloo, Ontario, N2L 3G1, Canada

Mastrangelo, Christina, Prof. Dr., University of Virginia, Department of Systems Engineering, Charlottesville, VA 22903, U.S.A.
e-mail: cm4m@watt.seas.virginia.edu

Melgaard, Henrik, Dr., Novo Nordisk A/S, Niels Steensens Vej 1, DK-2820 Gentofte, Denmark
e-mail: hmel@novo.dk

Nieckula, Jacek, Polish Academy of Sciences, Systems Research Institute, ul. Newelska 6, PL-01447 Warszawa, Poland
e-mail: nieckula@ibspan.waw.pl

Nishina, Ken, Prof. Dr., Nagoya Institute of Technology, Department of Systems Engineering, Gokiso-cho, Showa-ku, Nagoya 466-8555, Japan
e-mail: nishina@system.nitech.ac.jp

Ohta, Hiroshi, Prof. Dr., Osaka Prefecture University, Department of Industrial Engineering, College of Engineering, Sakai, Osaka 599-8531, Japan
e-mail: ohta@ie.osakafu-u.ac.jp

Ojima, Yoshikazu, Prof. Dr., Science University of Tokyo, Department of Industrial Administration, 2641, Yamazaki, Noda, Chiba, 278-8510, Japan
e-mail: ojima@ojim01.ia.noda.sut.ac.jp

Okuhara, Masao, Suwa Junior College, Science University of Tokyo, 391-0292, Japan

Pohlen, Michael F., University of Delaware, Department of Business Administration, Newark, DE 19716, U.S.A.

Porter, Joseph M., Proctor & Gamble Company, Cincinnati, OH 45202, U.S.A.

Reynolds, Marion, Jr., Prof. Dr., Virginia Polytechnic Institute and State University, Department of Statistics, Blacksburg, VA 24061-0439, U.S.A.
e-mail: mrr@vt.edu

Ryan, Thomas P., Prof. Dr., Case Western Reserve University, Department of Statistics, Cleveland, OH 44106-7054, U.S.A.
e-mail: tpr@po.cwru.edu

Saniga, Erwin M., Prof. Dr., University of Delaware, Department of Business Administration, Newark, DE 19716, U.S.A.
e-mail: sanigae@be.udel.edu

Schmid, Wolfgang, Prof. Dr., Department of Statistics, Europe University Viadrina, Postfach 776, D-15207 Frankfurt(Oder), Germany
e-mail: schmid@euv-frankfurt-o.de

Schöne, Alexander, Department of Statistics, Europe University Viadrina, Postfach 776, D-15207 Frankfurt(Oder), Germany

Shore, Haim, Prof. Dr., Ben Gurion University of the Negev, Department of Industrial Engineering and Management, Faculty of Engineering Science, P.O. Box 653, Beer-Shewa 84105, Israel
e-mail: shor@bgumail.bgu.ac.il

Sparks, D. Ross, CSIRO, Locked Bag 17, North Ryde, NSW 2113, Australia
e-mail: ross.sparks@cmis.csiro.au

Steiner, Stefan, Prof. Dr., University of Waterloo, Department of Statistics and Actuarial Science, Waterloo, Ontario, N2L 3G1, Canada
e-mail: shsteine@uwaterloo.ca

Stoumbos, Zachary, Prof. Dr., Department of Management Science and Information Systems, Faculty of Management, Rutgers University, 180 University Avenue, Newark, NJ 07102-1895, U.S.A.
e-mail: stoumbos@andromeda.rutgers.edu

Thyregod, Poul, Prof. Dr., The Technical University of Denmark, IMM, Building 321, DK-2800 Lyngby, Denmark
e-mail: pt@imm.dtu.dk

Tsui, Kwok-Leung, Prof. Dr., Georgia Institute of Technology, School of Industrial and Systems Engineering, Atlanta, GA 30332-0205, U.S.A.
e-mail:ktsui@isye.gatech.edu

Vännman, Kerstin, Prof. Dr., Lulea University of Technology, Division of Quality Technology & Statistics, S-97187 Lulea, Sweden
e-mail: kerstin.vannman@ies.luth.se

Vining, Geoffrey, Prof. Dr., Virginia Polytechnic Institute and State University, Department of Statistics, Blacksburg, VA 24061-0439, U.S.A.
e-mail: vining@stat.ufl.edu

Wallgren, Erik, ESA, University of Orebro, S-70182 Orebro, Sweden
e-mail: erik.wallgren@hoe.se

Wilrich, Peter-Th., Prof. Dr., Freie Universität Berlin, Institut f. Statistik u. Ökonometrie, Garystraße 21, D-14195 Berlin, Germany
e-mail: wilrich@wiwiss.fu-berlin.de

Woodall, William H., Prof. Dr., University of Alabama, Department of Management Science and Statistics, P.O. Box 870226, Tuscaloosa, AL 35487-0226, U.S.A.
e-mail: wwoodall@cba.ua.edu

Part 1

Sampling Inspection

Methodological Foundations of Statistical Lot Inspection

Rainer Göb

Institute for Applied Mathematics and Statistics, University of Würzburg, Germany

Abstract. Statistical lot inspection is an application of the theory of statistical hypothesis testing in the industrial problem of deciding on hypotheses on the quality of finite lots by means of random samples from the lot. Both the systematic industrial implementation and the theoretical investigation of statistical lot inspection trace back to the work of H. F. Dodge at Bell Telephone Laboratories in the 1920s. Investigations on statistical lot inspection have received enormous interest in academic literature. However, rigorous methodological foundations are still missing. In particular, there is no generally accepted model for statistical lot inspection under continuous quality characteristics.

This paper presents an elementary model for statistical lot inspection, with special emphasis on the topic of continuous quality characteristics. Under this model, the paper develops most powerful tests for the most important lot quality indicators under continuous quality characteristics: lot mean, average square deviation from target, lot minimum.

1 Introduction: Statistical Lot Inspection

Statistical lot inspection is an application of the theory of statistical hypothesis testing in the industrial problem of deciding on hypotheses on the quality of finite lots by means of random samples from the lot. Both the systematic industrial implementation and the theoretical investigation of statistical lot inspection trace back to the work of H. F. Dodge at Bell Telephone Laboratories in the 1920s, see *Dodge* (1969). In the following decades statistical lot inspection received enormous interest by industry and by adademic literature, see the bibliographies *Wetherill & Chiu* (1975) and *von Collani* (1990a). Over a long period, elements of statistical lot inspection constituted the major industrial application of statistical quality control and the preferred subject of scientific investigations in statistical quality control.

Interest in statistical lot inspection has been decreasing both in application and in research since the 1970s. Considering the reasons for this change of paradigms we have to distinguish between *systematic* reasons resulting from trends and necessities of industrial progress and *accidental* reasons resulting from the appearance of statistical lot inspection in application and research.

From a systematic point of view the following observation is important. During the first half of the 20th century the big industrial companies were primarily interested in mass manufacturing under relatively low quality standards; the cost of raw material, energy, and labour used to be relatively small. Correspondingly, industrial quality control followed the ***inspection principle***, i.e., the principle of inspecting finished product with the objective of detecting and sorting out intolerable items or lots. From the 1960s on industry had to face new challenges: global competition, increasing customer requirements, increasing complexity of products, increasing cost of raw material, energy, and labour, stricter legal obligations. Under these constraints the inspection principle of early modern mass production had to be replaced by a principle of ***defects prevention*** for all stages of a product's life cycle: product planning, product design, production design, manufacture, storage, shipping, service and maintenance. The new trends in industrial quality control were encouraged and reflected by publications on business management. Among these, Feigenbaum's *Total Quality Control* and Demings's *Out of the Crisis* can be considered as milestones. For statistical quality control as a scientific discipline the new trends had two major consequences. The extended scope of quality control opened new fields for application of statistics. On the other hand, along with the shift from inspection to prevention, the classical lot inspection schemes lost a good deal of their reputation. Rigorous opinions go as far as to recommend total discarding of statistical lot inspection schemes, see *Deming* (1986), pp. 430-432 for instance. Correspondingly the interest of scientific investigation shifted from lot inspection schemes to topics like statistical design of experiments, statistical process control, statistical theory of reliability and maintenance.

Let us consider the way in which the properties of statistical sampling schemes recommended by standards or research contributed to a growing distrust against statistical lot inspection. The deficiencies of the standards, above all the most popular MIL-STD series have been exposed and criticized extensively by several authors, see *Hald* (1981) for instance. However, many of the alternatives offered by research are in some sense too complicated to be implemented on shopfloor, e.g. procedures based on sequential sampling or Bayes schemes. These problems have been discussed by numerous contributions.

Another aspect has remained nearly undetected and undiscussed: The poverty of statistical lot inspection schemes with respect to item quality characteristics and with respect to the ways of measuring lot quality by suitable *lot quality indicators*. From the beginning, the simple discrete conforming/nonconforming scheme with *lot proportion nonconforming* (proportion of nonconforming items in the lot) as the corresponding lot quality indicator has been dominating in research, industrial standards and industrial application. The methods of statistical lot inspection for the control of lot proportion nonconforming are known as *sampling by attributes* and *sampling by variables for lot proportion nonconforming*. These are, by far, the preferred subjects of academic literature on statistical lot inspection. Among the monographs specialized on lot inspection, *Hald* (1981) is exclusively concerned with sampling by attributes. *Schilling* (1982) dedicates some space to topics like reliability sampling, but the major part of the book is on sampling by attributes and sampling by variables for lot proportion nonconforming. The same situation holds for general monographs on statistical quality control. Among these, an exceptional case is *Burr* (1976) where statistical lot inspection for the control of the lot mean under continuous item quality characteristics is considered in some detail.

Certainly the elementary conforming/nonconforming scheme is the most important way of measuring item quality. By a final consumer product quality is always assessed as conforming or nonconforming. However, the nearly exclusive role of this scheme in research and standards does not reflect the reality and necessities of industrial practice. Most of the so-called ***secondary*** technical quality characteristics, see *Ishikawa* (1985), which are considered by engineers, are continuous quality characteristics like weight, volume, acceleration, tensile strength, measures of purity, lifetime etc. The diversity and complexity of modern industrial products and modern industrial environments often requires to consider such continuous item quality characteristics and corresponding lot quality indicators like lot mean, lot minimum, average square deviation per item from a target in the lot. However, there is now generally accepted stochastic model for statistical lot inspection under continuous item quality characteristics and the corresponding lot quality indicators. An early approach by *Thyregod* (1973) has widely been ignored. The approaches of statistical quality control manuals only adapt process models to lot inspection by means of the ***asymptoticity argument*** which asserts that a "large" lot can be treated like a process. This argument only appeals to intuition and has never been presented in a mathematically sound version.

The present paper gives a survey of the contributions of *Göb* (1995), *Göb* (1996a), *Göb* (1996b), *Göb* (1996c), *Göb* (1997a), *Göb* (1997b), *Göb* (1997c), to the problem of statistical lot inspection under continuous item quality characteristics. Paragraph 2 introduces a stochastic model of statistical lot inspection, with special emphasis on continuous item quality characteristics. To describe statistical lot inspection in the framework of parametric statistics, this model considers the values of lot quality indicators as parameters of conditional distributions which can be determined from assumptions on the process behind the lots. Under this parametric model, paragraphs 3, 4, 5 and 6 investigate most powerful tests for the important lot quality indicators lot mean, lot proportion nonconforming, average square deviation from target per item in the lot, lot minimum. Paragraph 7 sketches the economic design (choice of sample size and critical region) of the tests developed in the preceding paragraphs. We conclude in paragraph 8 with some remarks on possible topics of future research.

2 An Elementary Model of Statistical Lot Inspection

The model exposed below was introduced by *Göb* (1996a). An early precursor of the underlying approach is *Thyregod* (1973). We outline the basic features of the model for the case of continuous item quality indicators. The reader is referred to *Göb* (1996a) for a more general and more detailed exposition.

2.1 The Item Quality Indicator

We consider discrete items of a certain type of product. The quality of each item i is expressed by a random continuous *item quality indicator* $\xi_i \in \mathbb{R}$. In simple cases the item quality indicator coincides with a one-dimensional continuous quality characteristic. In other cases the item quality indicator may also be a function of several quality characteristics, e.g. a weighted sum.

We consider a sequence 1,2,3,... of items as an output of a process of manufacture, storage, transport or other handling. Following the usual parametric model of statistical process control we assume $\xi_1, \xi_2, \xi_3, \ldots$ to be an independent family, each ξ_i with a distribution function $F_{\boldsymbol{\vartheta}_i}$ from a parametric family $(F_{\boldsymbol{\vartheta}})_{\boldsymbol{\vartheta}\in\Theta}$. Since we are interested in continuous item quality indicators we suppose a Lebesgue density $f_{\boldsymbol{\vartheta}}$ for each $F_{\boldsymbol{\vartheta}}$.

In SPC (statistical process control), a popular instance of the above model is the family of densities of normal distribution indexed by the mean μ and the variance σ^2, i.e.,

$$f_{(\mu,\sigma^2)} \quad = \quad \frac{1}{\sqrt{2\pi}\sigma} \exp\left(\frac{-(x-\mu)^2}{2\sigma^2}\right) \tag{1}$$

with $\boldsymbol{\vartheta} = (\mu, \sigma^2) \in \Theta \subset \mathbb{R} \times (0; +\infty)$.

Recently, SPC models are tending to investigate a wider range of distribution models, in particular lifetime distributions. An important representative from this area is the two-parametric exponential distribution defined by the densities

$$f_{(\lambda,d)} \quad = \quad \begin{cases} 0, & \text{if } x \leq d, \\ \\ \lambda \exp\Big(-\lambda(x-d)\Big), & \text{if } x > d, \end{cases} \tag{2}$$

with $\boldsymbol{\vartheta} = (\lambda, d) \in \Theta \subset (0; +\infty) \times \mathbb{R}$, where λ is the *scale parameter* and d is the *location* or *threshold parameter*.

From a stochastic point of view the quality level of the process considered at the output of item i is determined by the distribution parameter $\boldsymbol{\vartheta}_i$. Thus we use $\boldsymbol{\vartheta}_i$ as the *process quality indicator*. Basically, the major part of SPC methods can be considered as instances of statistical hypotheses on the actual value $\boldsymbol{\vartheta}_i$ of the process quality indicator.

2.2 The Lot Quality Indicator

A lot of size N is considered as a subsequence of items $K+1, ..., K+N$ of the process considered in paragraph 2.1, above. For simplicity's sake we use the index $K = 0$ and we consider the lot represented by items $1, ..., N$. We assume that the process behind the lots is stable enough to guarantee that the process quality indicator remains unchanged at least for the items in the lot. In our notation this amounts to assume that the quality indicators $\xi_1, .., \xi_N$ of the items $1, ..., N$ in the lot have the same density $f_{\boldsymbol{\vartheta}}$ with fixed $\boldsymbol{\vartheta} \in \Theta$.

In any economically and technically reasonable case lot quality cannot be identified directly with the vector $\boldsymbol{\xi} = (\xi_1, .., \xi_N)$ of quality indicators of the items in the lot, but only with a function $T(\boldsymbol{\xi})$ thereof, the so-called *lot quality indicator*. We restrict attention to the simple case of a one-dimensional function $T\colon \mathbb{R}^N \to \mathbb{R}$. For this case we list some important lot quality indicators which are reasonable under continuous item quality indicators.

The *lot mean* or *lot average* (arithmetic mean of the item quality indicators in the lot):

$$T(\xi_1, .., \xi_N) \quad = \quad \frac{1}{N}\sum_{i=1}^{N} \xi_i. \tag{3}$$

The *lot minimum* (minimum item quality indicator in the lot) respectively the *lot maximum* (maximum item quality indicator in the lot):

$$T(\xi_1, .., \xi_N) \; = \; \min\{\xi_1, .., \xi_N\}, \text{ respectively } T(\xi_1, .., \xi_N) \; = \; \max\{\xi_1, .., \xi_N\}. \tag{4}$$

The *lot average square deviation from target a* (average square deviation from target a per item in the lot):

$$T(\xi_1, .., \xi_N) \quad = \quad \frac{1}{N}\sum_{i=1}^{N} (\xi_i - a)^2. \tag{5}$$

The familiar lot quality indicator *lot proportion nonconforming* (proportion of nonconforming items in the lot) of acceptance sampling by attributes can also introduced in a model based on continuous item quality indicators. The continuous item quality ξ_i indicators are transformed into a conforming/nonconforming scheme by comparison of ξ_i with a *specification range* $R \subset \mathbb{R}$: Item i is counted as *conforming*, if $\xi_i \in R$, and as *nonconforming*, if $\xi_i \in \mathbb{R} \setminus R$. Then lot proportion nonconforming can be expressed as

$$T(\xi_1, .., \xi_N) \; = \; \frac{1}{N}\Big|\{i \,|\, 1 \le i \le N,\, \xi_i \in \mathbb{R} \setminus R\}\Big|. \tag{6}$$

However, we shall see in paragraph 4, below, that from the point of view of statistical methodology this lot quality indicator does not define a scheme of statistical lot inspection under continuous item quality indicators in a strict sense.

2.3 Statistical Lot Inspection

Let us assume a model with a specificied lot quality indicator $T(\boldsymbol{\xi})$. Then from the point of view of quality all lots with identical value $T(\boldsymbol{\xi}) = z$ of the lot quality indicator are equivalent. The value z of the lot quality indicator is the crucial value focused on by lot inspection. Statistical lot inspection can be considered as an application of the theory of hypothesis testing to the problem of testing hypotheses about the the value z of the lot quality indicator. To apply the parametric theory of statistical hypothesis testing we need a parametric model analogous to the ones considered by SPC. We observe that for lots with coincident value z of the lot quality indicator random variation of the item quality indicators $\xi_1, ..., \xi_N$ is restricted by $T(\boldsymbol{\xi}) = z$. The corresponding probability law is defined by the joint conditional distribution function

$$F_{\boldsymbol{\vartheta}|z}(x_1, ..., x_N) = \mathrm{P}_{\boldsymbol{\vartheta}}\Big(\xi_1 \le x_1, ..., \xi_N \le x_N \,|\, T(\boldsymbol{\xi}) = z\Big) \tag{7}$$

of the item quality indicators $\xi_1, \ldots, \xi_N$ under $T(\boldsymbol{\xi}) = z$. We assume that the lot quality indicator $T(\boldsymbol{\xi})$ is sufficient for a component $(\vartheta_{k+1}, ..., \vartheta_m)$ of $\boldsymbol{\vartheta} = (\vartheta_1, ..., \vartheta_m)$. Thus

$$F_{\boldsymbol{\vartheta}|z} = F_{\boldsymbol{\vartheta}'|z} \quad \text{with} \quad \boldsymbol{\vartheta}' = (\vartheta_1, ..., \vartheta_k)\,. \tag{8}$$

Using the conditional distribution the problem of testing hypotheses on the lot quality indicator can be considered in the framework of parametric statistics as the problem of testing the parameter z of the family $(F_{\boldsymbol{\vartheta}'|z})_{\boldsymbol{\vartheta}' \in \Theta', z \in Z}$ with $\boldsymbol{\vartheta}' = (\vartheta_1, ..., \vartheta_k)$ as unspecified nuisance parameters, where $Z \subset \mathbb{R}$ is the range of $T\colon \mathbb{R}^N \to \mathbb{R}$.

By this approach we can give a clear account of the analogy and the difference of statistical process inspection and statistical lot inspection:

- Statistical process inspection uses random samples from a production process for a statistical test for the hypothesis H: "$\boldsymbol{\vartheta}$ satisfactory" against the alternative K: "$\boldsymbol{\vartheta}$ unsatisfactory" on the process quality indicator $\boldsymbol{\vartheta}$.
- Statistical lot inspection uses random samples from the lot for a statistical test for the hypothesis H: "z satisfactory" against the alternative K: "z unsatisfactory" on the lot quality indicator z which coincides with the parameter z of the conditional distribution (8).

The statistical tests of lot inspection are often called ***sampling plans***. Acceptance of hypothesis H is identified with ***acceptance of the lot***, rejection of hypothesis H is identified with ***rejection of the lot***. These decision alternatives have various interpretations in industrial practice, e.g., reselling or processing of the items in case of acceptance, resending to the producer, screening, sorting, reworking in case of rejection.

From a methodological point of view the status of the process parameter $\boldsymbol{\vartheta}$ and the lot parameter z is essentially different. In most cases, direct observation of the process parameter $\boldsymbol{\vartheta}$ is impossible. For instance, hypotheses on the parameter $\boldsymbol{\vartheta} = (\mu, \sigma^2)$ of a normal distribution can only be examined by statistical methods. The lot parameter, however, satisfies the functional relation $T(\boldsymbol{\xi}) = z$. Thus, complete knowledge of the realizations $x_1, ..., x_N$ of the lot variables $\xi_1, ..., \xi_N$ entails complete knowledge of the lot parameter z. If the observations $\xi_1, ..., \xi_N$ are reliable any hypothesis on z can be verified in a strict sense. On shopfloor, however, 100 percent inspection of the lot mostly is impossible because of economic, organizational and technical reasons. In this case information on the lot is restricted to a random sample $(\xi_{\nu_1}, ..., \xi_{\nu_n})$ of size $n < N$ and we have to use statistical methods (estimation or testing) for investigations on the lot quality indicator z. For process inspection, application of statistics is a methodological necessity, whereas in lot inspection application of statistics results from practical restrictions.

2.4 Most Powerful Lot Tests

If we consider tests on hypotheses on the lot quality indicator z based on random samples from the lot, the optimality principles of the classical theory of hypothesis testing apply without any difference. Thus we are interested in uniformly powerful (UMP) or uniformly most powerful unbiased (UMPU) tests for hypotheses on z, based on random samples from the lot.

It is easy to verify, see *Göb* (1996a), that the conditional distribution of the item quality indicators $\xi_{\nu_1}, ..., \xi_{\nu_n}$ in a sample of items $\nu_1, ..., \nu_n$ taken without replacement from the lot equals the conditional distribution of the quality indicators $\xi_1, ..., \xi_n$ of the first n items $1, ..., n$ in the lot. From a stochastic point of view it is thus justified to consider $\xi_1, ..., \xi_n$ as the sample variables.

In the following four paragraphs 3, 4, 5 and 6 we consider some special instances of most powerful tests for the lot quality indicator under continuous item quality indicators in the framework extablished by section 2.3. We shall also consider the asymptoticity argument sketched in the introduction 1. This means that we shall investigate the test statistics of the most powerful lot tests and their distributions for lot size N tending to infinity in order to find out whether process models can really be applied to statistical lot inspection in case of large lots in an asymptotic sense.

3 Tests of Significance for the Lot Mean

In this paragraph we consider the lot mean $T(\boldsymbol{\xi}) = \frac{1}{N} \sum_{i=1}^{N} \xi_i$, compare (3), as the lot quality indicator. It is important to distinguish the areas of application of the two lot quality indicators lot mean and lot proportion nonconforming (6) which is based on a transformation of the underlying continuous item quality indicators to a conforming/nonconforming scheme by means of specification ranges. Let us first consider two illustrative examples.

Example 3.1 The diameter of balls to be used in ball-bearings of certain width has to satisfy strict upper and lower specification limits, otherwise the ball bearing won't work. Here it is indispensable to convert continuous scale measuring (diameter) into a conforming/nonconforming rating and to use the lot proportion nonconforming (6) as the lot quality indicator. •

Example 3.2 Adipine acid is shipped in containers to a department of a textile company for production of nylon. The contents of the containers are successively fed into the nylon manufacturing process. The output of nylon is an increasing continuous function of the input substances. Under these assumptions the definition of conforming/nonconforming containers makes no sense and it is reasonable to use the lot mean (3) as the lot quality indicator. •

These two examples are typical for the different areas of application of lot proportion nonconforming and lot mean as lot quality indicators. These areas of application correspond more or less to two different industrial segments: The *parts* industry and the *substance* industry.

In discrete parts manufacturing interest is in reproducing individual items, e.g. ball bearings, screws, resistors) for individual usage with quality characteristics as close as possible to targets or in tight specification ranges. It is natural that such items are rated in a conforming/nonconforming scheme, and that the quality of lots consisting of such items is rated by lot proportion nonconforming.

Substance industries, chemical industries for instance, are concerned with types of product (continuous substances, liquids, powders) where the distinction of individual items is not inherent in the product but determined by requirements of transport or storage, e.g. in tanks or containers which can count as items. The quality characteristics which are used to rate these arbitrary items are often measures of quantity or purity of substance. In the corresponding lot quality indicator these item quality characteristics are merged into a total measure of quantity or purity which depends on the sum of the individual measure values. Typically, this is the case if the entire lot is fed as a process component into a further process to produce another continuous substance, or if a final user is interested in the total quantity or purity of substance in the lot, for instance in the case of artificial fertilizers or plant protectants in agriculture. For such cases the lot mean obviously is the adequate measure of lot quality.

Let us now start to investigate statistical lot inspection for the lot mean. In the unconditional process distribution we assume each ξ_i to be distributed according to a normal distribution with density $f_{(\mu,\sigma^2)}$ defined by (1). The process parameter is $\boldsymbol{\vartheta} = (\mu, \sigma) \in \mathbb{R} \times (0; +\infty)$, where μ is called the *process mean* and σ^2 is called the process variance. Since $T(\boldsymbol{\xi}) = \frac{1}{N} \sum_{i=1}^{N} \xi_i$ is sufficient for the process mean μ the conditional distribution of the sample variables $\xi_1, ..., \xi_n$ under $T(\boldsymbol{\xi}) = z$ depends on the variance σ^2 and the lot mean z alone.

We shall consider the problems of testing the following one-sided and two-sided hypotheses H_i against alternatives K_i:

$$H_1 : z \leq z_0, \quad K_1 : z > z_0, \tag{9}$$

$$H_2 : z \geq z_0, \quad K_2 : z < z_0, \tag{10}$$

$$H_3 : z = z_0, \quad K_3 : z \neq z_0. \tag{11}$$

3.1 Tests for the Lot Mean under Known Process Variance

We consider the problem of testing hypotheses on the lot mean z under known process variance $\sigma^2 = \sigma_0^2$. In a modern industrial environment with stable and controlled processes of manufacturing and transport, and with exchange of information among supplier and customer, the assumption of a known process variance is satisfied in many cases. *Göb* (1996a) shows that the UMP respectively UMPU tests for the above problems (9), (10), (11), under this assumption are based on the statistic

$$t(\xi_1, ..., \xi_n) \quad = \quad \frac{\bar{\xi} - z_0}{\sigma} \sqrt{\frac{Nn}{N-n}} \qquad \text{where} \quad \bar{\xi} = \frac{1}{n} \sum_{i=1}^{n} \xi_i. \tag{12}$$

The conditional distribution of $t(\xi_1, ..., \xi_n)$ under $T(\boldsymbol{\xi}) = \frac{1}{N} \sum_{i=1}^{N} \xi_i = z$ is the normal distribution $\mathcal{N}\left(\frac{z-z_0}{\sigma}\sqrt{\frac{Nn}{N-n}}, 1\right)$. We obtain the following UMP and UMPU level α tests. Let ϕ: $\mathbb{R} \to \mathbb{R}$ be the distribution function of the normal distribution $\mathcal{N}(0, 1)$.

The test

$$\text{Reject iff} \quad \frac{\bar{\xi} - z_0}{\sigma} \sqrt{\frac{Nn}{N-n}} \;\geq\; c \tag{13}$$

with critical value c determined by

$$\alpha \;\stackrel{!}{=}\; 1 - \phi(c)$$

is UMP at level α for $H_1 : z \leq z_0$ against $K_1 : z > z_0$.

The test

$$\text{Reject iff} \quad \frac{\bar{\xi} - z_0}{\sigma} \sqrt{\frac{Nn}{N-n}} \;\leq\; c \tag{14}$$

with critical value c determined by

$$\alpha \;\stackrel{!}{=}\; \phi(c)$$

is UMP at level α for $H_2 : z \geq z_0$ against $K_2 : z < z_0$.

The test

$$\text{Reject iff} \quad \left| \frac{\bar{\xi} - z_0}{\sigma} \sqrt{\frac{Nn}{N-n}} \right| \;\geq\; c \tag{15}$$

with critical value $c > 0$ determined by

$$\alpha \;\stackrel{!}{=}\; \phi(-c) + 1 - \phi(c) \;=\; 2 \cdot \phi(-c)$$

is UMPU at level α for $H_3 : z = z_0$ against $K_3 : z \neq z_0$.

3.2 Tests for the Lot Mean under Unknown Process Variance

We also have to provide tests for the case of unknown process variance, i.e. for σ^2 ranging over $(0; +\infty)$. For this case, UMP and UMPU tests for the problems (9), (10), (11) are investigated by *Göb* (1996b). The tests use the statistic

$$t(\xi_1, \ldots, \xi_n) \;=\; \frac{\sqrt{\frac{Nn}{N-n}}(\bar{\xi} - z_0)}{\sqrt{\frac{1}{n-1} \sum\limits_{i=1}^{n} \left(\xi_i - \bar{\xi}\right)^2}} \,. \tag{16}$$

The conditional distribution of $t(\xi_1, \ldots, \xi_n)$ under $T(\boldsymbol{\xi}) = \frac{1}{N} \sum\limits_{i=1}^{N} \xi_i = z$ is the noncentral t-distribution with noncentrality parameter $\delta = \frac{z - z_0}{\sigma} \sqrt{\frac{Nn}{N-n}}$ and with $n-1$ degrees of freedom, i.e., under $z = z_0$ the central t-distribution with $n-1$ degrees of freedom. We obtain the following UMP and UMPU level α tests. Let $\mathrm{T}_{n-1}\colon \mathbb{R} \to \mathbb{R}$ be the distribution function of the central t-distribution with $n-1$ degrees of freedom.

The test

$$\textit{Reject iff} \quad \frac{\sqrt{\frac{Nn}{N-n}}(\bar{\xi} - z_0)}{\sqrt{\frac{1}{n-1}\sum_{i=1}^{n}\left(\xi_i - \bar{\xi}\right)^2}} \geq c \tag{17}$$

with critical value c determined by

$$\alpha \overset{!}{=} 1 - \mathrm{T}_{n-1}(c)$$

is UMPU at level α for $H_1 : z \leq z_0$ against $K_1 : z > z_0$.

The test

$$\textit{Reject iff} \quad \frac{\sqrt{\frac{Nn}{N-n}}(\bar{\xi} - z_0)}{\sqrt{\frac{1}{n-1}\sum_{i=1}^{n}\left(\xi_i - \bar{\xi}\right)^2}} \leq c \tag{18}$$

with critical value c determined by

$$\alpha \overset{!}{=} \mathrm{T}_{n-1}$$

is UMPU at level α for $H_2 : z \geq z_0$ against $K_2 : z < z_0$.

The test

$$\textit{Reject iff} \quad \left|\frac{\sqrt{\frac{Nn}{N-n}}(\bar{\xi} - z_0)}{\sqrt{\frac{1}{n-1}\sum_{i=1}^{n}\left(\xi_i - \bar{\xi}\right)^2}}\right| \geq c \tag{19}$$

with critical value $c > 0$ determined by

$$\alpha \overset{!}{=} \mathrm{T}_{n-1}(-c) + 1 - \mathrm{T}_{n-1}(c) = 2 \cdot \mathrm{T}_{n-1}(-c)$$

is UMPU at level α for $H_3 : z = z_0$ against $K_3 : z \neq z_0$.

3.3 Asymptotics of Tests for the Lot Mean

Let us now discuss the asymptoticity argument. In the present case this argument asserts that the well-known most powerful tests for the process mean (Gauss and t-test) can be used to test the mean of large lots.

From the formula given by *Göb* (1996b) for the joint conditional distribution of the sample variables $\xi_1, ..., \xi_n$ under $T(\boldsymbol{\xi}) = \frac{1}{N}\sum_{i=1}^{N} \xi_i$ it is easy to see that $\xi_1, ..., \xi_n$ are asymptotically i.i.d. with normal distribution $\mathcal{N}(z, \sigma^2)$ for large lot size N. Thus it is justified to use the process model for large lots. *Göb* (1996b) discusses the limiting values for the test statistics (12) and (16) and the respective distributions for lot size N tending to infinity. For the case of known process variance $\sigma^2 = \sigma_0^2$ the limiting value of the test statistic (12) is the test statistic of the Gauss test which is most powerful under the process model. For the case of unknown process variance $\sigma^2 \in (0; +\infty)$ the limiting value of the test statistic (16) is the test statistic of the t-test which is most powerful under the process model. Thus for inspection for the lot mean the asymptoticity argument can be justified in a strict sense from the point of view of the lot model of paragraph 2.3.

4 Sampling by Variables for Lot Proportion Nonconforming

Sampling by variables for lot proportion nonconforming (sampling by measurement for lot proportion nonconforming) is presented in most textbooks on statistical quality control. It is also suggested by industrial standards like MIL-STD-414, ANSI/ASQC Z1.9 or ISO 3951. In statistical literature, sampling by variables for lot proportion nonconforming traces back to the contributions by *Jennett & Welch* (1939) and by *Romig* (1939).

The models of sampling by variables for lot proportion nonconforming consider continuous item quality indicators ξ_i with normal distribution $\mathcal{N}(\mu, \sigma^2)$. These continuous type data are converted into a conforming/ nonconforming scheme by comparing ξ_i with a specification range R. This leads to lot proportion nonconforming in the form (6) as the lot quality indicator. However, the statistic used for testing hypotheses on lot proportion nonconforming is the sample mean

$$\bar{\xi} = \frac{1}{n}\sum_{i=1}^{n} \xi_i$$

which is a continuous function of the unconverted continuous scale data.

The above approach seems strange from intuitive reasoning. The lot quality indicator (6) and the test statistic $\bar{\xi}$ do not seem to be well-matched since obviously the sample average measure value does not convey suitable information on the number of nonconforming items in the lot. From intuitive reasoning the only consistent approach is to apply the conversion to the binary conforming/nonconforming scheme also in the test statistic, i.e., to use the proportion

$$t(\xi_1,...,\xi_n) \quad = \quad \frac{1}{n}\left|\{i \,|\, 1 \le i \le n,\, \xi_i \in \mathbb{R} \setminus R\}\right| \tag{20}$$

of nonconforming items in the sample. This amounts to using the test suggested by acceptance sampling by attributes.

In fact, the intuitively obvious incompatibility among lot proportion nonconforming and the test statistic $\bar{\xi}$ leads to absurdities, as even admitted by proponents of sampling by variables for lot proportion nonconforming, see ***Duncan***, p. 256, ***Grant***, p. 475, ***Schilling***, p. 224, for instance. The following phenomena occur, see ***Göb*** (1996a) for details:

- Rejection although the actual sample contains no nonconforming items.
- And even worse: Possible rejection of lots containing 100 percent conforming items, occasionally even with high probability as demonstrated by ***von Collani*** (1990b) and by ***Hryniewicz*** (1991).
- In case of two-sided specification limits: Possible acceptance of lots containing 100 percent nonconforming items under 100 percent inspection.

Proponents of sampling by variables for lot proportion nonconforming try to defend against such criticism by the argument of ***discriminating power***. A typical statement of this argument is the following, see ***Burr*** (1976), page 323: "In general a measurement provides more information about the part or assembly, than does a mere determination that it lies within or outside limits. It is thus often possible to obtain the same discriminating power on a smaller sample size n, when using measurements than when using attributes." This type of argument is advocated unanimously in standard textbooks of statistical quality control, see for instance ***Duncan*** (1986), page 255, ***Grant*** (1964), page 442, ***Schilling*** (1982), page 222/223, ***Wadsworth et al.*** (1986), page 545.

From the point of view of stochastic methodology the above argument has to be interpreted as asserting that at a given sample size n the test using the sample average measure value $\bar{\xi}$ is more powerful than the test suggested by acceptance sampling by attributes which uses the proportion (20) of nonconforming items in the sample as the test statistic. However, the latter assertion is refuted by ***Göb*** (1996a). This paper shows that for testing the hypothesis $H: z \le z_0$ (lot proportion nonconforming sufficiently small) against the alternative $K: z > z_0$ (lot proportion nonconforming too large) under a prescribed level of significance the test of acceptance sampling by attributes based on the statistic (20) is UMP and, consequently, more powerful than the test based on the sample average $\bar{\xi}$ as suggested by sampling by variables for lot proportion nonconforming. The latter result holds for any underlying family $(f_\vartheta)_{\vartheta\in\Theta}$ of density functions. This result refutes the above argument of discriminating power and confirms all intuitive doubts on sampling by measurement for lot proportion nonconforming. Attributes sampling always provides the best (most powerful) test for lot proportion nonconforming.

Proponents of sampling by measurement for lot proportion nonconforming try to defend their approach by the asymptoticity argument which states that, inspite of the facts pointed out above, sampling by measurement for lot proportion nonconforming is adequate in case of large lot size N. This is hard to believe from an intuitive point of view since the lack of intuitive plausibility of the measurement test statistic $\bar{\xi}$ for the control of lot proportion nonconforming does not vanish for large lot size.

For a rigorous discussion of the asymptoticity argument, *Göb* (1996a) considers a limiting process analogous to the one which leads from hypergeometric to binomial distribution: lot size N tends to infinity and lot proportion nonconforming z_N indexed by the lot size N tends to a value q in $[0;1]$. *Göb* (1996a) shows that the asymptotically UMP test for $H\colon z \leq z_0$ (lot proportion nonconforming sufficiently small) against $K\colon z > z_0$ (lot proportion nonconforming too large) under a prescribed level of significance is again the test of acceptance sampling by attributes based on the statistic (20). The assumption of large lot size N provides no arguments in defence of sampling by measurement for lot proportion nonconforming against attributes sampling.

5 Tests of Significance for the Average Square Deviation from Target

Under the assumption of continuous item quality indicators the lot quality indicators lot proportion nonconforming (6) and lot mean (3) stand for two competitive and mutually exclusive lot quality models. Under (6) each item is rated separately as conforming or nonconforming. The continuous item quality indicators ξ_i are absorbed in a discrete binary scheme. Large and small deviations from specifications cannot be distinguished. The lot mean (3) is a continuous strictly increasing function of each ξ_i. The original continuous scale information does not get lost. (3) does not treat the items as completely isolated units, small values ξ_i can be compensated by large values ξ_j.

Some applications, however, require a concept of lot quality settled between the extremes (3) and (6). Consider the following example.

Example 5.1 Packages of a composite plant-protective agent are shipped to a wine-growers' cooperative. For protection of one acre of cultivated area one package of agent is dissolved in 20 liters of water, the resulting fluid is spread out over the area. For proper application a package is supposed to contain 100 cm^3 of agent. Deviations from the target value 100 cm^3 in both directions continuously diminish the grape output of the wine yard. If the package contains less than 100 cm^3 the plants are attacked by noxious parasites and pests. If the package contains more than 100 cm^3 the vines and the ecological balance of the wine yard are damaged by an excess of chemical substances. •

Neither of the lot quality indicators (3) and (6) fits to the requirements of example 5.1. As in (6) the individual deviations of the item quality indicators from target should accumulate, but the deviation measure should be a continuous and not a step function as (6). On the other hand the continuous lot quality indicator (3) obviously is inadequate since mutual compensation of values above and below target should not be admitted.

Presumably, examples like 5.1 form the intuitive background of sampling by variables for lot proportion nonconforming. This approach reflects the need for a concept of lot quality which combines the aspect of specifications with the aspect of continuous scale measuring. However, as shown in paragraph 4, above, the solution suggested by variables sampling for lot proportion nonconforming is inadequate.

Let us try to establish a sound model to cope with situations as given by example 5.1. An appropriate lot quality indicator should be a continuous strictly increasing function of each's package deviation from the target value $a = 100$. This requirement is met by the ***average square deviation from target a per item in the lot*** (***lot average square deviation from target a***) $T(\boldsymbol{\xi}) = \frac{1}{N}\sum_{i=1}^{N}(\xi_i - a)^2$ introduced by (5). Clearly the interest is to have lots with $z = T(\boldsymbol{\xi})$ as small as possible. From an economic point of view the lot quality indicator $T(\boldsymbol{\xi}) = \frac{1}{N}\sum_{i=1}^{N}(\xi_i - a)^2$ makes sense in the following situation. For each item i the cost incurred from the deviation of ξ_i from target a is a quadratic function $L \cdot (\xi_i - a)^2$ of the deviation $\xi_i - a$, where $L > 0$. The income from a lot of size N is I_N. Then the profit incurred from a lot of size N is

$$I_N - \sum_{i=1}^{N} L \cdot (\xi_i - a)^2 \quad = \quad I_N - N \cdot L \cdot T(\xi_1, ..., \xi_N).$$

The study of statistical methods under this type of economic objective functions has been propagated by Taguchi and his scholars, see *Taguchi et al.* (1989), for instance.

Let us consider the lot quality indicator $T(\boldsymbol{\xi})$ in the case where the unconditional process distribution of the item quality indicators $\xi_1, ..., \xi_N$ is a normal distribution $\mathcal{N}(\mu, \sigma^2)$. The process parameter is $\boldsymbol{\vartheta} = (\mu, \sigma) \in$

$\mathbb{R} \times (0; +\infty)$, where μ is called the *process mean* and σ^2 is called the process variance.

As mentioned in the introduction, the industry of the first half of the 20th century reduced quality control to the quality control of finished product (product quality control) and statistical quality control to statistical lot inspection. Nowadays main emphasis is on the control of processes which lead to finished product in the customer's incoming material department: above all the control of manufacturing processes but also the control of storage and transport. In a modern quality control system satisfactory average product quality is achieved by process control under suitable application of statistics. However, even if stable expected quality is guaranteed item quality still varies at random around the mean. Thus lot inspection in product control is still advisable so as to prevent unwelcome large quality variation of items in a lot.

In our model we can reflect this relation among process control and product control in the following way. Expected item quality is determined by the process mean $\mu = E[\xi_i]$. Process control and its statistical tools keep the process mean on target. Thus for lot inspection in product control the process mean μ can be assumed to be known and fixed. Under reasonable circumstances we then have to assume that this known and fixed item quality mean coincides with the intended target value a. Inserting $a = \mu$ in (5) we finally obtain the lot quality indicator

$$T(\xi_1, \dots, \xi_N) = \frac{1}{N} \sum_{i=1}^{N} (\xi_i - \mu)^2 . \tag{21}$$

Let us now consider the lot quality indicator in the framework of paragraph 2.3. As well-known, $T(\boldsymbol{\xi}) = \frac{1}{N} \sum_{i=1}^{N} (\xi_i - \mu)^2$ is sufficient for the process variance σ^2. Hence the conditional distribution of the sample variables $\xi_1, \dots, \xi_n$ under $T(\boldsymbol{\xi}) = z$ depends on the process mean μ and the lot average square deviation z alone.

We shall consider the problems of testing the following one-sided and two-sided hypotheses H_i against alternatives K_i:

$$H_1 \colon\ z \le z_0, \quad K_1 \colon\ z > z_0, \tag{22}$$

$$H_2 \colon\ z \ge z_0, \quad K_2 \colon\ z < z_0, \tag{23}$$

$$H_3 \colon\ z \le z_1 \text{ or } z \ge z_2, \quad K_3 \colon\ z_1 < z < z_2, \tag{24}$$

$$H_4 \colon\ z_1 \le z \le z_2, \quad K_4 \colon\ z < z_1 \text{ or } z > z_2, \tag{25}$$

$$H_5\colon\ z = z_0, \quad K_5\colon\ z \neq z_0. \tag{26}$$

Clearly industrial quality control is mainly interested in the schemes (22) and (23) where z_0 is interpreted as a break even value for lot quality. In special cases, however, schemes (24), (25) and (26) are also interesting. For instance, assume that in ordinary course of business a consumer reckons upon lots of normal satisfactory quality with $z_1 \leq T(\boldsymbol{\xi}) \leq z_2$. In such a case she will use scheme (25) to sort out lots of intolerable quality level with $T(\boldsymbol{\xi}) > z_2$ as well as lots of extremely good quality level $T(\boldsymbol{\xi}) < z_1$ which may be used for special purposes.

Corresponding to intuition, the UMP and UMPU tests developed by *Göb* (1997) are all based on the statistic

$$t(\xi_1, ..., \xi_n) \quad = \quad \sum_{i=1}^{n} (\xi_i - \mu)^2 \,. \tag{27}$$

The conditional distribution of $\frac{1}{Nz}t(\xi_1, ..., \xi_n)$ is a beta-distribution of order $\left(\frac{n}{2}, \frac{N-n}{2}\right)$, hence the conditional of $t(\xi_1, ..., \xi_n)$ is

$$\mathrm{P}_\mu\left(t(\xi_1, ..., \xi_n) \leq y \,\middle|\, T(\boldsymbol{\xi}) = z\right) \quad = \quad \mathrm{B}_{\frac{n}{2}, \frac{N-n}{2}}\left(\frac{y}{Nz}\right) \tag{28}$$

where $\mathrm{B}_{\frac{n}{2}, \frac{N-n}{2}}$ is the distribution function of beta-distribution of order $\left(\frac{n}{2}, \frac{N-n}{2}\right)$.

In detail, *Göb* (1997) develops the following UMP and UMPU tests.

The test

$$\textit{Reject iff} \quad \sum_{i=1}^{n} (\xi_i - \mu)^2 \geq c \tag{29}$$

with critical value c determined by

$$\alpha \quad \stackrel{!}{=} \quad 1 - \mathrm{B}_{\frac{n}{2}, \frac{N-n}{2}}\left(\frac{c}{Nz_0}\right)$$

is UMP at level α for $H_1 : z \leq z_0$ against $K_1 : z > z_0$.

The test

$$\textit{Reject iff} \quad \sum_{i=1}^{n} (\xi_i - \mu)^2 \leq c \tag{30}$$

with critical value c determined by

$$\alpha \quad \stackrel{!}{=} \quad \mathrm{B}_{\frac{n}{2}, \frac{N-n}{2}}\left(\frac{c}{Nz_0}\right)$$

is UMP at level α for $H_2 : z \geq z_0$ against $K_2 : z < z_0$.

The test

$$\textit{Reject iff} \quad c_1 \leq \sum_{i=1}^{n} (\xi_i - \mu)^2 \leq c_2 \tag{31}$$

with critical values c_1, c_2 determined by

$$\mathrm{B}_{\frac{n}{2},\frac{N-n}{2}}\left(\frac{c_2}{Nz_1}\right) - \mathrm{B}_{\frac{n}{2},\frac{N-n}{2}}\left(\frac{c_1}{Nz_1}\right) \overset{!}{=} \alpha \overset{!}{=} \mathrm{B}_{\frac{n}{2},\frac{N-n}{2}}\left(\frac{c_2}{Nz_2}\right) - \mathrm{B}_{\frac{n}{2},\frac{N-n}{2}}\left(\frac{c_1}{Nz_2}\right)$$

is UMP at level α for $H_3 : z \leq z_1 \text{ or } z \geq z_2$ against $K_3 : z_1 < z < z_2$.

The test

$$\textit{Reject iff} \quad \sum_{i=1}^{n} (\xi_i - \mu)^2 \leq c_1 \quad \text{or} \quad \sum_{i=1}^{n} (\xi_i - \mu)^2 \geq c_2 \tag{32}$$

with critical values c_1, c_2 determined by

$$\begin{aligned} \mathrm{B}_{\frac{n}{2},\frac{N-n}{2}}\left(\frac{c_1}{Nz_1}\right) + 1 - \mathrm{B}_{\frac{n}{2},\frac{N-n}{2}}\left(\frac{c_2}{Nz_1}\right) &\overset{!}{=} \alpha \\ \overset{!}{=} \mathrm{B}_{\frac{n}{2},\frac{N-n}{2}}\left(\frac{c_1}{Nz_2}\right) + 1 - \mathrm{B}_{\frac{n}{2},\frac{N-n}{2}}\left(\frac{c_2}{Nz_2}\right) \end{aligned}$$

is UMPU at level α for $H_4 : z_1 \leq z \leq z_2$ against $K_4 : z > z_1 \text{ or } z > z_2$.

The test

$$\textit{Reject iff} \quad \sum_{i=1}^{n} (\xi_i - \mu)^2 \leq c_1 \quad \text{or} \quad \sum_{i=1}^{n} (\xi_i - \mu)^2 \geq c_2 \tag{33}$$

with critical values c_1, c_2 determined by

$$\mathrm{B}_{\frac{n}{2},\frac{N-n}{2}}\left(\frac{c_1}{Nz_0}\right) + 1 - \mathrm{B}_{\frac{n}{2},\frac{N-n}{2}}\left(\frac{c_2}{Nz_0}\right) \overset{!}{=} \alpha$$

and

$$\left(\frac{c_1}{c_2}\right)^{\frac{n}{2}} \overset{!}{=} \left(\frac{Nz_0 - c_2}{Nz_0 - c_1}\right)^{\frac{N-n}{2}-1}$$

is UMP at level α for $H_5 : z = z_0$ against $K_5 : z \neq z_0$.

6 Tests of Significance for the Lot Minimum

The lot minimum $T(\boldsymbol{\xi}) = \min\{\xi_1, ..., \xi_N\}$ is a reasonable lot quality indicator in particular for lifetime data ξ_i. Consider the case where $\xi_1, ..., \xi_N$ are the lifetimes of technical devices or parts to be implemented in a serial system which collapses with the first breakdown of one of the elements $1, ..., N$. Then the quality of the lot $1, ..., N$ is evidently determined by $T(\boldsymbol{\xi}) = \min\{\xi_1, ..., \xi_N\}$.

Göb (1995) considers a model with the lot quality indicator $T(\boldsymbol{\xi}) = \min\{\xi_1, ..., \xi_N\}$ where the process distribution of the item quality indicators is the two-parametric exponential distribution given by the density (2). Here the process parameter is $\boldsymbol{\vartheta} = (\lambda, d) \in (0; +\infty) \times \mathbb{R}$, where λ is the *scale parameter* and d is the *location* or *threshold parameter*. From its interpretation as a lower limit of the process variables d may aslo be referred to as the *process minimum*. As well-known, $T(\boldsymbol{\xi}) = \min\{\xi_1, ..., \xi_N\}$ is sufficient for d, see *Bain & Engelhardt* (1991), p. 155, for instance. Hence the conditional distribution of the sample variables $\xi_1, ..., \xi_n$ under $T(\boldsymbol{\xi}) = z$ depends on the process scale parameter λ and the lot minimum z.

The following phenomenon distinguishes the investigation of the present model from the investigations of the models of paragraphs 3, 4 and 5: The conditional distribution of the sample variables $\xi_1, ..., \xi_n$ is neither absolutely continous nor discrete but a mixture of an absolutely continuous and a discrete component. Thus the investigation of uniformly most powerful tests is considerably complicated. Locally most powerful tests can be determined without difficulty. For the one-sided problem $H_1 : z \geq z_0$ against $K_1 : z < z_0$ *Göb* (1995) develops a test whose power uniformly comes close to the maximum power possible. For the problem $H_2 : z \leq z_0$ against $K_2 : z > z_0$ *Göb* (1995) shows that at under a reasonably small significance level a UMP or UMPU test for does not exist. The paper develops an intuitively reasonable test for this problem. For details the reader is referred to *Göb* (1995).

7 The Economic Design of Tests in Statistical Lot Inspection

The model of paragraph 2 for statistical lot inspection under continuous item quality indicators requires parametric assumptions on the process of manufacture or transport behind the lot. This underlying process model leads to a natural approach to the economic design of tests (sampling plans) for statistical lot inspection.

We consider tests for decision on acceptance or rejection of the lots. For each sample size n let a test statistic t_n be fixed. Then the test (sampling plan) is determined by the pair (n, S) of the sample size n and the critical region S. Consider any economic loss function $G(N, n, S, z)$ which quantifies the expected economic cosnequences of decision on acceptance/rejection on lots of size N with value z of the lot quality indicator by means of the sampling plan (n, S). Let

$$\Gamma \quad = \quad \{H_{\vartheta} \,|\, \vartheta \in \Theta\} \tag{34}$$

be the set of possible distribution functions H_{ϑ} of the random lot quality indicator $T(\boldsymbol{\xi}) = T(\xi_1, ..., \xi_N)$ in the parametric process model of paragraph 2.1. For fixed $\vartheta \in \Theta$ the long run economic consequences incurred from deciding on lots of size N from the process by means of the sampling plan (n, S) is quantified by

$$E_{\vartheta}\Big[G\Big(N, n, S, T(\boldsymbol{\xi})\Big)\Big] \quad = \quad \int_{\mathbb{R}} G(N, n, S, z)\, dH_{\vartheta}(z)\,. \tag{35}$$

It is reasonable to design the sampling plan so as to protect the consumer under the worst possible case in the long run. Thus sampling plans (n, S) applied to lots of size N are evaluated by the objective function

$$Z_N(n, S) = \sup_{\vartheta \in \Theta} \int_{\mathbb{R}} G(N, n, S, z)\, dH_{\vartheta}(z) =_{(35)} \sup_{H \in \Gamma} \int_{\mathbb{R}} G(N, n, S, z)\, dH(z)\,, \tag{36}$$

and the economically optimum sampling $(n^{\star}, S^{\star})$ is determined as the sampling plan which minimizes $Z_N(n, S)$. This approach to the design of sampling plans is an instance of the Γ-minimax approach suggested by *Krumbholz* (1982): Incomplete knowledge on the distribution of lot quality is represented by a set Γ of admissible distributions of lot quality, and the sampling plan is chosen so as to give protection against the worst case possible under Γ. Thus we may use the term ***Γ-minimax sampling plan*** for the optimum sampling plan $(n^{\star}, S^{\star})$.

The Γ-minimax design for UMP respectively UMPU tests under the model of paragraph 2 and under a linear cost model is investigated by the following papers: *Göb* (1997b) and *Göb* (1997c) investigate the Γ-minimax design of the tests for the lot mean under known process variance presented in paragraph 3.1. *Göb* (1996c) investigates the Γ-minimax design of the tests for the average square deviation from target presented in paragraph 5.

8 Conclusion

The results presented by paragraphs 3, 4, 5, and 6 show that the model introduced by paragraph 2 for statistical lot inspection under continuous item quality indicators can successively be applied to develop most powerful lot tests. Paragraph 7 shows that the model also provides a reasonable approach to the economic design of sampling plans under continuous item quality indicators. However, the model can also be applied in case of discrete item quality indicators, see *Göb* (1992a), *Göb* (1992b), *Göb* (1996a).

We have introduced the model for a problem of statistical quality control. Obviously, however, the problem of statistical inference on a characteristic $z = T(\xi_1, ..., \xi_N)$ of a finite set of items $1, ..., N$ with corresponding individual characteristics $\xi_1, ..., \xi_N$ also appears in many other contexts, e.g., in biosciences.

In the area of continuous item quality indicators the following topics remain for further research:

(i) Investigation of further lot quality indicators. In particular, lot quality indicators with a finite number of points of discontinuity.
(ii) Investigation of further parametric process models.
(iii) Further investigation in the economic design of tests for lot inspection.

References

1. Bain, L.J., and Engelhardt, M. (1991) Statistical Analysis of Reliability and Life-Testing Models. Marcel Dekker, New York/Basel/Hong Kong
2. Burr, I.W. (1976) Statistical Quality Control Methods. Marcel Dekker, New York, Basel
3. von Collani, E. (1990a) Wirtschaftliche Qualitätskontrolle - eine Übersicht über einige neue Ergebnisse. OR Spektrum **12**, 1–23
4. von Collani, E. (1990b) ANSI/ASQC Z1.4 versus ANSI/ASQC Z1.9. Economic Quality Control **5**, 60–64
5. Deming, W.E. (1986) Out of the Crisis. Massachusetts Institute of Technology, Cambridge, Massachusetts
6. Dodge, H.F. (1969) Notes on the Evolution of Acceptance Sampling Plans. Parts I,II,III. Journal of Quality Technology **1**, 77–88, 155–162, 225–232
7. Duncan, A.J. (1986) Quality Control and Industrial Statistics. Fifth ed. Richard D. Irwin, Homewood, Illinois
8. Feigenbaum, A.V. (1961) Total Quality Control. Mc Graw-Hill, New York
9. Göb, R. (1992a) Some Interesting Sampling Distributions and Their Applications. Proceedings of the Second Würzburg-Umeå Conference in Statistics, Würzburg 1992, 99–104
10. Göb, R. (1992b) Type A Operating Characteristic Functions of Defects Sampling Plans. Metrika **39**, 139–153
11. Göb, R. (1995) Tests for the Minimum of Finite Lots Generated by a Process under Two-Parameter Exponential Distribution. Research Reports of the Würzburg Research Group on Quality Control **56**
12. Göb, R. (1996a) An Elementary Model of Statistical Lot Inspection and its Application to Sampling by Variables. Metrika **44**, 135–163
13. Göb, R. (1996b) Tests of Significance for the Mean of a Finite Lot. Metrika **44**, 223–238

14. Göb, R. (1996c) Economic Γ-Minimax Acceptance Sampling Plans for the Average Square Deviation from Target in Finite Lots. Economic Quality Control **11**(1), 23–52
15. Göb, R. (1997a) Tests of Significance for the Average Square Deviation from Target in a Finite Lot. Metrika **45**, 131–169
16. Göb, R. (1997b) Economic Inspection for the Mean of a Finite Lot. Research Reports of the Würzburg Research Group on Quality Control **77**, Revised version accepted for publication in Statistics & Decisions
17. Göb, R. (1997c) An Economic Γ-Minimax Decision Scheme for Inspecting the Lot Mean. Economic Quality Control **12**(4), 227–253
18. Grant, E.L. (1964) Statistical Quality Control. McGraw-Hill, New York, Toronto, London
19. Hald, A. (1981) Statistical Theory of Sampling Inspection by Attributes. John Wiley, New York
20. Hryniewicz, O. (1991) A Note on E. von Collanis Paper. "ANSI/ASQC Z1.4 versus ANSI/ASQC Z1.9". Economic Quality Control **6**, 16–18
21. Ishikawa, K. (1985) What is Total Quality Control? The Japanese Way. Prentice-Hall, Englewood Cliffs, New Jersey
22. Jennett, W.J., and Welch, B.L. (1939) The Control of Proportion Defective as Judged by a Single Quality Characteristic Varying on a Continuous Scale. Supplement to the Journal of the Royal Statistical Society **6**, 80–88
23. Krumbholz, W. (1982) Die Bestimmung einfacher Attributprüfpläne unter Berücksichtigung von unvollständiger Vorinformation. Allgemeines Statistisches Archiv **66**, 240-253
24. Romig, H.G. (1939) Allowable Average in Sampling Inspection. Ph.D. Thesis, Columbia University, New York
25. Schilling, E.G. (1982) Acceptance Sampling in Quality Control. Marcel Dekker, New York/Basel
26. Taguchi, T., Elsayed, E.A., and Hsiang, T.C. (1989) Quality Engineering in Production Systems. McGraw-Hill, New York
27. Thyregod, P. (1973) On Exchangeable Prior Information in Sampling Finite Populations. In Bulletin of the International Institute, Proceedings of the 39th Session, **3**, 530–535
28. Wadsworth, H.M., Stephens, K.S., and Godfrey, A.B. (1986) Modern Methods for Quality Control and Improvement. John Wiley, New York
29. Wetherill, G.B., and Chiu, W.K. (1975) A Review of Acceptance Sampling Schemes with Emphasis on the Economic Aspect. International Statistical Review **43**, 191–209

Credit–based Accept–zero Sampling Schemes for the Control of Outgoing Quality

David H. Baillie[1] and Chris A.J. Klaassen[2]

[1] Baillie Associates, Chesham, Bucks, England
[2] Institute for Business and Industrial Statistics, University of Amsterdam, The Netherlands

Abstract. A general procedure is presented for switching between accept-zero attributes or variables sampling plans to provide acceptance sampling schemes with a specified limit on the (suitably defined) average outgoing quality (AOQ). The switching procedure is based on credit, defined as the total number of items accepted since the last rejection. The limit on the AOQ is maintained by mandatory 100% inspection of lots that are rejected when the credit is zero, with acceptance of all inspected items found to be conforming in such lots.

The general procedure is developed and the application of the procedure to three situations is considered. Denoting the smallest value in a sample by $X_{(1)}$, a single, lower specification limit by L and a constant by c, the method is applied to:

(a) sampling by attributes, with acceptance criterion $X_{(1)} \geq L$;
(b) sampling by variables from a normal distribution with unknown process mean μ and known process standard deviation σ, with acceptance criterion $X_{(1)} \geq L + c\sigma$;
(c) sampling by variables from a normal distribution with unknown process mean μ and unknown process standard deviation σ, with acceptance criterion $X_{(1)} \geq L + cs$, where s is the sample standard deviation.

Finally it is shown that the guarantee on the upper limit to the AOQ remains valid when *all* rejected lots are submitted to 100% inspection.

1 Theory and main result

Consider a sequence of lots of sizes $N_1, N_2 \ldots$ with quality determined by a sequence of parameters $\theta_1, \theta_2 \ldots$. For a lot of size N with parameter $\theta \in (-\infty, \infty)$ a sample of size n results in $X_1, \ldots, X_n$, and the process quality q_θ is defined here by

$$q_\theta = P_\theta(X_1 < L). \tag{1}$$

Define the credit, K, as the total number of items accepted since the last rejection. A test procedure (still to be chosen, depending on N, n, L and K, but quite arbitrary for the moment except that rejection will always take place if nonconforming items are found in the sample) results in an acceptance probability

$$p_\theta = p_\theta(N, n, L, K). \tag{2}$$

We will assume that q_θ and p_θ are differentiable in θ with derivatives $\dot{q}_\theta$ and $\dot{p}_\theta$ respectively. The sample size n will depend on the credit K and, before the l-th lot in the sequence is sampled, the credit will be equal to $K_l = \sum_{i=1}^{l-1} N_i$ if no rejection has occurred.

Let the random variable J denote the index of the first lot in the sequence that is rejected. Assuming independence between lots we have

$$P(J=j) = \left[\prod_{l=1}^{j-1} p_{\theta_l}(N_l, n_l, L, K_l)\right] \{1 - p_{\theta_j}(N_j, n_j, L, K_j)\}. \tag{3}$$

Consider the inspection strategy for which all rejected lots except for the first in a sequence are returned to the producer *without* further inspection. If the first lot in the sequence (i.e. when $K = 0$) is rejected on the basis of the sample results, the lot is submitted to 100% inspection; conforming items found in the sampling and 100% inspection of the first lot are allowed into the marketplace, while the nonconforming items are returned to the producer. This strategy for inspection and for credit updating is represented schematically in the flowchart below.

At the time of the first rejection the expected proportion of defectives in the market equals zero for $J = 1$ and

$$\frac{\sum_{j=1}^{J-1}(N_j - n_j)q_{\theta_j}}{\sum_{j=1}^{J-1} N_j} \quad \text{if} \quad J > 1. \tag{4}$$

We might call the expectation of this over J the *expected outgoing quality at rejection.* As 100% inspection is only applied to the first lot in a sequence, we denote this expectation by E_f . Writing ρ_i for $\prod_{l=1}^{i} p_{\theta_l}$ with $\rho_0 = 1$, it is found that

$$\begin{aligned}
E_f &= \sum_{i=2}^{\infty} \frac{\sum_{j=1}^{i-1}(N_j - n_j)q_{\theta_j}}{\sum_{h=1}^{i-1} N_h} \left(\prod_{l=1}^{i-1} p_{\theta_l}\right)(1 - p_{\theta_i}) \\
&= \sum_{i=2}^{\infty} \frac{\sum_{j=1}^{i-1}(N_j - n_j)q_{\theta_j}}{\sum_{h=1}^{i-1} N_h}(\rho_{i-1} - \rho_i) \\
&= \sum_{i=1}^{\infty} \frac{\sum_{j=1}^{i}(N_j - n_j)q_{\theta_j}}{\sum_{h=1}^{i} N_h}(\rho_i - \rho_{i+1}) \\
&= \sum_{j=1}^{\infty}\sum_{i=j}^{\infty} \frac{(N_j - n_j)q_{\theta_j}}{\sum_{h=1}^{i} N_h}(\rho_i - \rho_{i+1}) \\
&\leq \sum_{j=1}^{\infty} \frac{(N_j - n_j)q_{\theta_j}}{\sum_{h=1}^{j} N_h} \sum_{i=j}^{\infty}(\rho_i - \rho_{i+1}) \\
&\leq \sum_{j=1}^{\infty} \frac{(N_j - n_j)}{K_j + N_j} q_{\theta_j}\rho_j.
\end{aligned} \tag{5}$$

Conditions and a test procedure are required such that $E_f \leq \pi_0$. It is evident that sufficient conditions for

$$\sum_{j=1}^{\infty} \frac{(N_j - n_j)}{K_j + N_j} q_{\theta_j} \rho_j \leq \pi_0 \tag{6}$$

are that

$$\sum_{j=1}^{k} \frac{(N_j - n_j)}{K_j + N_j} \frac{q_{\theta_j} \rho_j}{1 - \rho_k} \leq \pi_0, \quad \forall k. \tag{7}$$

Viewing the left hand side of (7) as a function Ψ_k of $\theta_1, \ldots, \theta_k$ it can be seen that

$$\Psi_k(\theta_1, \ldots, \theta_k) = \Psi_{k-1}(\theta_1, \ldots, \theta_{k-1}) \frac{1 - \rho_{k-1}}{1 - \rho_k} + \frac{N_k - n_k}{K_k + N_k} q_{\theta_k} \frac{\rho_k}{1 - \rho_k}. \tag{8}$$

If q_θ and p_θ are defined such that $p_{-\infty} = 0,\ q_{-\infty} = 1,\ p_\infty = 1$ and $q_\infty = 0$, it follows from (8) that

$$\Psi_k(\theta_1, \ldots, \theta_{k-1}, -\infty) = \Psi_{k-1}(\theta_1, \ldots, \theta_{k-1})\{1 - \rho_{k-1}\} \tag{9}$$

and

$$\Psi_k(\theta_1, \ldots, \theta_{k-1}, \infty) = \Psi_{k-1}(\theta_1, \ldots, \theta_{k-1}). \tag{10}$$

Consequently, in order to be able to use induction, it suffices to find conditions for (7) to hold for $k = 1$ and for $\sup_{\theta_k} \Psi_k(\theta_1, \ldots, \theta_k) \leq \pi_0$ for $k > 1$. Since $\Psi_k(\theta_1, \ldots, \theta_k)$ is differentiable in θ_k we determine the stationary points of $\Psi_k(\theta_1, \ldots, \theta_k)$ by solving $\Psi_k'(\theta_k) = \frac{\partial}{\partial \theta_k} \Psi_k(\theta_1, \ldots, \theta_k) = 0$. Writing $\Psi_k(\theta_k)$ for $\Psi_k(\theta_1, \ldots, \theta_k)$ we have

$$\Psi_k'(\theta_k) = \Psi_k(\theta_k) \frac{\dot{p}_{\theta_k} \rho_{k-1}}{1 - \rho_k} + \frac{N_k - n_k}{K_k + N_k} \frac{\{\dot{q}_{\theta_k} p_{\theta_k} + q_{\theta_k} \dot{p}_{\theta_k}\} \rho_{k-1}}{1 - \rho_k}. \tag{11}$$

At these stationary points θ_k we have $\Psi_k(\theta_k) \leq \pi_0$ if and only if

$$-\frac{N_k - n_k}{K_k + N_k} \left(\frac{\dot{q}_{\theta_k} p_{\theta_k}}{\dot{p}_{\theta_k}} + q_{\theta_k} \right) \leq \pi_0. \tag{12}$$

By induction, for (7) to hold for all k, it suffices that for all θ

$$\frac{q_\theta p_\theta}{1 - p_\theta} \leq \frac{N_1}{N_1 - n_1} \pi_0 \quad \text{for } k = 1, K_1 = 0, \tag{13}$$

and

$$-q_\theta - \frac{\dot{q}_\theta p_\theta}{\dot{p}_\theta} \leq \frac{K_k + N_k}{N_k - n_k} \pi_0 \quad \text{for } k > 1. \tag{14}$$

We summarize our main result as

Theorem 1 *Let in the above set–up the quality q_θ of a lot and the acceptance probability p_θ be differentiable in θ with derivatives $\dot{q}_\theta$ and $\dot{p}_\theta$ respectively. Furthermore, assume*

$$q_{-\infty} = 1,\ p_{-\infty} = 0,\ q_{\infty} = 0,\ p_{\infty} = 1. \tag{15}$$

Fix π_0, $0 \le \pi_0 \le 1$. If for each lot size N and credit value K, the sample size n and the test procedure are chosen such that (13) and (14) hold for all θ, then

$$E_f \le \pi_0 \tag{16}$$

is valid, i.e. the expected outgoing quality at rejection E_f does not exceed the limit π_0.

The sampling schemes implied by this theorem have inherent switching rules, since good past performance results in high credit and hence (cf. (14)) in small sample sizes. The switching rules in classical schemes, e.g. in ISO 2859 Part 1 (1989), seem to lack such a firm basis.

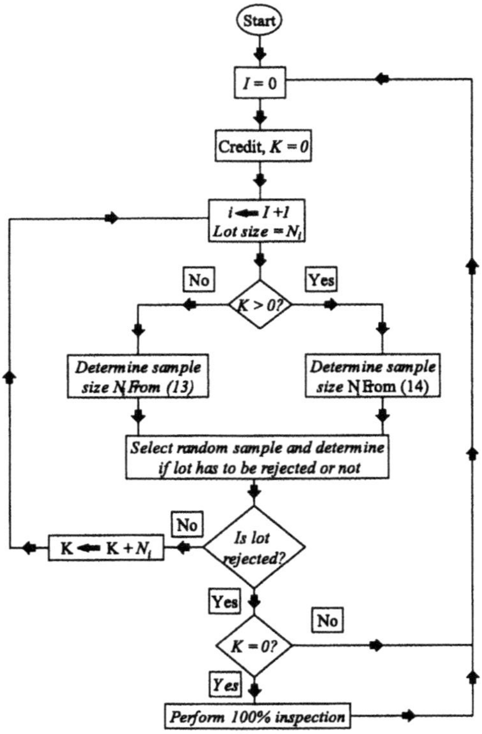

2 Application to sampling by attributes from a process

Suppose that $q_\theta = 1-\theta$ represents the fraction nonconforming in a population and sampling is by attributes with acceptance number zero. Then for a sample of size n, the probability of acceptance is

$$p_\theta = \theta^n. \tag{17}$$

We have $\theta \in (0,1)$ here and the following analogue of (15):

$$q_0 = 1,\ p_0 = 0,\ q_1 = 0,\ p_1 = 1. \tag{18}$$

With this change Theorem 1 is still valid. The left hand side of (13) equals

$$\frac{(1-\theta)\theta^{n_1}}{1-\theta^{n_1}} = \frac{1}{\theta^{-1}+\theta^{-2}+\ldots+\theta^{-n_1}} \tag{19}$$

and is bounded above by $1/n_1$, which is the limit as $\theta \uparrow 1$. Consequently, (13) always holds if

$$n_1 \geq \frac{N_1}{N_1\pi_0+1}. \tag{20}$$

For (14) we require

$$\sup_{0<\theta<1}\left\{\theta - 1 + \frac{\theta^{n_k}}{n_k\theta^{n_k-1}}\right\} = \frac{1}{n_k} \leq \frac{K_k+N_k}{N_k-n_k}\,\pi_0, \tag{21}$$

i.e.

$$n_k \geq \frac{N_k}{(K_k+N_k)\pi_0+1}, \tag{22}$$

As $K_1 = 0$, (20) is of the same form as (22). Hence if (22) is satisfied for all k, the expected outgoing quality at rejection will not exceed π_0. Klaassen (1996a, b) derived a similar result for a somewhat different definition of E_f and for the hypergeometric acceptance probability instead of the binomial probability in (17).

Example 2.1

Suppose that the required AOQL is $\pi_0 = 1.5\%$ and the first lot submitted in a sequence is of size $N = 201$. Sampling is by attributes and testing is nondestructive. Credit K is zero, so the appropriate sample size is found from (22) to be the smallest value of n such that

$$n \geq \frac{N}{(K+N)\pi_0+1} = \frac{201}{(0+201)0.015+1} = 50.06, \tag{23}$$

i.e. $n = 51$. Fifty-one items are selected at random from the lot and inspected. None are found to be nonconforming, so the credit is increased to $K = 201$.

Suppose that the second lot in the sequence is of size 192. From (22) it is seen that the second sample size must satisfy

$$n \geq \frac{N}{(K+N)\pi_0+1} = \frac{192}{(201+192)0.015+1} = 27.85, \qquad (24)$$

i.e. $n = 28$. A random sample of 28 items is accordingly selected from the second lot and inspected. One nonconforming item is found in the sample. The lot is therefore considered to be non-acceptable, and the credit K is reset to zero. As the credit was greater than zero when the sample size for this lot was determined, it is not necessary for the purpose of guaranteeing the AOQL to subject the lot to 100% inspection. By prior agreement between the supplier and the consumer, all conforming items found in the sample are accepted by the consumer, the remaining items being returned to the supplier with any nonconforming items found in the sample identified. Accordingly, the consumer accepts 27 items from the second lot, and returns the other 165 to the supplier with the nonconforming item identified.

If the sample from the second lot had yielded no nonconforming items, then the credit would have increased from $K = 201$ to $K = 201 + 192 = 393$.

3 Application to the variables acceptance criterion $X_{(1)} \geq L + c_n\sigma$ for σ known, μ unknown and $c_n \geq 0$

For sampling by variables, a sample of size n is taken and the sample values $X_1, \ldots, X_n$ are ordered so that $X_{(1)} \leq \ldots \leq X_{(n)}$. The lot is accepted if $X_{(1)} \geq L + c_n\sigma$. It is assumed that X is normally distributed with known standard deviation σ but with unknown mean μ.

Suppose that the parameter $\theta = \mu$. The fraction nonconforming is given by

$$q_\theta = \Phi\{(L-\theta)/\sigma\} \qquad (25)$$

where $\Phi(.)$ is the standard normal distribution function. The probability of acceptance is given by

$$p_\theta = \Phi^n\{-(L-\theta)/\sigma - c_n\}. \qquad (26)$$

To simplify the following algebra, without loss of generality we set $L = 0$ and $\sigma = 1$. Thus

$$q_\theta = \Phi(-\theta) \quad \text{and} \quad p_\theta = \Phi^n(\theta - c_n). \qquad (27)$$

The derivatives of these functions are

$$\dot{q}_\theta = -\phi(-\theta) \quad \text{and} \quad \dot{p}_\theta = n\Phi^{n-1}(\theta - c_n)\phi(\theta - c_n). \qquad (28)$$

Substituting in (13) it is found that the inequality for $k = 1$ is

$$\frac{\Phi(-\theta)\Phi^{n_1}(\theta - c_{n_1})}{1 - \Phi^{n_1}(\theta - c_{n_1})} \leq \frac{N_1\pi_0}{N_1 - n_1}. \qquad (29)$$

As θ tends to minus infinity, the left hand side of (29) clearly tends to zero. As θ tends to plus infinity, both the numerator and denominator tend to zero. Using l'Hôpital's rule, we find that the left hand side tends to

$$\frac{1}{n_1} \lim_{\theta \to \infty} \exp(\tfrac{1}{2}c_{n_1}^2 - \theta c_{n_1}) = 0. \tag{30}$$

A tedious argument shows that the left-hand side of (29) has a unique stationary point, namely a maximum. To find it we differentiate the left-hand side of (29) and equate the derivative to zero:

$$\begin{aligned} &\{1 - \Phi^{n_1}(\theta - c_{n_1})\}\{\Phi(-\theta) n_1 \ Phi^{n_1 - 1}(\theta - c_{n_1})\phi(\theta - c_{n_1}) \\ &\quad - \Phi^{n_1}(\theta - c_{n_1})\phi(-\theta)\} \\ &= \Phi(-\theta)\Phi^{n_1}(\theta - c_{n_1})(-n_1)\Phi^{n_1 - 1}(\theta - c_{n_1})\phi(\theta - c_{n_1}), \end{aligned} \tag{31}$$

which gives, on rearranging,

$$1 - \Phi^{n_1}(\theta - c_{n_1}) = \frac{n_1 \Phi(-\theta)\phi(\theta - c_{n_1})}{\Phi(\theta - c_{n_1})\phi(-\theta)}. \tag{32}$$

Denote the solution in θ to (32) by θ_1^* . Substituting in (29), we see that the required sample size n_1 must satisfy

$$\frac{1}{n_1} \exp(\tfrac{1}{2}c_{n_1}^2 - \theta_1^* c_{n_1})\Phi^{n_1 + 1}(\theta_1^* - c_{n_1}) \le \frac{N_1 \pi_0}{N_1 - n_1}. \tag{33}$$

We now consider the case where $k > 1$. Substituting (27) and (28) in (14) we find that n_k must satisfy

$$-\Phi(-\theta) + \frac{\phi(\theta)\Phi(\theta - c_{n_k})}{n_k \phi(\theta - c_{n_k})} \le \frac{K_k + N_k}{N_k - n_k} \pi_0, \quad \text{for } k > 1. \tag{34}$$

The left hand side of (34) can be simplified to

$$-\Phi(-\theta) + \frac{1}{n_k} \exp(\tfrac{1}{2}c_{n_k}^2 - \theta c_{n_k})\Phi(\theta - c_{n_k}). \tag{35}$$

Expression (35) tends to -1 as θ tends to minus infinity, and to zero as θ tends to infinity. For (34) to provide a meaningful constraint on n_k , it is evident that (35) must exceed zero for some intermediate value of θ . In fact, it may be shown that (35) has a unique maximum attained at the unique stationary point. The derivative of (35) is

$$\phi(\theta) + \frac{1}{n_k} \exp(\tfrac{1}{2}c_{n_k}^2 - \theta c_{n_k})\{\phi(\theta - c_{n_k}) - c_{n_k}\Phi(\theta - c_{n_k})\}. \tag{36}$$

Equating this to zero and rearranging, we see that the stationary value of (35) occurs when

$$\begin{aligned}\frac{1}{n_k}&\exp(\tfrac{1}{2}c_{n_k}^2 - \theta c_{n_k})\Phi(\theta - c_{n_k}) \\ &= \frac{1}{c_{n_k}}\{\phi(\theta) + \frac{1}{n_k}\exp(\tfrac{1}{2}c_{n_k}^2 - \theta c_{n_k})\phi(\theta - c_{n_k})\} \\ &= \frac{(n_k + 1)\phi(\theta)}{n_k c_{n_k}}.\end{aligned} \tag{37}$$

Substituting from (37) in (35), we see that at its stationary value (35) is equal to

$$-\Phi(-\theta) + \frac{(n_k + 1)\phi(\theta)}{n_k c_{n_k}}. \tag{38}$$

Denote the solution in θ to (37) by θ_k^{**}. Then from (34) and (38) it is seen that for $k > 1$ the required sample size n_k must satisfy

$$-\Phi(-\theta_k^{**}) + \frac{(n_k + 1)\phi(\theta_k^{**})}{n_k c_{n_k}} \leq \frac{K_k + N_k}{N_k - n_k}\,\pi_0. \tag{39}$$

Example 3.1

The required AOQL is $\pi_0 = 0.5\%$, credit K is 2215 and the distribution of the quality characteristic is known to be normal with process standard deviation $\sigma = 0.024$. A lower specification limit L is given as 0.75, and c has been chosen to be 0.25. The next lot is of size 100. The acceptance criterion for an accept-zero plan by variables is therefore

$$\text{Accept if } X_{(1)} \geq L + c\sigma\text{; reject otherwise,}$$

i.e.

$$\text{Accept if } X_{(1)} \geq 0.756\text{; reject otherwise.}$$

What sample size is required for an accept-zero plan by variables?

Before implementing a sampling by variables accept-zero scheme, it is advisable to prepare a table of the suprema of the left hand sides of (13) and (14) for the given value of c, for a range of values of n. For the present example, this means calculating the suprema of the left hand sides of (29) and (34) for $c = 0.25$. The calculated values are tabulated below for sample sizes 2 to 25.

As $K > 0$, we need to find the smallest value of n that satisfies (34) for all θ. As the sample size takes the value 2, 3, 4, 5, ... the right hand side of (34) takes the values 0.1181, 0.1193, 0.1206, 0.1218 ... Comparing these with corresponding elements of table 1, we see that $0.277726 > 0.1181$, $0.179446 > 0.1193$, $0.131489 > 0.1206$, but $0.103271 < 0.1218$. Hence the smallest sample size from the next lot that guarantees the AOQL for known process standard deviation is 5.

Table 1. Suprema of left hand sides of (29) and (34) for $c = 0.25$

n	$K = 0$	$K > 0$	n	$K = 0$	$K > 0$	n	$K = 0$	$K > 0$
2	0.270121	0.277726	10	0.047556	0.048732	18	0.025202	0.025795
3	0.174669	0.179446	11	0.042900	0.043952	19	0.023774	0.024331
4	0.128064	0.131489	12	0.039049	0.040000	20	0.022494	0.023019
5	0.100629	0.103271	13	0.035814	0.036680	21	0.021341	0.021837
6	0.082626	0.084761	14	0.033058	0.033853	22	0.020297	0.020766
7	0.069938	0.071722	15	0.030684	0.031418	23	0.019347	0.019793
8	0.060534	0.062060	16	0.028619	0.029299	24	0.018479	0.018904
9	0.053294	0.054624	17	0.026806	0.027440	25	0.017684	0.018089

4 Application to the variables acceptance criterion $X_{(1)} \geq L + c_n s$ for both σ and μ unknown and $c_n > 0$

Denoting the sample standard deviation by s, we now consider the problem of applying the acceptance criterion $X_{(1)} \geq L + c_n s$ to normally distributed data when both σ and μ are unknown; as before, c_n is assumed to be greater than zero.

The probability of acceptance is

$$\begin{aligned} p &= P(X_{(1)} \geq L + cs|\mu, \sigma) \\ &= P(X_{(1)} \geq \mu + \sigma\Phi^{-1}(q) + cs) \\ &= P\Big(\frac{X_{(1)} - \mu}{\sigma} \geq \Phi^{-1}(q) + cs/\sigma\Big) \\ &= P(Z_{(1)} \geq \Phi^{-1}(q) + cs'), \end{aligned} \tag{40}$$

where $Z_{(1)}$ is the smallest of n standard normal variates and s' is their sample standard deviation. It is evident that p is a single-valued function of the quality q and n, i.e. p is independent of the particular values of μ and σ that result in a fraction nonconforming of q. To take advantage of this, set $\theta = q$ in (13) and (14). The revised inequalities are

$$\frac{qp_q}{1 - p_q} \leq \frac{N_1}{N_1 - n_1}\pi_0 \quad \text{for } k = 1, K_1 = 0, \tag{41}$$

and

$$-q - \frac{p_q}{\dot{p}_q} \leq \frac{K_k + N_k}{N_k - n_k}\pi_0 \quad \text{for } k > 1. \tag{42}$$

Except for sample size 2, for which accurate results have been obtained by numerical integration, it is relatively difficult to determine the suprema of the left hand sides of (41) and (42). The table below gives the suprema for $c = 0.25$ and sample sizes from 2 to 16. For the sample sizes 3 to 16, the values were determined approximately by a search procedure over θ using 10 million simulations at each value.

Table 2. Approximate suprema of left hand sides of (41) and (42) for $c = 0.25$

n	$K=0$	$K>0$	n	$K=0$	$K>0$	n	$K=0$	$K>0$
2	0.26523	0.29466	7	0.065	0.070	12	0.037	0.038
3	0.166	0.182	8	0.057	0.060	13	0.034	0.035
4	0.121	0.131	9	0.050	0.053	14	0.031	0.032
5	0.095	0.102	10	0.045	0.047	15	0.029	0.030
6	0.077	0.083	11	0.041	0.042	16	0.027	0.028

5 The case where all rejected lots are subjected to 100% inspection

Consider the sampling strategy where all rejected lots are subjected to 100% inspection; conforming items found during sampling and 100% inspection are allowed into the market, but all nonconforming items found are returned to the producer. As before, credit is reset to zero whenever a lot is rejected, and a new sequence is started.

This strategy would be appropriate when the value of the workmanship in each item is large in comparison with the cost of testing it.

At time J within a sequence, the probability of a bad bargain resulting from that sequence, i.e. the proportion of nonconforming items entering the market from that sequence, equals

$$\frac{\sum_{j=1}^{J-1} D_j}{\left(\sum_{j=1}^{J} N_j\right) - D_J} \quad \text{for } J = 1, 2, \ldots. \tag{43}$$

where D_j is the number of nonconforming items in the j-th lot of the sequence. As 100% inspection is applied to only the *last* lot in the sequence, we denote the expectation of this by E_l. Note that this and the previous strategy both lead to a zero fraction nonconforming in the marketplace when $J = 1$, for which the two strategies are identical, and that for $J > 1$ we have

$$\frac{\sum_{j=1}^{J-1} D_j}{\left(\sum_{j=1}^{J} N_j\right) - D_J} \le \frac{\sum_{j=1}^{J-1} D_j}{\sum_{j=1}^{J-1} N_j}. \tag{44}$$

Hence

$$\begin{aligned} E_l &= \sum_{i=2}^{\infty} \left(\frac{\sum_{j=1}^{i-1} D_j}{\left(\sum_{j=1}^{i} N_j\right) - D_i} (\rho_{i-1} - \rho_i) \right) \\ &\le \sum_{i=2}^{\infty} \left(\frac{\sum_{j=1}^{i-1} D_j}{\sum_{j=1}^{i-1} N_j} (\rho_{i-1} - \rho_i) \right) = E_f. \end{aligned} \tag{45}$$

Thus the sampling strategy that guarantees that $E_f \leq \pi_0$ will also guarantee that $E_l \leq \pi_0$. Sampling plans for the former strategy may therefore also be used for the latter strategy.

References

1. ISO 2859 (1989) Sampling procedures for inspection by attributes. International Organization for Standardization
2. C.A.J. Klaassen (1996a) A Bonus-Malus system in statistical sampling for testing jewelry on fineness. In Proc. IPMI Seminar on Precious Metal Sampling and Analysis, San Antonio, Texas. International Precious Metals Institute, Allentown, Pensylvania
3. C.A.J. Klaassen (1996b) Bonus-Malus in acceptance sampling on attributes. Technical report, submitted to Technometrics

Acceptance Sampling Plans by Attributes with Fuzzy Risks and Quality Levels

Przemysław Grzegorzewski[1,2]

[1] Systems Research Institute, Polish Academy of Sciences, Warsaw, Poland
[2] Institute of Mathematics, Warsaw University of Technology, Warsaw, Poland

Abstract. A method for designing single sampling plans by attributes with relaxed producer's and consumer's risks and quality levels is suggested. Since small deviations of the parameter (risk, quality level) from its target value are very often of no importance this soft method for designing sampling plans is more flexible than the conventional one.

1 Introduction

A single acceptance sample plan by attributes is an ordered pair (n, c), where n is a sample size and c is an acceptance number. This plan depends on four parameters: producer's quality level θ_1, consumer's quality level θ_2, producer's risk α and consumer's risk β. Various economic and technological factors must be taken into account in defining these parameters, which makes it difficult for producers and consumers to uniquely specify their quality levels and risks. Since n and c must be integers it may happen that a desired plan does not exist or very large samples are needlessly required if the producer and consumer's standpoints are too rigid.

To relax this rigidity of the conventional design plans with relaxed risks and quality levels are considered. Single sampling plans by attributes with relaxed risks were first discussed by Ohta and Ichihashi (1988) and Kanagawa and Ohta (1990). They have reformulated plan designing problem as a fuzzy mathematical programming one. Then Tamaki, Kanagawa and Ohta (1991) tried to exercise the modal theory of Dubois and Prade for designing plans with relaxed θ_1, θ_2, α and β.

In the present paper we propose a new method of designing sampling plans by attributes. This method is based on the generalized concept of probabilities of type I and type II errors for fuzzy hypothesis testing problem, suggested by Arnold (1996). We consider situations with relaxed both risks and quality levels. This method is more flexible than conventional methods, since small deviations of the requirement (risk or quality level) from its target value are very often of no importance. By relaxing risks and quality levels one can reduce a sample size. The trade-off problem between the reduction of sample size and the relaxation of the requirements on risks and quality was solved by the *maximum satisfaction* optimality criterion.

2 Sampling plans by attributes

Suppose that a lot of size N has been submitted for inspection. We observe a random sample of n items $X_1, \ldots, X_n$ taken from the lot described as follows

$$X_i = \begin{cases} 0 & \text{if the item is conforming} \\ 1 & \text{if the item is nonconforming} \end{cases}, \quad i = 1, \ldots, n. \tag{1}$$

The observed total number of nonconforming items in the sample is

$$d = \sum_{i=1}^{n} X_i. \tag{2}$$

Let c denote a critical number, called the acceptance number. The decision criteria are as follows

- if $d \leq c$ then accept the lot
- if $d > c$ then reject the lot. (3)

Thus any single acceptance sampling plan by attributes may be described by an ordered pair (n, c). Acceptance sampling doesn't reduce to single-sampling only. There are double-sampling, multiple-sampling and sequential-sampling plans, sampling schemes and other techniques. For more details concerning acceptance sampling we refer the reader to Schilling (1982). However, further on by a sampling plan (or a plan, for short) we will understand a single acceptance sampling plan by attributes.

The procedure of a single-sampling plan by attributes is equivalent to a hypothesis testing for the critical quality level θ_0. Namely, we consider a random sample $X_1, \ldots, X_n$ from the hypergeometric distribution, i.e. $X_i \sim Hy(N, D, n)$, where D denotes the number of nonconforming items in the lot. If the ratio $\frac{n}{N}$ is small (say, $\frac{n}{N} \leq 0.1$) then we can use the binomial approximation, i.e. $X_i \sim Bin(n, \theta)$, where $\theta = \frac{D}{N}$ is the proportion of nonconforming items (also called, fraction nonconforming). Thus our sampling plan is equivalent to testing hypothesis $H : \theta \leq \theta_0$ against the alternative hypothesis $K : \theta > \theta_0$.

As we have mentioned above such a test/plan may be described by an ordered pair (n, c). The problem of designing a plan is to find such two numbers n and c that certain requirements concerning that plan are fulfilled. Traditionally we specify four parameters: producer's quality level θ_1, consumer's quality level θ_2, producer's risk α and consumer's risk β (where $\beta < 1 - \alpha$). Then assuming that binomial sampling is appropriate the sample size n and the acceptance number c ($c \leq n$) are the solutions of

$$\begin{cases} P(\theta_1) = 1 - \alpha \\ P(\theta_2) = \beta, \end{cases} \tag{4}$$

where $P(\theta)$ denotes the probability that a lot of quality θ (i.e. a lot with fraction θ nonconforming) will be accepted.

Usually (4) cannot be realized since both n and c must be integers and a desired plan does not exist. Therefore a following requirements are adopted instead of (4):

$$\begin{cases} P(\theta_1) \geq 1 - \alpha \\ P(\theta_2) \leq \beta. \end{cases} \tag{5}$$

There are often many plans (n, c) which satisfy (5). Hence we need a criterion for choosing optimal sampling plan. The most popular optimality criterion is that based on minimizing the sample size. We say that a plan (n^*, c^*) satisfying (5) is $n-$*optimal* if $n^* \leq n$ for all plans (n, c) satisfying (5). The problem of determining optimal sampling plans was considered by Guenther (1973), Hald (1967, 1977, 1981), Jeach (1980) and Stephens (1978).

3 Sampling plans with relaxed risks

Applying (5) we always get a solution, however very large samples are often required if the producer and consumer's standpoints concerning risks and quality levels are too rigid. Fujino and Okuno (1981) reformulated the problem of designing sampling plans. They called such a plan (n, c) optimal if it minimizes

$$\max \left\{ \frac{|1 - \alpha - P(\theta_1)|}{\alpha}, \frac{|P(\theta_2) - \beta|}{\beta} \right\}. \tag{6}$$

Adopting their minimax criterion we relax the conditions on quality levels and risks, however the sample size can be reduced. In practice it is not essential that risks and quality levels are exact - it suffice that they are close, in some sense, to the desired values. Thus we face two important problems:

(a) how to describe these relax conditions on risks and quality levels?
(b) how to design the optimal plan for relaxed risks and quality levels?

Single sampling plans by attributes with relaxed risks were first discussed by Ohta and Ichihashi (1988). They have reformulated plan designing problem as a fuzzy mathematical programming one. Ohta and Ichihashi considered generalized conditions (4)

$$\begin{cases} P(\theta_1) \cong 1 - \alpha \\ P(\theta_2) \cong \beta, \end{cases} \tag{7}$$

where the symbol $\cong$ stands for an approximate relation treated as fuzzy equality associated with fuzzy numbers corresponding to expressions of the type: "about α", "approximately between α_1 and α_2", etc.

Let $\mu_A(\alpha)$ and $\mu_B(\beta)$ describe a grade of satisfaction with a sampling plan with actual producer's risk α and actual consumer's risk β, respectively. Thus, more precisely, we can consider two fuzzy sets A and B with membership functions $\mu_A, \mu_B : [0, 1] \to [0, 1]$, respectively, for modeling risks, where

1 represents full satisfaction while 0 corresponds to the complete lack of satisfaction. Examples of producer's risk are given in Figure 1. Function given in Figure 1a describes risk corresponding to the statement: "about α", while function given in Figure 1b describes risk corresponding to the statement: "rather smaller than α but surely not greater than α_2" or "approximately smaller than α". Note that function given in Figure 1c corresponds to the traditional model (4): producer's risk is equal to α, while function given in Figure 1d corresponds to the model (5): the risk is smaller than α.

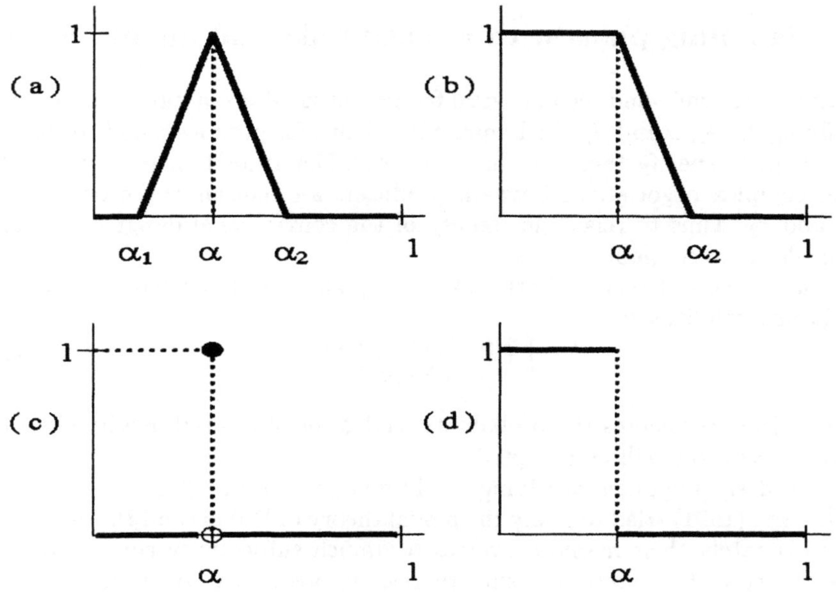

Fig. 1. Examples of membership functions corresponding to producers' risk

Now the problem of designing optimal sampling plan satisfying (7) reduces to finding such (n, c) that maximize

$$\min_{\gamma \in [0,1]} \{\mu_A(\gamma), \mu_B(\gamma)\} . \tag{8}$$

In a special case of

$$\mu_A(\gamma) = \begin{cases} \frac{\gamma}{\alpha} & \text{if } \gamma \in [0, \alpha) \\ 2 - \frac{\gamma}{\alpha} & \text{if } \gamma \in [\alpha, 2\alpha] \\ 0 & \text{if } \gamma \in (2\alpha, 1] \end{cases} \qquad \mu_B(\gamma) = \begin{cases} \frac{\gamma}{\beta} & \text{if } \gamma \in [0, \beta) \\ 2 - \frac{\gamma}{\beta} & \text{if } \gamma \in [\beta, 2\beta] \\ 0 & \text{if } \gamma \in (2\beta, 1] \end{cases} \tag{9}$$

problem (7) and (8) defined by Ohta and Ichihashi reduces to problem (6) suggested by Fujino and Okuno.

Recently Kanagawa and Ohta (1990) considered generalized conditions (5)

$$\begin{cases} P(\theta_1) \approx\geq 1-\alpha \\ P(\theta_2) \approx\leq \beta, \end{cases} \tag{10}$$

where the symbols $\approx\geq$ and $\approx\leq$ stand for approximate inequalities treated as fuzzy inequalities associated with fuzzy numbers corresponding to expressions of the type: "rather greater than α", "slightly less than β", etc. (see Figure 1b).

4 Sampling plans with relaxed risks and quality levels

Various economic and technological factors must be taken into account while defining θ_1, θ_2, α and β, which makes it difficult for producers and consumers to uniquely specify these four main factors. Determination involves careful and complex negotiations between producers and consumers, especially for θ_1 and θ_2. Thus to relax the rigidity of the conventional design let us also consider relaxed quality levels.

In the case of relaxed both risks and quality levels we have to consider following conditions:

$$\begin{cases} P(\sim\theta_1) \approx\geq 1-\alpha \\ P(\sim\theta_2) \approx\leq \beta, \end{cases} \tag{11}$$

where $P(\sim\theta)$ denotes the probability that a lot of relaxed quality close (in some sense) to θ will be accepted.

For designing plans satisfying conditions given above Tamaki, Kanagawa and Ohta (1991) tried to apply the modal theory of Dubois and Prade (1983). Unfortunately their method involves too much subjectivity connected with the choice of the modal measure (necessity, possibility, etc.) and determination of additional limits for these measures. Here we propose a different method which utilize Arnold's approach to fuzzy hypotheses testing.

Suppose that we consider fuzzy null hypothesis H against fuzzy alternative K described by their membership functions $\mu_H : \Theta \to [0,1]$ and $\mu_K : \Theta \to [0,1]$, respectively. These membership functions give the degree to which θ belongs to each hypothesis. In particular, $\mu_H(\theta) = 1$ means that θ completely belongs to H, whereas $\mu_H(\theta) = 0$ indicates that θ is not at all a member of H. Examples of such fuzzy hypotheses are given in Figures 2 and 3.

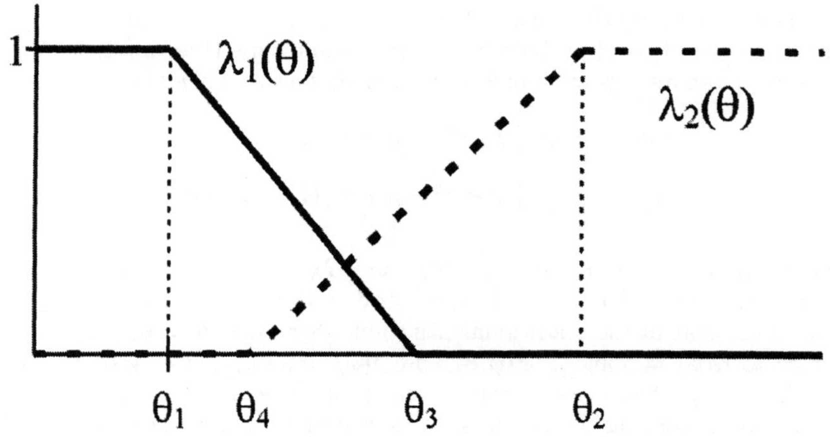

Fig. 2. Membership functions of fuzzy hypotheses $H : \mu_H(\theta) = \lambda_1(\theta)$ against $K : \mu_K(\theta) = \lambda_2(\theta)$

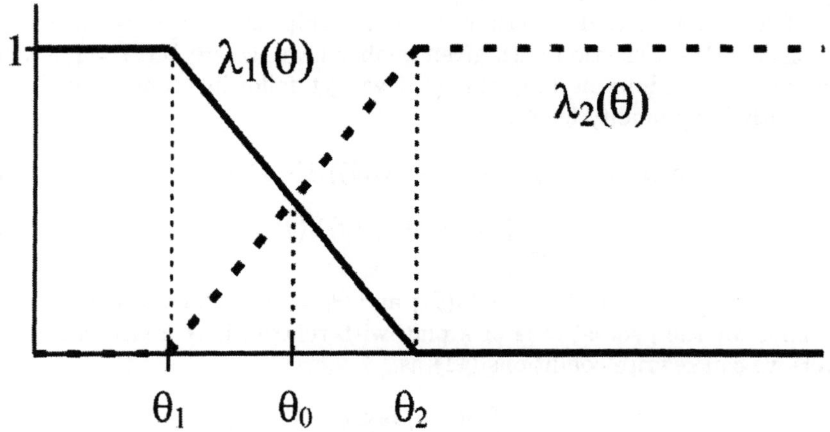

Fig. 3. Membership functions of fuzzy hypotheses $H : \mu_H(\theta) = \lambda_1(\theta)$ against $K : \neg H$, i.e. $\mu_K(\theta) = 1 - \mu_H(\theta) = 1 - \lambda_1(\theta) = \lambda_2(\theta) \quad \forall \theta \in \Theta$

In order to verify such fuzzily formulated hypotheses Arnold (1996) generalized the definition of probabilities of type I and type II errors. Let $M(\phi, \cdot) : \Theta \to [0,1]$ denote a power function of a given test ϕ. Then the quantities

$$\alpha(\phi) = \sup_{\theta \in \Theta_H} \{(\mu_H(\theta) - \mu_K(\theta))M(\phi, \theta)\} \tag{12}$$

$$\beta(\phi) = \sup_{\theta \in \Theta_K} \{(\mu_K(\theta) - \mu_H(\theta))(1 - M(\phi, \theta))\}, \tag{13}$$

where $\Theta_H = \{\theta \in \Theta : \mu_H(\theta) > \mu_K(\theta)\}$ and $\Theta_K = \{\theta \in \Theta : \mu_H(\theta) < \mu_K(\theta)\}$, are generalized probabilities of type I and type II errors, respectively. It is easily seen that in the traditional situation where $H \cap K = \emptyset$, (12) and (13) reduce to $\alpha(\phi) = \sup_{\theta \in H} M(\phi, \theta)$ and $\beta(\phi) = \sup_{\theta \in K}(1 - M(\phi, \theta))$ which are classical probabilities of type I and type II errors, respectively. Hence proceeding according to the classical approach to hypotheses testing a test ϕ is said to be a test on significance level α ($\alpha \in [0,1]$) if the inequality $\alpha(\phi) \leq \alpha$ holds.

Now let us go back to designing sampling plans. Suppose we have two functions $\lambda_1, \lambda_2 : [0,1] \to [0,1]$, where $\lambda_1(\theta_1)$ describes a grade of satisfaction with a sampling plan with actual producer's quality level θ_1 while $\lambda_2(\theta_2)$ is a grade of dissatisfaction with a plan with actual consumer's quality level θ_2. Of course, we can also consider a membership function $\widetilde{\lambda}_2$ corresponding to consumer's satisfaction with given quality level, where $\widetilde{\lambda}_2(\theta) = 1 - \lambda_2(\theta)$. According to definitions (12) and (13) we get following conditions for the plan with fuzzy quality levels:

$$\alpha(n, c) = \sup_{\theta \in \Theta_1} \{(\lambda_1(\theta) - \lambda_2(\theta))(1 - P(\theta))\} \leq \alpha \tag{14}$$

$$\beta(n, c) = \sup_{\theta \in \Theta_2} \{(\lambda_2(\theta) - \lambda_1(\theta))P(\theta)\} \leq \beta, \tag{15}$$

where $\Theta_1 = \{\theta \in [0,1] : \lambda_1(\theta) > \lambda_2(\theta)\}$ and $\Theta_2 = \{\theta \in [0,1] : \lambda_1(\theta) < \lambda_2(\theta)\}$.

Thus for the general case of a plan with relaxed both risks and quality levels we can rewrite conditions (11) as

$$\begin{cases} \alpha(n, c) \approx\leq \alpha \\ \beta(n, c) \approx\leq \beta. \end{cases} \tag{16}$$

Since we may get many plans satisfying (16) we need an optimality criterion for choosing the best plan. We suggest to use for that purpose a following function

$$S(n, c) = \min \{\mu_A(\alpha(n, c)), \mu_B(\beta(n, c)), \xi(n)\}, \tag{17}$$

where $\xi : N \to [0,1]$. Here, as before, $\mu_A(\alpha)$ and $\mu_B(\beta)$ describe a grade of satisfaction with a sampling plan (n, c) with producer's risk $\alpha(n, c)$ and consumer's risk $\beta(n, c)$, while $\xi(n)$ describes a grade of satisfaction with a sample size n. This function, called a *satisfaction function*, seems to be more appropriate than (8) because it takes into account not only risks but a sample size as well. Such a function was also used by Kanagawa and Ohta (1990). Hence now our optimality criterion is based on maximizing satisfaction function $S(n,c)$. We say that a plan (n^*,c^*) satisfying (16) is *$MS-$optimal* if $S(n^*,c^*) \geq S(n,c)$ for all plans (n,c) satisfying (16).

Theoretically one may choose any functions μ_A, μ_B and ξ. However, in practice, these functions should fulfill a few regularity conditions. For example it is clear that both producer and consumer would be satisfied with a plan with no risk and their grades of satisfaction would decrease if their risks α and β increase. On the contrary the grade of producer's satisfaction $\lambda_1(\theta)$ would decrease while the grade of consumer's dissatisfaction $\lambda_2(\theta)$ would increase as θ increases. The grade of satisfaction with a sample size would decrease as n increases. Moreover, there exist such an integer n_0 that the inspection with a sample of size greater than $n_0$would be unprofitable.

One may ask whether for given functions μ_A, μ_B, λ_1, λ_2 and ξ (even regular) $MS-$optimal sampling plan exist. The answer is positive.

Proposition 1. *Let μ_A, μ_B, $\lambda_1 : [0,1] \to [0,1]$ denote nonincreasing functions such that $\mu_A(0) = \mu_B(0) = \lambda_1(0) = 1$ and $\mu_A(1) = \mu_B(1) = \lambda_1(1) = 0$. Let $\lambda_2 : [0,1] \to [0,1]$ be a nondecreasing function such that $\lambda_2(0) = 0$ and $\lambda_2(1) = 1$. Suppose $\xi : N \to [0,1]$ is a nonincreasing function such that $\xi(1) = 1$ and exists a n_0 such that $\xi(n_0) = 0$. Then there exist $MS-$optimal single sampling plan by attributes satisfying (16).*

Below, instead of an easy proof, we will show a practical algorithm for designing a desired plan.

Algorithm for designing $MS-$optimal single-sampling plan by attributes

STEP 0: Initialize n $(n = 1, 2, \ldots)$ and $S(n^*,c^*) = 0$.
STEP 1: Set an acceptance number $c := 0$.
STEP 2: Find

$$\begin{cases} \alpha(n,c) = \sup\limits_{\theta \in \Theta_1} \left\{ (\lambda_1(\theta) - \lambda_2(\theta)) \left(1 - \sum\limits_{k=0}^{c} \binom{n}{k} \theta^k (1-\theta)^{n-k} \right) \right\} \\ \beta(n,c) = \sup\limits_{\theta \in \Theta_2} \left\{ (\lambda_2(\theta) - \lambda_1(\theta)) \sum\limits_{k=0}^{c} \binom{n}{k} \theta^k (1-\theta)^{n-k} \right\} \end{cases} \tag{18}$$

STEP 3: Find

$$S(n,c) = \min \left\{ \mu_A(\alpha(n,c)), \mu_B(\beta(n,c)), \xi(n) \right\}.$$

STEP 4: If $S(n,c) > S(n^*,c^*)$ then $(n^*,c^*) := (n,c)$.
STEP 5: If $c < n$ then $c := c+1$; goto STEP 2.
STEP 6: If $n < n_0$ then $n := n+1$; goto STEP 1.
STEP 7: Stop.

Remark 1 We'll have, at worst, $\frac{1}{2}(1+n_0)n_0$ iterations.

Remark 2 If we consider hypotheses $H : \mu_H(\theta) = \lambda_1(\theta)$ against $K : \neg H$, i.e. $\mu_K(\theta) = 1 - \mu_H(\theta) = 1 - \lambda_1(\theta) = \lambda_2(\theta) \quad \forall\theta \in [0,1]$ (see Figure 3), where $\theta_2 = \sup\{\mathrm{supp}(\lambda_1)\}$, $\theta_1 = \inf\{\mathrm{supp}(\lambda_2)\}$ and $\lambda_1(\theta_0) = \lambda_2(\theta_0) = 0.5$ then (18) reduce to

$$\begin{cases} \alpha(n,c) = \sup\limits_{\theta_1 \le \theta < \theta_0} \left\{ \frac{\theta_0-\theta}{\theta_0-\theta_1} \left(1 - \sum\limits_{k=0}^{c} \binom{n}{k} \theta^k (1-\theta)^{n-k} \right) \right\} \\ \beta(n,c) = \sup\limits_{\theta_0 < \theta \le \theta_2} \left\{ \frac{\theta-\theta_0}{\theta_2-\theta_0} \sum\limits_{k=0}^{c} \binom{n}{k} \theta^k (1-\theta)^{n-k} \right\}. \end{cases} \tag{19}$$

5 Numerical example

Suppose we are interested in designing a sampling plan by attributes with following parameters: producer's risk $\alpha = 5\%$, consumer's risk $\beta = 10\%$, producer's quality level 2% and consumer's quality level 9%. It can be found easily that the *n-optimal* plan satisfying (5) is: $n = 87$ and $c = 4$ (with actual risks 0.0307 and 0.09884, respectively).

Now let us admit a slight relaxation of plan parameters and consider a sampling plan by attributes with a following risks: "α rather smaller than 5% but surely not greater than 8%" and "β rather smaller than 10% but surely not greater than 20%". To model this fuzzy risks we use membership functions as in Fig. 1b.

Moreover, let $\xi(n)$ - describing a grade of satisfaction with a sample size n - have a following form:

$$\xi(n) = \begin{cases} 1 & \text{if} \quad n \le n_1 \\ 1 - \sqrt{\frac{n-n_1}{n_0-n_1}} & \text{if} \quad n_1 < n \le n_0 \\ 0 & \text{if} \quad n > n_0, \end{cases} \tag{20}$$

where $n_1 < n_0$. In our study we consider functions with n_1 equal to 50 or 30 and n_0 equal to 300, 200 and 100.

To model fuzzy quality levels we use membership functions as in Fig. 2 with different θ_i, $i = 1,2,3,4$ chosen in such a way that they correspond to relaxed levels mentioned above. Thus, e.g., a membership function with $\theta_1 = 0.02$ and $\theta_3 = 0.025$ one may interpret as: "producer's quality level is rather smaller than 2% but surely not greater than 2,5%", while a membership function with $\theta_4 = 0.09$ and $\theta_2 = 0.095$ corresponds to situation where the consumer is satisfied with a plan with quality level smaller than 9% but he is completely dissatisfied with a plan with quality level 9,5%. Then the $MS-$optimal single sampling plan by attributes satisfying (16) is $n = 69$ and $c = 3$ (for $n_1 = 50$ and $n_0 = 300$, 200 and 100). If we consider other relaxation of quality levels, i.e. $\theta_1 = 0.02$, $\theta_3 = 0.03$ and $\theta_4 = 0.09$ and $\theta_2 = 0.1$ then we get $n = 65$ and $c = 3$ (also for $n_1 = 50$ and $n_0 = 300, 200$ and 100). For other function $\xi(n)$, i.e. for $n_1 = 30$ and $n_0 = 100$, we obtain $n = 64$, $c = 3$ and $n = 61$, $c = 3$, respectively.

6 Conclusions

In the present paper we have proposed a method for designing single sampling plans by attributes with fuzzy producer's and consumer's risks and quality levels. Since small deviations of the parameter (risk, quality level) from its target value are very often of no importance our soft method for designing sampling plans is more flexible than the conventional one. By relaxing risks and quality levels one can reduce sample size. The trade-off problem between the reduction of sample size and the relaxation of the requirements on risks and quality levels was solved by, so called, $MS-$optimality criterion taking into account the grade of satisfaction with the plan of given risks and sample size. Our soft method for designing sampling plans is also general: it may be applied for determining more complicated acceptance sampling plans by attributes (e.g. double-sampling plans, multiple-sampling plans) as well as for designing acceptance plans by variables.

References

1. Arnold B. F. (1996) An Approach to Fuzzy Hypothesis Testing. Metrika **44**, 119–126
2. Dubois D., Prade H. (1983) Ranking Fuzzy Numbers in the Setting of Possibility Theory. Inform. Sci. **30**, 183–224
3. Fujino Y., Okuno T. (1981) On Single Sampling Plans by Attributes with Given Producer's and Consumer's Risk. J. Jap. Soc. Quality Control **11**, 169–177
4. Guenther W. C. (1973) A Sample Size Formula for the Hypergeometric. J. Quality Technology **5**, 167–170
5. Hald A. (1967) The Determination of Single Sampling Attribute Plans with Given Producer's and Consumer's Risk. Technometrics **9**, 401–415
6. Hald A. (1977) A Note on the Determination of Single Sampling Attribute Plans of Given Strength. Technometrics **19**, 211–212
7. Hald A. (1981) Statistical Theory of Sampling Inspection by Attributes. University Press, London
8. Jeach J. L. (1980) Determination of Acceptance Numbers and Sample Size for Attribute Sampling Plans. J. Quality Technology **12**, 187–190
9. Kanagawa A., Ohta H. (1990) A Design for Single Sampling Attribute Plan Based on Fuzzy Sets Theory. Fuzzy Sets and Systems **37**, 173–181
10. Ohta H., Ichihashi H. (1988) Determination of Single–Sampling Attribute Plans Based on Membership Functions. Int. J. Prod. Res. **26**, 1477-1485
11. Schilling E. G. (1982), Acceptance Sampling in Quality Control. Dekker, New York

12. Stephens L. (1978) A Closed Form Solution for Single Sample Acceptance Sampling Plans. J. Quality Technology **10**, 159–163
13. Tamaki F., Kanagawa A., Ohta H. (1991) A Fuzzy Design of Sampling Inspection Plans by Attributes. Japanese Journal of Fuzzy Theory and Systems **3**, 315–327

Acceptance Sampling by Variables under Measurement Uncertainty

Henrik Melgaard[1] and Poul Thyregod[2]

[1] Novo Nordisk A/S, Gentofte, Denmark
[2] Department of Mathematical Modelling, Technical University of Denmark, Lyngby, Denmark

Abstract. The paper discusses the influence from measurement uncertainty in case of acceptance sampling by variables. Two kinds of measurement uncertainty are considered in this paper, one is the pure random noise changing from one item to the other. The other type of measurement error considered is a constant error on all items in a lot, but where the error varies from lot to lot. The largest deterioration on the operating characteristics is seen from the low frequency measurement error type.

1 Introduction

It is a general idea - at least among statisticians - that sampling by variables is a more efficient procedure for acceptance sampling, than sampling by attributes. This idea is supported by the fact that under a correct model, a parametric estimator of the fraction nonconforming is more efficient than the nonparametric estimator based upon the crude count of nonconforming items in the sample.

However, in recent years various arguments have been suggested questioning the superiority of acceptance sampling plans by variables. v.Collani [7] has indicated some pitfalls when the theory is used as a basis for sentencing individual lots.

Other authors [6] have discussed the non-robustness towards the distributional assumptions.

In industrial praxis, an other important issue is the presence of measurement errors. Therefore, in the present paper, we shall discuss the implications of using acceptance sampling by variables under measurement error.

The theory for acceptance sampling by variables under a normal distribution dates back to the papers by Lieberman and Resnikoff [4] and [3] . The current ISO standard with sampling plans, ISO 3951, [5], is based upon this original idea of basing the test upon a minimum variance, unbiased estimate of the fraction nonconforming in the process. A description of the theory may be found in [8].

Other authors [2], [1] have suggested other principles for approaching the statistical testing problems. In the present paper we shall, however, limit ourselves to the plans provided in ISO 3951. The principal results are valid for likelihood-ratio based tests as well as for the tests based upon the minimum variance unbiased estimate.

In [9], Owen and Chou, consider the effect of measurement error and a constant offset error on the OC curves of the one-sided plans. In the present paper, however, these considerations are extended to the double-sided plans in the case of measurement error and a random 'laboratory' bias.

2 OC curves

2.1 No measurement error

We shall assume that items in a lot are produced from a process in control. Let X denote the quantity of interest. We shall assume that the distribution of X over the items in a lot may be described by i.i.d random variables that follow a normal distribution with mean $\mathsf{E}[X] = \mu$ and $\mathsf{V}[X] = \sigma^2$. Under the assumption of simple random sampling from the lot, it follows that the values $X_1, X_2, \ldots, X_n$ of the quantity of interest in the sample likewise are i.i.d and normally distributed with the same distribution as X.

Let the specification limits for X be U and L, respectively. Then the expected fraction of nonconforming items is given by

$$p \equiv \Phi\Big(\frac{L-\mu}{\sigma}\Big) + \Phi\Big(\frac{\mu - U}{\sigma}\Big) \tag{1}$$

Without loss of generality we may choose the limits to be $L = -1$ and $U = 1$, thus

$$p \equiv \Phi\Big(\frac{-1-\mu}{\sigma}\Big) + \Phi\Big(\frac{\mu - 1}{\sigma}\Big) \tag{2}$$

2.2 Independent measurement error acting on all items

Assume that the quantity of interest cannot be measured without measurement error.

Let $Y = Y(X)$ denote the measurement result for an item with value X. Assume that

$$Y_i = X_i + E_i \tag{3}$$

where $E_1, \ldots, E_n$ are i.i.d and normally distributed with $\mathsf{E}[E_i] = 0$ and $\mathsf{V}[E_i] = \sigma_e^2$.

Let as usual

$$\overline{Y} = \sum Y_i/n \tag{4}$$

and

$$s_y^2 = \sum (Y_i - \overline{Y})^2 \Big/ (n-1) \tag{5}$$

Then $\overline{Y}$ follows a normal distribution with $\mathsf{E}[\overline{Y}] = \mu$ and $\mathsf{V}[\overline{Y}] = (\sigma^2 + \sigma_e^2)/n$ and s_y^2/σ^2 follows a $(1 + \sigma_e^2/\sigma^2)\chi^2(f)/f$ distribution with $f = n - 1$ and $\overline{Y}$ and s_y^2 are independent.

2.3 Same measurement error on all items

Assume instead that the major part of the measurement error is a random "laboratory" or instrument bias that affects all measurements in the same manner, i.e.

$$Y_i = X_i + B \tag{6}$$

where B varies from lot to lot as i.i.d and normally distributed variables with $\mathsf{E}[B] = 0$ and $\mathsf{V}[B] = \sigma_B^2$.

Under this assumption we have that

$\overline{Y}$ follows a normal distribution with $\mathsf{E}[\overline{Y}] = \mu$ and $\mathsf{V}[\overline{Y}] = \sigma_B^2 + \sigma^2/n$ and s_y^2/σ^2 follows a $\chi^2(f)/f$ distribution with $f = n-1$ and $\overline{Y}$ and s_y^2 are independent.

Thus, in this case the uncertainty of the estimate, $\overline{Y}$, of the position, is heavily influenced by the measurement uncertainty, but the estimate, s_y^2 of the process spread is not affected.

However, unless σ_B^2 is known, the estimate s_y^2 does not provide enough information about the uncertainty (sampling error) of $\overline{Y}$.

2.4 Both types of measurement error

In practice, both types of measurement error will often be present. Thus,

$$Y_i = X_i + B + E_i \tag{7}$$

where the assumptions on B and E_i are as before.

In this case $\overline{Y}$ follows a normal distribution with $\mathsf{E}[\overline{Y}] = \mu$ and $\mathsf{V}[\overline{Y}] = \sigma_B^2 + (\sigma^2 + \sigma_e^2)/n$ and s_y^2/σ^2 follows a $(1 + \sigma_e^2/\sigma^2)\chi^2(f)/f$ distribution with $f = n-1$ and $\overline{Y}$ and s_y^2 are independent. Thus, as well the uncertainty of the estimate, $\overline{Y}$, of the position, as the estimate s_y^2 of the process spread is affected.

The sampling plans considered in this study are the ones from ISO 3951, based on the work of Lieberman and Resnikoff [4]. From the given choice of (n, k) we are able to find the critical p

$$p_c = B_{n/2-1}\left(\frac{1}{2}\left(1 - k\frac{\sqrt{n}}{n-1}\right)\right) \tag{8}$$

where B_r denote the distribution function of the symmetrical beta distribution with parameter r. The maximum value of the sample standard deviation is found from

$$s_{y,max} = \frac{\sqrt{n}}{n-1}/(1 - 2B_{n/2-1}^{-1}(p_c/2)) \tag{9}$$

The OC-curve is determined by numerical integration using the IMSL-subroutine DQDAGS. Although the area of integration is a region in the $(\bar{Y}, s_y)$ plane, the integral is reduced to a one-dimensional integral of the function

F over the interval of admissible values of the χ^2 distributed variable $t = (n-1)s_y^2/(\sigma^2+\sigma_e^2)$. The OC-curve is given by

$$OC_{\mu,\sigma} = \int_0^{t_m} (\Phi(\frac{\mu+\bar{Y}_{max}(t)}{\sqrt{\sigma_B^2+(\sigma^2+\sigma_e^2)/n}}) - \Phi(\frac{\mu-\bar{Y}_{max}(t)}{\sqrt{\sigma_B^2+(\sigma^2+\sigma_e^2)/n}})g_{n-1}(t)dt \tag{10}$$

where $g_{n-1}(t)$ is the χ^2_{n-1} density function with $n-1$ degrees of freedom.

3 Approximate OC curves

For practical purposes it is useful to have the OC curves, coresponding to the one-sided plans, which are similar to the ones given in ISO 3951. These will be close to the double specification limits case, when $\sigma \ll \sigma_{max}$. For the one-sided case without measurement errors we have that the lot is accepted, i.e. $\hat{p} \leq p_c$ if and only if $\bar{Y} + ks_y \leq U$, where k is given by the relation $k = (n-1)(1-2B^{-1}_{n/2-1}(p_c))/\sqrt{n}$, therefore

$$\begin{aligned} OC_0(p) &= P\{\hat{p} \leq p_c\} = P\{\bar{Y} + ks_y \leq U\} \\ &= P\left\{\frac{U-\bar{Y}}{s_y} \geq k\right\} \\ &= P\left\{\frac{\frac{\sqrt{n}(U-\mu)}{\sigma} - \frac{\sqrt{n}(\bar{Y}-\mu)}{\sigma}}{s_y/\sigma} \geq k\sqrt{n}\right\} \\ &= P\{t_{n-1,\delta_0} \geq k\sqrt{n}\} \end{aligned} \tag{11}$$

where $\delta_0 = -\sqrt{n}z_p$ and $t_{\nu,\delta}$ is the non-central t distribution with ν degrees of freedom and noncentrality parameter δ, and $z_p = \Phi^{-1}(p)$, see [4].

By using the same sampling plan from ISO 3951, but in this case random measurement error is present, according to the model given in Section 2.2, the OC curve is given by

$$\begin{aligned} OC_1(p) &= P\left\{\frac{U-\bar{Y}}{s_y} \geq k\right\} \\ &= P\left\{\frac{\frac{\sqrt{n}(U-\mu)}{\sigma\sqrt{1+\sigma_e^2/\sigma^2}} - \frac{\sqrt{n}(\bar{Y}-\mu)}{\sigma\sqrt{1+\sigma_e^2/\sigma^2}}}{s_y/(\sigma\sqrt{1+\sigma_e^2/\sigma^2})} \geq k\sqrt{n}\right\} \\ &= P\{t_{n-1,\delta_1} \geq k\sqrt{n}\} \end{aligned} \tag{12}$$

where $\delta_1 = -\sqrt{n}z_p/\sqrt{1+\sigma_e^2/\sigma^2}$. One thing to remark is that for $\sigma_e/\sigma =$ constant the OC curve is not dependent of σ. Another thing to remark is that the OC curve in case of random measurement noise is always to the left of the 'noise free' OC curve. The change of the OC curve may be expressed as a monotonic transformation of the p-axes. If the proportion of defective in the noise-free case is called p_*, we have that $z_{p_*} = z_p/\sqrt{1+\sigma_e^2/\sigma^2}$ and hence the p in case of random noise is given by

$$p = \Phi(\Phi^{-1}(p_*)\sqrt{1+\sigma_e^2/\sigma^2}) \tag{13}$$

Now we return to the case of a systematic measurement error as described in Section 2.3. The OC curve is then given by

$$\begin{aligned} OC_2(p) &= P\left\{\frac{U-\bar{Y}}{s_y} \geq k\right\} \\ &= P\left\{\frac{\frac{\sqrt{n}(U-\mu)}{\sigma\sqrt{1+n\sigma_B^2/\sigma^2}} - \frac{\sqrt{n}(\bar{Y}-\mu)}{\sigma\sqrt{1+n\sigma_B^2/\sigma^2}}}{s_y/\sigma} \geq \frac{k\sqrt{n}}{\sqrt{1+n\sigma_B^2/\sigma^2}}\right\} \\ &= P\left\{t_{n-1,\delta_2} \geq \frac{k\sqrt{n}}{\sqrt{1+n\sigma_B^2/\sigma^2}}\right\} \end{aligned} \tag{14}$$

where $\delta_2 = -\sqrt{n}z_p/\sqrt{1+n\sigma_B^2/\sigma^2}$. It is remarked, as with the previous noise model, that if σ_B/σ = constant, the OC curve of this model is independent of σ. If we put

$$n_* = \frac{n}{1+n(\sigma_B/\sigma)^2} \tag{15}$$

in the equation for $OC_2(p)$ we see, that the equation is very close to the noise free one, but with a sample size of n_*. The only difference is that the degrees of freedom in the t distribution is not changed. This means, that approximately the OC curve in case of a systematic measurement error is similar to the OC curve of the noise free case, but with a reduced sample size. In the examples given in the previous section, we had a sample size of $n = 50$ and a measurement uncertainty of $\sigma_B/\sigma = 0.2$. Approximately this corresponds to the OC curve of the noise free case with a sample size of

$$n_* = \frac{50}{1+50(0.2)^2} = 50/3 \tag{16}$$

This fact is also revealed from the figures of the previous section.

Finally the OC curve in case of both types of measurement error is given by

$$\begin{aligned} OC_3(p) &= P\left\{\frac{U-\bar{Y}}{s_y} \geq k\right\} \\ &= P\left\{\frac{\frac{\sqrt{n}(U-\mu)}{\sqrt{n\sigma_B^2+\sigma^2+\sigma_e^2}} - \frac{\sqrt{n}(\bar{Y}-\mu)}{\sqrt{n\sigma_B^2+\sigma^2+\sigma_e^2}}}{s_y/(\sigma\sqrt{1+\sigma_e^2/\sigma^2})} \geq \frac{k\sqrt{n}\sqrt{\sigma^2+\sigma_e^2}}{\sqrt{n\sigma_B^2+\sigma^2+\sigma_e^2}}\right\} \\ &= P\left\{t_{n-1,\delta_3} \geq \frac{k\sqrt{n}\sqrt{\sigma^2+\sigma_e^2}}{\sqrt{n\sigma_B^2+\sigma^2+\sigma_e^2}}\right\} \end{aligned} \tag{17}$$

where $\delta_3 = -\sqrt{n}z_p/\sqrt{1+n\sigma_B^2/\sigma^2}$.

4 Examples from using ISO 3951

The OC-bands shown in this section are computed exact for combined double specification limits based on the estimate $\hat{p}(\bar{Y}, s_y)$ of Section 2. We have chosen two sampling plans from ISO 3951 to illustrate the consequenses of the two types of measurement error. The OC were evaluated for 10 values of p in the relevant range. For each value of p the OC was calculated at $\sigma_i = 0.1(i-1)\sigma_{max}(p)$, $i = 2, ..., 11$ and at $\sigma_1 = 10^{-5}\sigma_{max}(p)$, similar to the approach of [6].

For each of the sampling plans the different scenarios are illustrated. First the corresponding acceptance region is shown and then the OC-bands of the different measurement error models. The dashed OC-bands is for the model without measurement error. The levels of the measurement error were chosen to be

1. $\sigma_e = 0.2\sigma$
2. $\sigma_B = 0.2\sigma$
3. $\sigma_e = 0.1414\sigma$ and $\sigma_B = 0.1414\sigma$ so that $\sqrt{\sigma_e^2 + \sigma_B^2} = 0.2\sigma$

If the variances of the measurement errors are made relative to the variance of the process, the OC-bands become relativly narrow. If on the other hand the variance of the measurement error are chosen to be a fixed number, the OC become very dependent upon the variance of the process, hence the OC-bands become very broad. We believe that in practice there will be some prior knowledge about the process variation, and that the measurement system is chosen on the basis of this prior knowledge in order to have a prescribed rate of discrimination. Therefore we find it reasonable to present the OC-bands of the models with the size of the measurement errors chosen relative to the process variation. In the figures the acceptance region and OC-bands are given for the sampling plans corresponding to AQL = 1.0, n = 50 and AQL = 6.5 , n = 50.

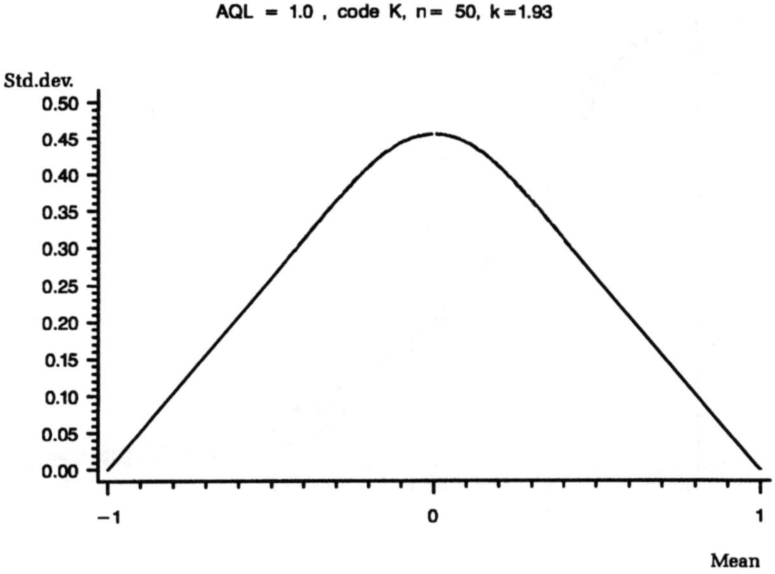

Fig. 1. Acceptance region for "s" method of ISO 3951

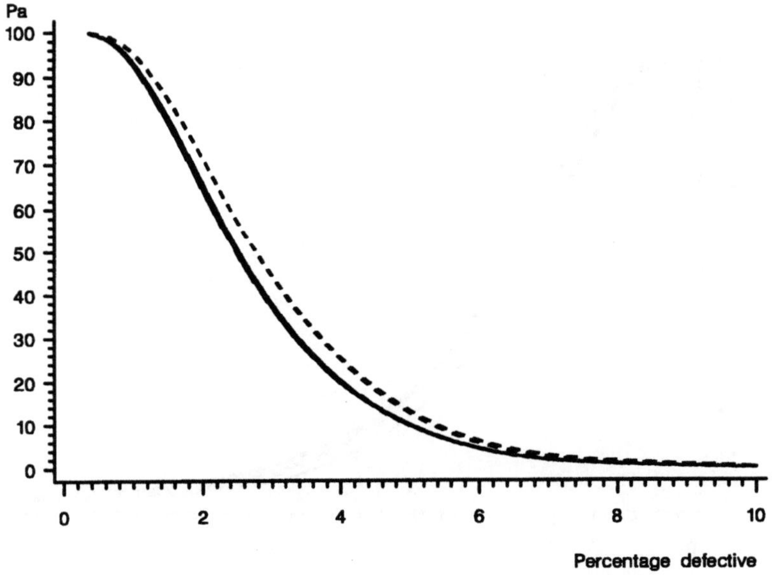

Fig. 2. OC-curves under high frequency measurement error, AQL=1.0 , code K, n=50, $\sigma_e = 0.2\sigma$, the dashed curve is for the model without measurement error

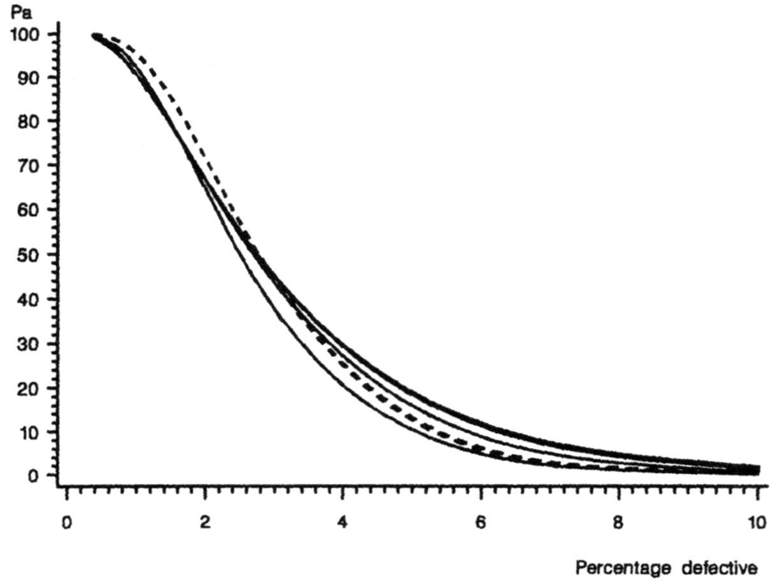

Fig. 3. OC-curves under low frequency measurement error, AQL=1.0 , code K, n=50, $\sigma_B = 0.2\sigma$, the dashed curve is for the model without measurement error

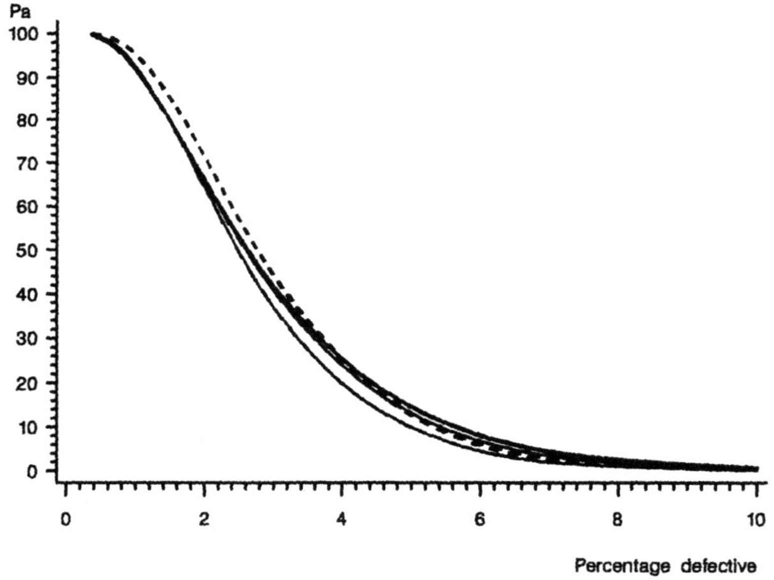

Fig. 4. OC-curves under combined measurement error, AQL=1.0 , code K, n=50, $\sigma_e = 0.14\sigma$, $\sigma_B = 0.14\sigma$, the dashed curve is for the model without measurement error

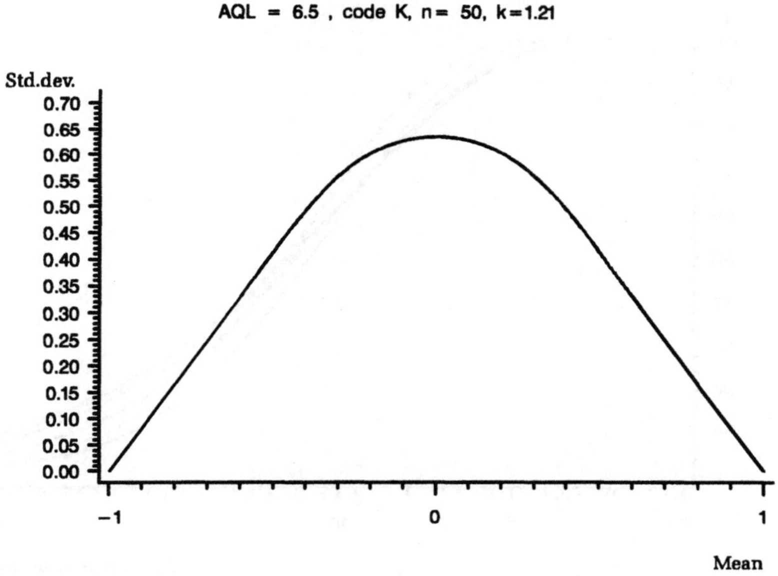

Fig. 5. Acceptance region for "s" method of ISO 3951

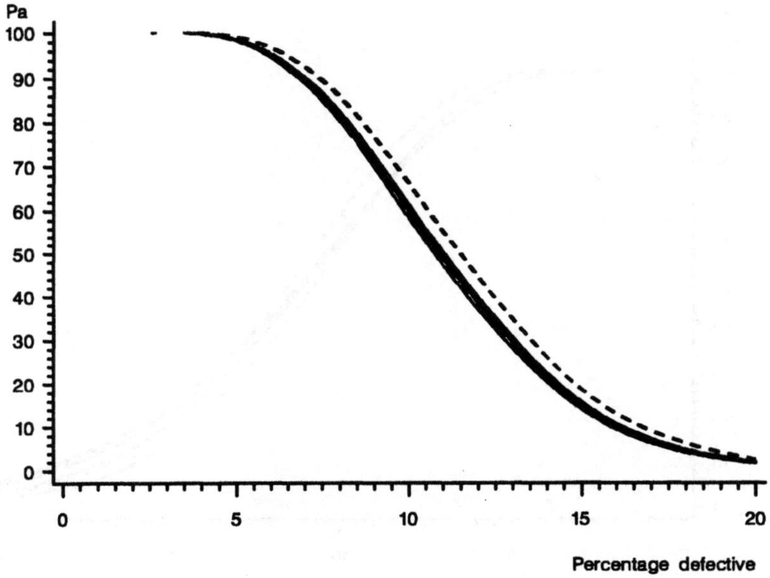

Fig. 6. OC-curves under high frequency measurement error, AQL=6.5 , code K, n=50, $\sigma_e = 0.2\sigma$, the dashed curve is for the model without measurement error

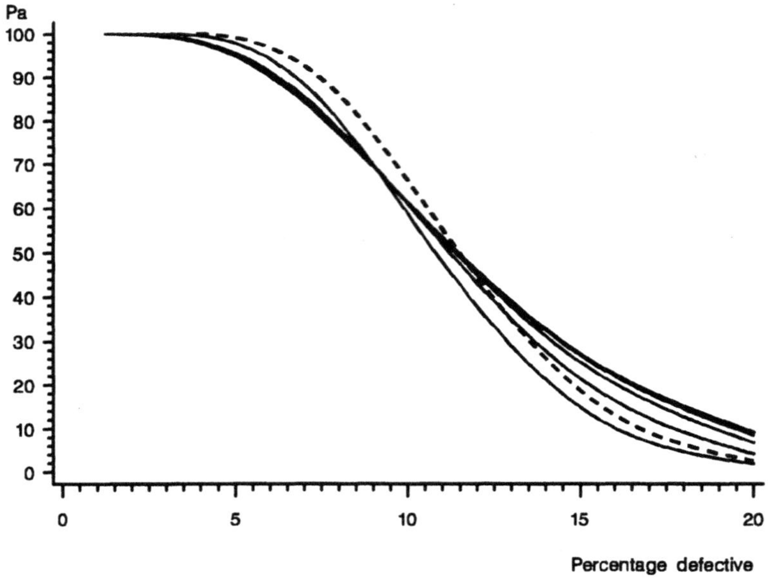

Fig. 7. OC-curves under low frequency measurement error, AQL=6.5 , code K, n=50, $\sigma_B = 0.2\sigma$, the dashed curve is for the model without measurement error

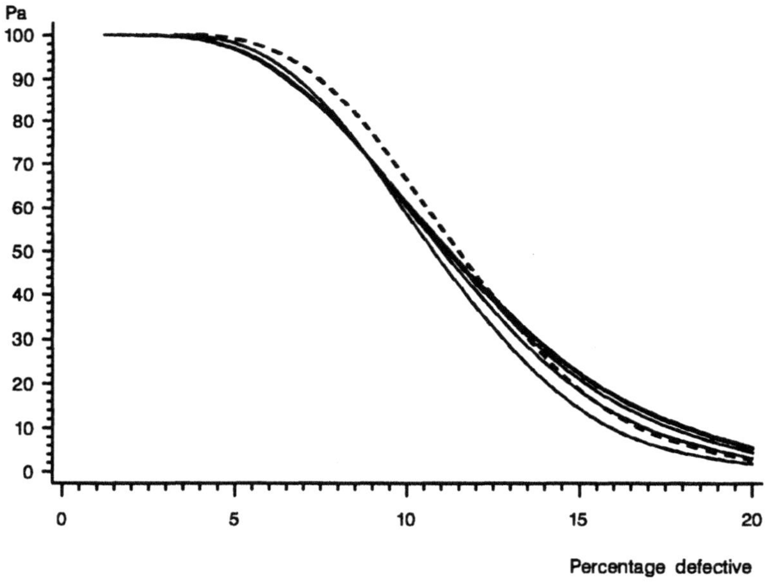

Fig. 8. OC-curves under combined measurement error, AQL=6.5 , code K, n=50, $\sigma_e = 0.14\sigma$, $\sigma_B = 0.14\sigma$, the dashed curve is for the model without measurement error

References

1. Lei, Ding-Hwa and Vardeman, Stephen B. (1998) The LRT Method of constructing a two-sided "variables" acceptance region and its comparison with other methods. Commun.Statist.-Theory Meth. **27**, 329-351
2. M.Bruhn-Suhr, W.Krumbholz and H.-J.Lenz (1992) On the design of Exact Single Sampling Plans by Variables. In: Frontiers in statistical Quality Control **4** H.-J.Lenz et al. eds., Physica Verlag, Heidelberg, 83-90
3. G.J.Resnikoff (1952)A new two-sided acceptance region for sampling by variables. Technical Report No. **8**, Applied Mathematics and Statistics Laboratory, Stanford University, Stanford, CA. 1952
4. G.J.Lieberman and G.J.Resnikoff (155) Sampling plans for Inspection by Variables, Journ. Amer.Statist.Assoc **50**, 457-516
5. ISO 3951 (1989) Sampling Procedures and charts for inspection by variables for percent nonconforming. International Organization for Standardization, Geneva
6. W.Kössler and H.-J.Lenz (1997) On the Non-Robustness of Maximum-likelihood Sampling Plans by Variables. in: Frontiers in statistical Quality Control **5**, H.-J.Lenz et al. eds., Physica Verlag, Heidelberg, 38-49
7. E.v.Collani (1992) The Pitfall of Acceptance Sampling by Variables. In: Frontiers in statistical Quality Control **4**, H.-J.Lenz et al. eds., Physica Verlag, Heidelberg, 91-99
8. E. G. Schilling (1982) Acceptance Sampling in Quality Control. Marcel Dekker, New York, 1982
9. D. B. Owen and Youn-min Chou (1983) Effect of Measurement Error and Instrument Bias on Operating Characteristics for Variables Sampling Plans, Journ. Qual. Techn. **15**, 107-117

Part 2

Statistical Process Control

Simultaneous Shewhart–Type Charts for the Mean and the Variance of a Time Series

Sven Knoth, Wolfgang Schmid, and Alexander Schöne

Department of Statistics, Europe University Viadrina, Frankfurt(Oder), Germany

Abstract. We introduce simultaneous control charts for the mean and the variance of a time series. Our schemes are extensions of the well-known Shewhart charts for independent variables. We consider a modified $\bar{X}$–S^2–chart, a modified $\bar{X}$–R–chart, a residual $\bar{X}$–S^2–chart and a residual $\bar{X}$–R–chart. A comparison of these schemes is made by means of the average run length. It turns out that residual schemes lead to better results. For nearly all parameter combinations the residual $\bar{X}$–S^2–chart is found to be the best control design.

1 Introduction

Control charts are widely applied in industry to monitor production processes. The most popular scheme for the mean is the $\bar{X}$–chart and for the variance the control charts based on the sample variance S^2 and the range R respectively. These methods were introduced by Shewhart already in the 1930s. A basic assumption for the application of these charts is that the variables of the underlying data set are independent and normally distributed.

In recent years control schemes for correlated data were introduced such as residual charts (Alwan and Roberts (1988), Harris and Ross (1991), Wardell et al. (1994), Vander Wiel (1996)) and modified charts (Vasilopoulos and Stamboulis (1978), Yashchin (1993), and Schmid (1995, 1997a,b)). Residual charts are based on the idea of transforming the data such that the obtained variables are independent. This can be achieved for some stochastic processes (e. g., ARMA processes), but only under the restriction that the parameters of the process are known. Then all classical schemes can be applied to the residuals. Modified control charts are straightforward extensions of the well-known procedures for independent samples. Therefore, they are more natural for a practitioner. In principle, the variance of the independent process is replaced by the variance of the time series. A disadvantage of these schemes is the fact that the critical values depend on the model parameters. However, as shown in Kramer and Schmid (1997), this effect is negligible for a small degree of correlation.

Control designs for the variance in the presence of autocorrelation were only discussed by a few authors in literature (e. g., MacGregor and Harris (1994), Amin et al. (1997)). Amin et al. (1997) investigated the behavior of the average run length (ARL) of the S^2– and the R–chart under correlation.

They assumed that the samples themselves are independent, but the observations within each sample follow an autoregressive process of order one with coefficient α. Thus their model differs from that of the above authors who mostly dealt with correlation between the samples. They showed that for certain sample sizes n the behavior of the in–control ARL of the S^2–chart is fairly robust, provided that the correlation structure is not too strong (e. g., $n = 6, 7, .., 10$, $0 \le \alpha \le 0.5$). The R–chart turns out to be more sensitive. Furthermore, they introduced modified S^2– and R–charts which can even be applied for larger correlations. Again, the S^2–chart provides better results than the R–chart.

Frequently, it is desirable to control the mean and variance at the same time. The most famous simultaneous control designs are the $\bar{X}$–S and the $\bar{X}$–S^2–chart respectively (see, e. g., Montgomery (1991)). Gan (1995) discussed omnibus EWMA charts (developed by Domangue/Patch (1991)) and combined EWMA and CUSUM schemes. An EWMA–$\bar{X}$–S–chart was analyzed by Stemann (1997) and Mittag/Stemann (1997). Further, Kanagawa and Arizono (1997) used a $\bar{X}$–S chart based on the Kullback–Leibler information.

In this paper simultaneous schemes for time series are introduced. We focus on generalizations of the $\bar{X}$–S^2 and the $\bar{X}$–R chart. Four different control schemes are considered: the modified $\bar{X}$–S^2–chart, the modified $\bar{X}$–R–chart, the residual $\bar{X}$–S^2–chart, and the residual $\bar{X}$–R–chart. We compare these charts for the same modeling as used by Amin et al. (1997) (i. e. independent samples). The out–of–control situation is described by a combined change point model which takes into account shifts in the mean as well as changes in the variance. First, in Section 2 it is discussed how the simultaneous $\bar{X}$–S^2–chart and $\bar{X}$–R–chart behave in the presence of autocorrelation. It shows that neither scheme can be applied to autocorrelated time series. Both the in–control ARL and the out–of–control ARL may vary dramatically even if the correlation is small. This is our motivation for introducing modified simultaneous control charts for time series (Section 3 and 4). The modified $\bar{X}$–S^2–chart is discussed in Section 3.1. For the calculation of its in–control ARL we make use of a result from Schöne and Schmid (1998) about the joint distribution of a quadratic and a linear form. The modified $\bar{X}$–R–chart is treated in Section 3.2. Residual charts are studied in Section 4. For the residual $\bar{X}$–S^2–chart a formula for its ARL is derived. A comparison of all charts is given in Section 5. The results of this study are of great interest. A clear ranking between the schemes under consideration can be made. For nearly all parameter combinations the residual $\bar{X}$–S^2–chart turns out to be the best design.

2 Simultaneous Control Charts

In the sequel, we denote the observed process by $\{X_{i,j}\}$. The indices i and j refer to the sample number and the number within the sample, respectively.

It is always assumed that the sample size n is fixed. The target process (in–control process) is denoted by $\{Y_{i,j}\}$. Let $E(Y_{i,j}) = \mu_0$ and $Var(Y_{i,j}) = \gamma_0$. Both quantities shall be known. We assume that the random vectors $(Y_{i,1}, Y_{i,2}, \ldots, Y_{i,n})'$, $i \geq 1$ are independent. On the other hand, within the samples the observations may be correlated. Note that in many cases the period between two samples is large in comparison with the time needed to take a sample. Thus, although the observations are autocorrelated, usually the degree of correlation between the samples can be neglected. This is the main idea behind the modeling of the target process.

In order to compare control schemes it is necessary to determine the relationship between the target process and the observed process. In the following we will make use of the combined change point model

$$X_{i,j} = \begin{cases} Y_{i,j} & \text{for} \quad i < q \\ \Delta\,(Y_{i,j} - \mu_0) + \mu_0 + a\,\sqrt{\gamma_0} & \text{for} \quad i \geq q \end{cases} . \tag{1}$$

The integer q stands for the unknown change point. Here the actual change takes place between two samples or at the first observed value of a sample. We get

$$E(X_{i,j}) = \begin{cases} \mu_0 & \text{for} \quad i < q \\ \mu_0 + a\,\sqrt{\gamma_0} & \text{for} \quad i \geq q \end{cases}$$

and

$$Var(X_{i,j}) = \begin{cases} \gamma_0 & \text{for} \quad i < q \\ \Delta^2\,\gamma_0 & \text{for} \quad i \geq q \end{cases} .$$

We say that the process is in–control if $a = 0$ and $\Delta = 1$, else the process is out–of–control. For $a \neq 0$ a shift in the mean and for $\Delta \neq 1$ a change in the variance can be observed provided that $i \geq q$.

Now, in most studies it is assumed that the change already arises at the beginning, i. e. that $q = 1$. For independent variables this is in principle no restriction since the Shewhart chart does not depend on q and the influence on the EWMA and the CUSUM scheme is small. Since in the present case the samples are again independent, we will always assume that $q = 1$. We write $P_{a,\Delta}$ and $E_{a,\Delta}$ to denote that the probability is taken with respect to the model (1). The notation P_0 and E_0 refers to the in–control situation ($a = 0$ and $\Delta = 1$).

2.1 The $\bar{\mathbf{X}}$–$\mathbf{S^2}$–Chart

First, we assume that the random variables $Y_{i,j}$, $j = 1, 2, \ldots, n$, $i \geq 1$ are independent and normally distributed. In order to construct the $\bar{X}$–S^2–chart we determine the mean and the empirical variance for each sample, i. e.

$$\bar{X}_i = \frac{1}{n}\sum_{j=1}^{n} X_{i,j} \quad \text{and} \quad S_i^2 = \frac{1}{n-1}\sum_{j=1}^{n}\left(X_{i,j} - \bar{X}_i\right)^2 , \tag{2}$$

where i stands for the sample number. It is concluded that the process is out–of–control at time i if at least one of the standardized statistics is large, i. e.

$$\frac{|\bar{X}_i - \mu_0|}{\sqrt{\gamma_0}} > x_c \quad \text{or} \quad \frac{S_i^2}{\gamma_0} > s_c \tag{3}$$

for some prespecified constants x_c and s_c. These quantities are called critical values. The index c refers to the classical situation (*iid*).

The run length τ of the simultaneous $\bar{X}$–S^2–chart is given by

$$\tau = \min\left\{\tau_1, \tau_2\right\} \tag{4}$$

with

$$\tau_1 = \inf\left\{i \in I\!N : \frac{\bar{X}_i - \mu_0}{\sqrt{\gamma_0}} \notin [-x_c, x_c]\right\} , \; \tau_2 \; = \; \inf\left\{i \in I\!N : \frac{S_i^2}{\gamma_0} \notin [0, s_c]\right\} .$$

τ is equal to the first value at which it is concluded that the process is out–of–control.

The probability that no alarm is given at time i is equal to

$$P(a, \Delta) := P_{a,\Delta}\left(\frac{|\bar{X}_i - \mu_0|}{\sqrt{\gamma_0}} \le x_c, \; \frac{S_i^2}{\gamma_0} \le s_c\right)$$

$$= \left[\Phi\left(\sqrt{n}\,\frac{x_c - a}{\Delta}\right) - \Phi\left(\sqrt{n}\,\frac{-x_c - a}{\Delta}\right)\right] \chi^2_{n-1}\left((n-1)\frac{s_c}{\Delta^2}\right), \tag{5}$$

where Φ stands for the standard normal distribution function and χ^2_{n-1} for the χ^2–distribution function with $n-1$ degrees of freedom. Since the samples are independent the ARL is given by

$$E_{a,\Delta}(\tau) = \frac{1}{1 - P(a, \Delta)} .$$

Now a problem arises in the choice of the critical values. Given a prespecified value ξ for the in–control ARL one equation $E_0(\tau) = \xi$ with two variables (x_c and s_c) has to be fulfilled. In order to obtain a unique solution we need an additional condition. In this context it is commonly demanded that for both univariate charts the probability of a correct decision in the in–control state is the same. This means that

$$P_0\left(\frac{|\bar{X}_i - \mu_0|}{\sqrt{\gamma_0}} \le x_c\right) = P_0\left(\frac{S_i^2}{\gamma_0} \le s_c\right) . \tag{6}$$

Thus we obtain that

$$x_c = \frac{z_{1-\beta/2}}{\sqrt{n}} \quad \text{and} \quad s_c = \frac{\chi^2_{n-1,1-\beta}}{n-1} \quad \text{with} \quad \beta = 1 - \sqrt{1-1/\xi}, \quad (7)$$

where z_β and $\chi^2_{n-1,\beta}$ denote the β quantile of the standard normal and the χ^2–distribution with $n-1$ degrees of freedom, respectively.

2.2 The $\bar{\text{X}}$–R–Chart

In this Section it is again assumed that the observations $Y_{i,j}$, $j = 1, 2, \ldots, n$, $i \geq 1$, are independent and normally distributed. Besides the control charts based on the empirical variance, the range chart (R–chart) is frequently applied to detect deviations in the variance structure. Let $X_{i,(1)} \leq X_{i,(2)} \leq \ldots \leq X_{i,(n)}$ denote the order statistics of $X_{i,1}, X_{i,2}, \ldots, X_{i,n}$. Then the range of the i–th sample is given by $R_i = X_{i,(n)} - X_{i,(1)}$. For normally distributed variables these schemes were analyzed in detail in literature (e. g., Montgomery (1991)). It has turned out that in most cases the charts based on the empirical variance lead to better results. However, for small n, say $n \leq 6$, the range chart possesses nearly the same efficiency as the S–chart. Therefore, the simpler range chart is often preferred in practice.

By combining it with the $\bar{X}$–chart we obtain the $\bar{X}$–R–chart. The process is decided to be out–of–control at time i if

$$\frac{|\bar{X}_i - \mu_0|}{\sqrt{\gamma_0}} > \tilde{x}_c \quad \text{or} \quad \frac{R_i}{\sqrt{\gamma_0}} > r_c \quad (8)$$

for some constants $\tilde{x}_c$ and r_c. The corresponding run length is equal to

$$\tilde{\tau} = \min\{\tilde{\tau}_1, \tilde{\tau}_2\}$$

with

$$\tilde{\tau}_1 = \inf\left\{i \in I\!N : \frac{\bar{X}_i - \mu_0}{\sqrt{\gamma_0}} \notin [-\tilde{x}_c, \tilde{x}_c]\right\}, \; \tilde{\tau}_2 = \inf\left\{i \in I\!N : \frac{R_i}{\sqrt{\gamma_0}} \notin [0, r_c]\right\}.$$

The computation of the distribution function of the sample range R_i turns out to be more difficult. By standardizing R_i with the in–control standard deviation $\sqrt{\gamma_0}$ we obtain the well–known relation

$$P_0\left(\frac{R_i}{\sqrt{\gamma_0}} \leq r\right) = Q(r) := \int_{-\infty}^{\infty} n \left[\Phi(x+r) - \Phi(x)\right]^{n-1} \phi(x)\, dx,$$

where ϕ denotes the density function of the standard normal distribution.

Since the variables $\bar{X}_i$ and R_i are independent the ARL can be shown to be

$$E_{a,\Delta}(\tilde{\tau}) = \frac{1}{1 - \tilde{P}(a,\Delta)}$$

with

$$\begin{aligned}\tilde{P}(a,\Delta) &:= P_{a,\Delta}\left(\frac{|\bar{X}_i - \mu_0|}{\sqrt{\gamma_0}} \le \tilde{x}_c,\ \frac{R_i}{\sqrt{\gamma_0}} \le r_c\right) \\ &= \left[\Phi\left(\sqrt{n}\,\frac{\tilde{x}_c - a}{\Delta}\right) - \Phi\left(\sqrt{n}\,\frac{-\tilde{x}_c - a}{\Delta}\right)\right] Q(r_c/\Delta)\,. \qquad (9)\end{aligned}$$

As already explained in Section 2.1, the critical values $\tilde{x}_c$ and r_c are chosen such that $E_0(\tilde{\tau}) = \xi$ under the restriction that

$$P_0\left(\frac{|\bar{X}_i - \mu_0|}{\sqrt{\gamma_0}} \le \tilde{x}_c\right) = P_0\left(\frac{R_i}{\sqrt{\gamma_0}} \le r_c\right). \qquad (10)$$

We get that

$$\tilde{x}_c = x_c \quad \text{and} \quad r_c = Q^{-1}(1-\beta) \qquad (11)$$

where Q^{-1} stands for the inverse of Q. For computing Q^{-1} a software package implementation is applied (function g01fmf() from the NAG® Fortran Library Mark 17).

2.3 Behavior under Autocorrelation

Here we want to study the influence of correlation on the performance of the $\bar{X}$–S^2–chart and the $\bar{X}$–R–chart. Suppose that the charts are applied under the false assumption of independence. However, in reality, they follow an autoregressive process of order one (AR(1)) with mean μ_0. Let

$$Y_{i,j} = \mu_0 + \alpha\,(Y_{i,j-1} - \mu_0) + \varepsilon_{i,j} \qquad (12)$$

with $|\alpha| < 1$, where the variables $\varepsilon_{i,j}$, $j \ge 1$, $i \ge 1$, are independent and normally distributed with expectation 0 and variance σ_0^2. Under these conditions there exists a weakly stationary solution of (12) with expectation μ_0 and variance $\gamma_0 = \sigma_0^2/(1-\alpha^2)$.

Now we choose the in–control ARL equal to 500. Thus (x_c, s_c) and $(\tilde{x}_c, r_c)$ are determined as given in (7) and (11). If the critical values are taken in this way, we speak about the classical $\bar{X}$–S^2–chart and the classical $\bar{X}$–R–chart. Note that in the presence of correlation the formula (5) is no longer valid since the statistics $\bar{X}_i$ and S_i^2 are not independent. The same holds for $\bar{X}_i$ and R_i.

We consider both charts with respect to three parameter combinations in the change point model (1). State I corresponds to the in-control case $a = 0$ and $\Delta = 1$. With II and III we denote two out-of-control states, $a = 0.5, \Delta = 1.1$ and $a = 1, \Delta = 1.2$. In Figure 1 the ARLs of both schemes are given for the sample size $n = 5$. The ARLs of the $\bar{X}$–S^2–chart are computed by using a series expansion as introduced by Schöne and Schmid (1998). We briefly describe this method in Section 3.1. On the contrary, the values for the $\bar{X}$–R–chart are determined by a Monte–Carlo simulation study based on 10^7 repetitions.

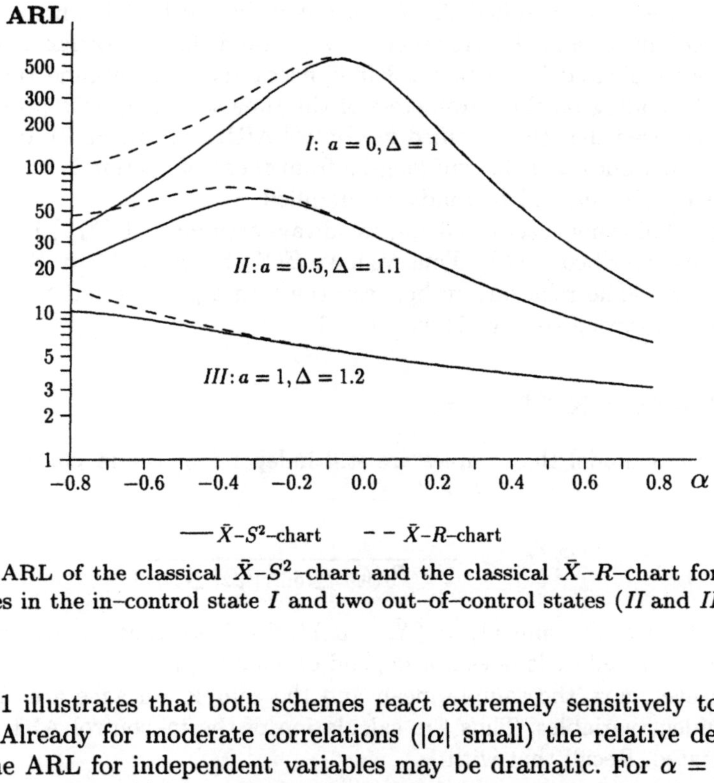

Fig. 1. ARL of the classical $\bar{X}$–S^2–chart and the classical $\bar{X}$–R–chart for AR(1) processes in the in-control state I and two out-of-control states (II and III)

Figure 1 illustrates that both schemes react extremely sensitively to correlation. Already for moderate correlations ($|\alpha|$ small) the relative deviation from the ARL for independent variables may be dramatic. For $\alpha = 0.1$ the relative error is 36 % for both charts in the in-control state. For the out-of-control combination it is 24 % ($\bar{X}$–S^2) and 25 % ($\bar{X}$–R) for state II, but only 7.7 % ($\bar{X}$–S^2) and 7.9 % ($\bar{X}$–R) for state III. Thus the errors are larger for the in-control process. It is interesting to observe that for a positive α the influence on both charts is nearly the same. But, in case of negative α the ARL of the $\bar{X}$–R–chart is always larger than the ARL of the $\bar{X}$–S^2–chart. For $\alpha = -0.1$ the relative deviations are 9.2 % (state I), 30.3 % (state II), 9 % (state III) for the $\bar{X}$–S^2–chart and 13.6 % (state I), 32.7 % (state II), 9.3 % (state III) for the $\bar{X}$–R–chart.

Choosing the critical values as in the *iid* case thus leads to dramatic deviations from the true ARL in the presence of autocorrelation. Thus it is necessary to determine the critical values as a function of the target process. This is the reason for introducing modified control schemes.

3 Modified Charts

Modified charts are direct extensions of the well–known control schemes for independent samples. Instead of the variance of an independent sample the variance of the underlying target process is used. Thus the modified $\bar{X}$–S^2–chart and the modified $\bar{X}$–R–chart coincide with the control schemes introduced in (3) and (8). However, now γ_0 stands for the variance of a time series. As indicated in Section 2.3 it is necessary to determine the critical values depending on the parameters of the time series. By this procedure it can be ensured that the specified in–control ARL is attained. To distinguish the critical values and the run lengths from those of Section 2 we made use of the subscript m, which stands for modified.

In the following (Sections 3 – 5) we always assume that $\{Y_{i,j}\}$ is an AR(1) process as described in (12). Thus we have $E(Y_{i,j}) = \mu_0$ and $Var(Y_{i,j}) = \gamma_0 = \sigma_0^2/(1-\alpha^2)$. The relationship between the target process and the observed process is always given by (1) with $q = 1$.

3.1 Modified $\bar{\mathbf{X}}$–$\mathbf{S^2}$–Chart

Since in our model the samples are still independent the in–control ARL is equal to

$$E_0(\tau_m) = \frac{1}{1 - P_0(Q_1 \le s_m, |Q_2| \le x_m)},$$

where $Q_1 = S_i^2/\gamma_0$ and $Q_2 = (\bar{X}_i - \mu_0)/\sqrt{\gamma_0}$. Note that $E_0(\tau_m)$ is only a function of n and α. It does not depend on σ_0 and μ_0.

However, now the sample mean and the sample variance are no longer independent variables. Thus the calculation of the in–control ARL is more complicated. Recall that with

$$\boldsymbol{X}_i = (X_{i,1}, X_{i,2}, \ldots, X_{i,n})^{\mathrm{T}}, \; \boldsymbol{h} = \frac{1}{n}\mathbf{1} \text{ and } \boldsymbol{A} = \frac{1}{n-1}\left(\boldsymbol{I} - \frac{1}{n}\mathbf{1}\mathbf{1}^{\mathrm{T}}\right)$$

the mean and the empirical variance can be written as follows

$$\bar{X}_i = \boldsymbol{h}^{\mathrm{T}}\boldsymbol{X}_i \quad \text{and} \quad S_i^2 = \boldsymbol{X}_i{}^{\mathrm{T}}\boldsymbol{A}\boldsymbol{X}_i\,,$$

where, as usual, $\boldsymbol{I}$ denotes the identity matrix and $\mathbf{1}$ the vector of ones. Schöne and Schmid (1998) derived a series expansion for the joint distribution function of a quadratic and a linear form in normal variables. It is shown that

$$P_0(Q_1 \le s, |Q_2| \le x) = \sum_{i,j=0}^{\infty} \frac{f_{i+1,j}}{\sqrt{2\pi}} \frac{4\beta}{2j+1} g_i^{(n+1)}(s)\, x^{2j+1} + \\ + \chi^2_{n-1}\left(\frac{s}{\beta}\right)\left[\Phi\left(\frac{x}{\sqrt{b_0}}\right) - \Phi\left(-\frac{x}{\sqrt{b_0}}\right)\right],$$

where the constants f_{ij} follow some iteration rule. β is equal to $(\lambda_1 + \lambda_{n-1})/2$ with λ_1 and λ_{n-1} as nonzero eigenvalues of $\mathbf{A}$ with the largest and the smallest magnitude respectively. The functions g_i are modified generalized Laguerre polynomials. χ^2_{n-1} and Φ stand again for the related probability distribution functions. By truncation the infinite sum at a suitable number of terms we obtain feasible approximations of the joint distribution function. For details we refer to Schöne and Schmid (1998).

This technique will be used to compute the critical values s_m and x_m. Similarly to the classical scheme we demand that the univariate charts behave symmetrically, i. e. for the univariate distribution functions holds

$$P_0(Q_1 \le s_m) = P_0(|Q_2| \le x_m)\,.$$

For more extensive tables of the critical values we refer to Knoth et al. (1998). In our comparison study all schemes are calibrated to fulfill an in–control ARL of 500. Here we focus on three parameter combinations which are summarized together with the critical values in Table 1.

Table 1. Critical values of the modified $\bar{X}$–S^2–chart (in–control ARL = 500)

n	a	x_m	s_m
5	0.3	1.869	4.249
10	0.5	1.678	3.303
10	-0.5	0.639	4.382

Again, the out–of–control situation is characterized by the probability that no alarm $P_m(a, \Delta)$. As in Schöne (1997) we use

$$P_m(a,\Delta) = \frac{1}{2}\left[\operatorname{sgn}(x_m - a)\, P_0\left(Q_1 \le \frac{s_m}{\Delta^2}, |Q_2| \le \left|\frac{x_m - a}{\Delta}\right|\right) + \\ + P_0\left(Q_1 \le \frac{s_m}{\Delta^2}, |Q_2| \le \frac{x_m + a}{\Delta}\right)\right].$$

3.2 Modified X̄–R–Chart

The in–control ARL of the modified $\bar{X}$–R–chart is given by

$$E_0(\tilde{r}_m) = \frac{1}{1 - P(Q_3 \le r_m, |Q_2| \le \tilde{x}_m)},$$

where $Q_3 = R_i/\sqrt{\gamma_0}$. This function exclusively depends on n and α.

As above the critical values are calculated such that $E_0(\tilde{r}_m) = \xi$ under the symmetry condition $P_0(Q_3 \le r_m) = P_0(|Q_2| \le \tilde{x}_m)$. The calculation of the joint distribution of Q_3 and $|Q_2|$ turns out to be complicated since in the present case the variables are not independent. Therefore, we make use of simulations. First, within a small simulation study we determine starting values for our algorithm. These values are found to be all less than 10. In the next step we improve these approximations. Therefore, we split the quadrangle $[0, 10]^2$ into 500^2 identical cells. At the center point of each cell we determine the empirical joint distribution function of $(Q_3, |Q_2|)$ (10^6 repetitions). Additionally, we derive the empirical marginal distribution functions of Q_3 and $|Q_2|$. In order to fulfill the in–control ARL and the symmetry condition we select a cell by iteration over the grid which provides errors smaller than prespecified bounds. After that, this cell is enlarged. The new length equals 0.2. Again this cell is partitioned into 500^2 subcells. In a new simulation study based on 10^8 repetitions the empirical distribution function is calculated at the center points of these subcells. Again we select by iteration a suitable cell with errors smaller than bounds which are stronger compared to the bounds of the latter iteration. The resulting critical values are listed in Table 2. Notice, that the critical values $\tilde{x}_m$ of the modified $\bar{X}$–S^2–chart and the $\bar{X}$–R–chart are nearly the same. Extensive tables of the critical values for various parameter combinations can be found in Knoth et al. (1998).

Table 2. Critical values of the modified $\bar{X}$–R–chart (in–control ARL = 500)

n	α	$\tilde{x}_m$	r_m
5	0.3	1.869	5.174
10	0.5	1.678	5.650
10	-0.5	0.640	6.668

4 Residual Charts

Residual charts are based on the consideration of the residuals of the time series. In most papers the deviation between the observed value and a predictor for this observation is measured. As in Harris and Ross (1991), Wardell et al. (1994), Kramer and Schmid (1997,98) we take the best linear predictor $\hat{X}_{i,j}$ for $X_{i,j}$ given $X_{i,j-1}, X_{i,j-2}, .., X_{i,1}$. Note that the predictor is calculated under the assumption that the process is in the in-control state. We get for an AR(1) process

$$\hat{X}_{i,1} = \mu_0 \quad \text{and} \quad \hat{X}_{i,j} = \alpha(X_{i,j-1} - \mu_0) + \mu_0$$

for $j \geq 2$. With respect to the relation (1) the residuals are given by

$$\begin{aligned} X_{i,1} - \hat{X}_{i,1} &= \Delta(Y_{i,1} - \mu_0) + a\sqrt{\gamma_0} \quad \text{and} \\ X_{i,j} - \hat{X}_{i,j} &= \Delta\varepsilon_{i,j} + a\sqrt{\gamma_0}(1-\alpha) \end{aligned}$$

for $j \geq 2$. Next we normalize the residuals such that they have the same variance. This leads to

$$\hat{\varepsilon}_{i,1} = \sqrt{1-\alpha^2}\left(X_{i,1} - \hat{X}_{i,1}\right) \quad \text{and} \quad \hat{\varepsilon}_{i,j} = X_{i,j} - \hat{X}_{i,j}\,,\ j \geq 2.$$

Since the variables $\{\varepsilon_{i,j}\}$ are independent and normally distributed these quantities are independent too, but not identically distributed. It follows that

$$\hat{\varepsilon}_{i,1} \sim \mathcal{N}\left(a\,\sigma_0, \Delta^2\sigma_0^2\right) \text{ and } \hat{\varepsilon}_{i,j} \sim \mathcal{N}\left(a\,\sigma_0\sqrt{\frac{1-\alpha}{1+\alpha}}, \Delta^2\sigma_0^2\right), j \geq 2. \quad (13)$$

The symbol $\mathcal{N}\left(\mu, \sigma^2\right)$ stands for the normal distribution function with expectation μ and variance σ^2.

Most authors have not discussed the influence of the starting problem on the ARL. They used $X_{i,1} - \mu_0 - \alpha(X_{i,0} - \mu_0)$ as first residual neglecting the fact that $X_{i,0}$ is not observable. Kramer and Schmid (1998) proposed the above procedure.

4.1 Residual $\bar{\mathbf{X}}$–$\mathbf{S}^2$–Chart

The procedure is similar to the classical $\bar{X}$–S^2–chart. We determine the mean and the empirical variance of the normalized residuals

$$\bar{\hat{\varepsilon}}_i = \frac{1}{n}\sum_{j=1}^{n}\hat{\varepsilon}_{i,j} \quad \text{and} \quad \hat{S}_i^2 = \frac{1}{n-1}\sum_{j=1}^{n}\left(\hat{\varepsilon}_{i,j} - \bar{\hat{\varepsilon}}_i\right)^2.$$

It is concluded that the process is out–of–control at time i if

$$\frac{|\bar{\hat{\varepsilon}}_i|}{\sigma_0} > x_r \quad \text{or} \quad \frac{\hat{S}_i^2}{\sigma_0^2} > s_r$$

with positive constants x_r and s_r. The run length of this control scheme is denoted by τ_r.

First we observe that

$$\frac{\bar{\hat{\varepsilon}}_i}{\sigma_0} \sim \mathcal{N}\left(\frac{a}{n}\left(1+(n-1)\sqrt{\frac{1-\alpha}{1+\alpha}}\right), \frac{\Delta^2}{n}\right).$$

Moreover, it follows from Corollary 5.1.3a of Mathai and Provost (1992) that

$$\frac{(n-1)}{\Delta^2\,\sigma_0^2}\hat{S}_i^2 \sim \chi^2_{n-1,\nu} \qquad \text{with} \quad \nu = \frac{n-1}{n}\left(\frac{a}{\Delta}\right)^2\left(1-\sqrt{\frac{1-\alpha}{1+\alpha}}\right)^2.$$

$\chi^2_{n-1,\nu}$ denotes the noncentral χ^2–distribution with $n-1$ degrees of freedom and noncentrality parameter ν. Consequently the ARL is equal to

$$E_{a,\Delta}(\tau_r) = \frac{1}{1-P_r(a,\Delta)} \tag{14}$$

with

$$P_r(a,\Delta) := \left\{\Phi\left(\left[x_r - \frac{a}{n}\left(1+(n-1)\sqrt{\frac{1-\alpha}{1+\alpha}}\right)\right]\frac{\sqrt{n}}{\Delta}\right) - \right.$$
$$\left. - \Phi\left(\left[-x_r - \frac{a}{n}\left(1+(n-1)\sqrt{\frac{1-\alpha}{1+\alpha}}\right)\right]\frac{\sqrt{n}}{\Delta}\right)\right\} \chi^2_{n-1,\nu}\left(\frac{n-1}{\Delta^2}s_r\right).$$

$P_r(a,\Delta)$ is equal to the probability that no alarm is given at time i. (14) shows that the in–control ARL of τ_r is the same as that of the $\bar{X}$–S^2–chart for independent variables. Consequently we choose $x_r = x_c$ and $s_r = s_c$. Note that of course the out–of–control behavior is different. This is due to the influence of the starting value.

Thus the critical values do not depend on the parameters of the target process. This is the great advantage of the residual charts.

4.2 Residual X̄–R–Chart

First we order the normalized residuals. This leads to $\hat{\varepsilon}_{i,(1)} \leq \hat{\varepsilon}_{i,(2)} \leq \ldots \leq \hat{\varepsilon}_{i,(n)}$. After that the sample range $\hat{R}_i = \hat{\varepsilon}_{i,(n)} - \hat{\varepsilon}_{i,(1)}$ is calculated. Then an alarm is given at time i if

$$\frac{|\bar{\hat{\varepsilon}}_i|}{\sigma_0} > \tilde{x}_r \quad \text{or} \quad \frac{\hat{R}_i}{\sigma_0} > r_r\,.$$

The run length of this scheme shall be $\tilde{\tau}_r$. Since for $a=0$ the variables $\{\hat{\varepsilon}_{i,j}\}$ are independent and identically normally distributed we obtain $E_{0,\Delta}(\tilde{\tau}_r) = E_{0,\Delta}(\tilde{\tau})$ provided that we take $\tilde{x}_r = x_c$ and $r_r = r_c$. In particular, the in–control ARLs of the residual $\bar{X}$–R–chart and the classical $\bar{X}$–R–chart are the same. Hence, we determine the critical values as in (11).

5 Comparison of the Control Schemes

In this Section the modified $\bar{X}$–S^2–chart ($m\bar{X}S^2$), the modified $\bar{X}$–R–chart ($m\bar{X}R$), the residual $\bar{X}$–S^2–chart ($r\bar{X}S^2$) and the residual $\bar{X}$–R–chart ($r\bar{X}R$) are compared with each other. The out–of–control situation is described by the combined change point model (1) with $q = 1$. The target process is assumed to be an AR(1) process with mean $\mu_0 = 0$ (cf. Section 2.3). Thus we have

$$X_{i,j} = \Delta Y_{i,j} + a\sqrt{\gamma_0} \quad \text{and} \quad Y_{i,j} = \alpha Y_{i,j-1} + \varepsilon_{i,j}$$

with $|\alpha| < 1$. The variables $\varepsilon_{i,j}$, $i, j \geq 1$ are assumed to be independent with $\varepsilon_{i,j} \sim \mathcal{N}(0,1)$. Consequently, the observations within a sample are autocorrelated, but the samples themselves are independent.

Here we present our results for the parameter combinations: $\alpha = 0.3$ and $n = 5$, $\alpha = 0.5$ and $n = 10$, $\alpha = -0.5$ and $n = 10$. The charts are analyzed for the following shifts and scale factors

$$a \in \{0, 0.25, 0.5, 0.75, 1, 1.25, 1.5, 2\} \quad \text{and}$$
$$\Delta \in \{1, 1.1, 1.2, 1.3, 1.4, 1.5, 1.75, 2\}.$$

As a measure of the performance of each chart the ARL is used. To compare the schemes each chart is calibrated such that the in–control ARL under the symmetry condition is approximately equal to 500. How this can be achieved has already been described above. Except for the residual $\bar{X}$–S^2–chart the out–of–control ARLs of all other schemes are estimated by simulations (10^8 repetitions). The results are summarized in Table 3 ($\alpha = 0.3$, $n = 5$), Table 4 ($\alpha = 0.5$, $n = 10$) and Table 5 ($\alpha = -0.5$, $n = 10$).

The results of our study are of great interest since an obvious ranking between the control schemes under consideration can be made. If no scale deviation is present (i. e. $\Delta = 1$) then all charts provide similar results for a positive coefficient α (see Table 3 and 4) while for $\alpha < 0$ (Table 5) the residual charts have smaller out–of–control ARLs. In this case the ARLs of the residual $\bar{X}$–S^2–chart and the residual $\bar{X}$–R–chart are of equal magnitude. The same observation is valid for the modified schemes. However, as already mentioned, they are a little bit worse. If exclusively a change in the scale occurs (i. e. $a = 0$) the residual $\bar{X}$–S^2–chart turns out to be the best one, followed by the residual $\bar{X}$–R–chart. The ARLs of the modified charts are seen to be larger. Within this class, the application of the modified $\bar{X}$–S^2–chart leads to better results. Nearly the same behavior can be observed for all other out–of–control situations treated in the tables. The residual $\bar{X}$–S^2–chart has always the smallest ARLs in the out–of–control state. Only for large shifts a the other charts are serious competitors. In second position, but with clear distance, we find the residual $\bar{X}$–R–chart. While in Table 3 and Table 5 the modified $\bar{X}$–S^2–chart provides better results than the modified $\bar{X}$–R–chart, the opposite is true in Table 4. However, the ARLs of the modified charts are much larger than that of the residual schemes. Thus we are able to give a clear recommendation to the practitioner: Use the residual $\bar{X}$–S^2–chart.

Note that these results are not contrary to those in Schmid (1995) and Kramer and Schmid (1998). In their studies the modified schemes were shown to be better than the residual charts for a positive coefficient. However, in these papers the sample size was equal to 1 and the variables between the samples follow a time series. Thus these results cannot be compared with the present ones since they are based on different assumptions.

6 Conclusions

In our study it is assumed that the samples are independent, but the variables within a sample may be correlated. It is quite interesting to observe the dominance of the residual schemes over the modified schemes. Exclusively for pure shifts all four schemes behave similarly. However for deviations in the scale, the performance of the residual charts is much better. For all processes under consideration the residual $\bar{X}$–S^2–chart provides the best results, followed by the residual $\bar{X}$–R–chart.

Table 3. ARLs of some simultaneous control schemes ($n = 5$, $\alpha = 0.3$, in–control ARL = 500)

		a								
		0.00	0.25	0.50	0.75	1.00	1.25	1.50	2.00	chart
Δ	1.00	500.42	305.35	111.44	39.40	15.64	7.21	3.87	1.69	$m\bar{X}S^2$
		500.00	304.48	111.06	39.31	15.63	7.21	3.87	1.69	$r\bar{X}S^2$
		499.04	305.17	111.29	39.38	15.63	7.20	3.86	1.69	$m\bar{X}R$
		500.58	305.06	111.45	39.38	15.63	7.21	3.87	1.69	$r\bar{X}R$
	1.10	154.48	113.45	55.74	24.87	11.70	6.10	3.57	1.71	$m\bar{X}S^2$
		143.76	107.33	54.05	24.48	11.60	6.07	3.56	1.71	$r\bar{X}S^2$
		156.19	114.47	55.96	24.90	11.70	6.10	3.57	1.71	$m\bar{X}R$
		150.70	111.34	55.10	24.69	11.65	6.08	3.57	1.71	$r\bar{X}R$
	1.20	61.91	50.73	30.85	16.68	9.07	5.26	3.32	1.72	$m\bar{X}S^2$
		54.93	45.87	28.90	16.06	8.88	5.19	3.29	1.72	$r\bar{X}S^2$
		63.32	51.62	31.17	16.77	9.10	5.26	3.32	1.72	$m\bar{X}R$
		59.38	48.99	30.11	16.42	8.99	5.23	3.31	1.72	$r\bar{X}R$
	1.30	30.10	26.34	18.55	11.68	7.19	4.58	3.09	1.72	$m\bar{X}S^2$
		26.00	23.12	16.86	10.97	6.91	4.46	3.04	1.71	$r\bar{X}S^2$
		31.00	27.03	18.89	11.80	7.23	4.59	3.09	1.72	$m\bar{X}R$
		28.61	25.17	17.92	11.39	7.07	4.52	3.06	1.72	$r\bar{X}R$
	1.40	16.92	15.43	12.01	8.48	5.79	4.00	2.86	1.71	$m\bar{X}S^2$
		14.45	13.34	10.70	7.80	5.47	3.85	2.79	1.70	$r\bar{X}S^2$
		17.53	15.93	12.30	8.62	5.85	4.02	2.88	1.72	$m\bar{X}R$
		16.03	14.67	11.52	8.22	5.66	3.93	2.83	1.70	$r\bar{X}R$
	1.50	10.64	9.97	8.30	6.38	4.73	3.50	2.65	1.69	$m\bar{X}S^2$
		9.08	8.58	7.32	5.79	4.41	3.33	2.56	1.66	$r\bar{X}S^2$
		11.07	10.33	8.55	6.52	4.80	3.54	2.66	1.70	$m\bar{X}R$
		10.08	9.46	7.94	6.16	4.60	3.43	2.61	1.70	$r\bar{X}R$
	1.75	4.70	4.55	4.16	3.64	3.09	2.59	2.17	1.60	$m\bar{X}S^2$
		4.08	3.97	3.68	3.27	2.83	2.42	2.06	1.55	$r\bar{X}S^2$
		4.90	4.74	4.31	3.74	3.16	2.63	2.20	1.61	$m\bar{X}R$
		4.48	4.35	3.99	3.50	2.99	2.52	2.13	1.58	$r\bar{X}R$
	2.00	2.83	2.79	2.66	2.47	2.25	2.02	1.82	1.48	$m\bar{X}S^2$
		2.52	2.49	2.39	2.24	2.07	1.89	1.72	1.43	$r\bar{X}S^2$
		2.94	2.89	2.75	2.54	2.30	2.07	1.85	1.50	$m\bar{X}R$
		2.73	2.68	2.56	2.39	2.18	1.97	1.78	1.46	$r\bar{X}R$

Table 4. ARLs of some simultaneous control schemes ($n = 10$, $\alpha = 0.5$, in–control ARL $= 500$)

		a								
		0.00	0.25	0.50	0.75	1.00	1.25	1.50	2.00	chart
Δ	1.00	500.15	275.62	87.41	28.28	10.78	4.96	2.75	1.36	$m\bar{X}S^2$
		500.00	274.80	87.16	28.26	10.79	4.97	2.75	1.36	$r\bar{X}S^2$
		501.12	275.73	87.40	28.28	10.78	4.97	2.75	1.36	$m\bar{X}R$
		499.37	274.96	87.34	28.28	10.79	4.97	2.75	1.36	$r\bar{X}R$
	1.10	146.80	102.39	45.71	18.92	8.55	4.43	2.64	1.39	$m\bar{X}S^2$
		112.70	84.26	41.53	18.13	8.38	4.39	2.63	1.39	$r\bar{X}S^2$
		141.63	99.80	45.22	18.84	8.53	4.42	2.64	1.39	$m\bar{X}R$
		131.37	94.42	43.95	18.57	8.48	4.41	2.64	1.39	$r\bar{X}R$
	1.20	55.98	44.80	25.79	13.20	6.94	3.98	2.55	1.42	$m\bar{X}S^2$
		35.36	30.48	20.24	11.56	6.47	3.83	2.49	1.41	$r\bar{X}S^2$
		52.79	42.75	25.11	13.03	6.89	3.97	2.54	1.42	$m\bar{X}R$
		45.99	38.10	23.34	12.50	6.74	3.92	2.52	1.42	$r\bar{X}R$
	1.30	26.02	22.56	15.49	9.44	5.67	3.58	2.44	1.44	$m\bar{X}S^2$
		14.71	13.54	10.63	7.40	4.90	3.28	2.31	1.41	$r\bar{X}S^2$
		24.28	21.26	14.87	9.22	5.60	3.55	2.42	1.44	$m\bar{X}R$
		20.22	18.05	13.19	8.52	5.34	3.45	2.38	1.43	$r\bar{X}R$
	1.40	14.08	12.81	9.89	6.89	4.65	3.20	2.31	1.45	$m\bar{X}S^2$
		7.62	7.24	6.23	4.93	3.71	2.76	2.10	1.40	$r\bar{X}S^2$
		13.13	12.03	9.43	6.67	4.56	3.16	2.29	1.44	$m\bar{X}R$
		10.63	9.89	8.05	5.96	4.22	3.00	2.22	1.42	$r\bar{X}R$
	1.50	8.58	8.04	6.71	5.16	3.82	2.84	2.16	1.44	$m\bar{X}S^2$
		4.65	4.50	4.08	3.49	2.87	2.32	1.89	1.36	$r\bar{X}S^2$
		8.05	7.58	6.39	4.98	3.73	2.79	2.14	1.44	$m\bar{X}R$
		6.42	6.12	5.33	4.32	3.36	2.60	2.04	1.40	$r\bar{X}R$
	1.75	3.61	3.51	3.24	2.87	2.48	2.11	1.80	1.38	$m\bar{X}S^2$
		2.16	2.14	2.06	1.93	1.78	1.63	1.48	1.25	$r\bar{X}S^2$
		3.47	3.38	3.13	2.79	2.42	2.07	1.78	1.37	$m\bar{X}R$
		2.79	2.73	2.58	2.36	2.11	1.86	1.64	1.32	$r\bar{X}R$
	2.00	2.15	2.12	2.04	1.93	1.79	1.65	1.51	1.29	$m\bar{X}S^2$
		1.48	1.47	1.45	1.41	1.36	1.31	1.25	1.15	$r\bar{X}S^2$
		2.11	2.08	2.01	1.90	1.77	1.63	1.50	1.28	$m\bar{X}R$
		1.76	1.75	1.70	1.63	1.55	1.46	1.37	1.21	$r\bar{X}R$

Table 5. ARLs of some simultaneous control schemes ($n = 10$, $\alpha = -0.5$, in–control ARL = 500)

Δ		a: 0.00	0.25	0.50	0.75	1.00	1.25	1.50	2.00	chart
	1.00	497.75	42.29	4.20	1.40	1.03	1.01	1.00	1.00	$m\bar{X}S^2$
		500.00	40.12	3.95	1.35	1.03	1.00	1.00	1.00	$r\bar{X}S^2$
		500.79	42.62	4.22	1.40	1.03	1.00	1.00	1.00	$m\bar{X}R$
		497.82	40.12	3.95	1.35	1.03	1.00	1.00	1.00	$r\bar{X}R$
	1.10	146.88	26.12	3.84	1.43	1.05	1.00	1.00	1.00	$m\bar{X}S^2$
		112.70	23.78	3.61	1.38	1.04	1.00	1.00	1.00	$r\bar{X}S^2$
		150.11	26.35	3.85	1.43	1.05	1.00	1.00	1.00	$m\bar{X}R$
		131.30	24.53	3.63	1.38	1.04	1.00	1.00	1.00	$r\bar{X}R$
	1.20	56.35	17.01	3.52	1.46	1.06	1.00	1.00	1.00	$m\bar{X}S^2$
		35.36	14.05	3.27	1.40	1.05	1.00	1.00	1.00	$r\bar{X}S^2$
		58.26	17.23	3.54	1.46	1.06	1.00	1.00	1.00	$m\bar{X}R$
		45.99	15.44	3.32	1.41	1.05	1.00	1.00	1.00	$r\bar{X}R$
	1.30	26.35	11.48	3.23	1.47	1.08	1.01	1.01	1.00	$m\bar{X}S^2$
		14.71	8.45	2.90	1.41	1.07	1.01	1.00	1.00	$r\bar{X}S^2$
		27.51	11.71	3.25	1.48	1.08	1.01	1.00	1.00	$m\bar{X}R$
		20.22	9.94	3.01	1.42	1.07	1.01	1.00	1.00	$r\bar{X}R$
Δ	1.40	14.34	8.02	2.94	1.48	1.10	1.01	1.00	1.00	$m\bar{X}S^2$
		7.62	5.39	2.52	1.40	1.08	1.01	1.00	1.00	$r\bar{X}S^2$
		15.06	8.24	2.96	1.48	1.10	1.01	1.00	1.00	$m\bar{X}R$
		10.63	6.65	2.70	1.42	1.08	1.01	1.00	1.00	$r\bar{X}R$
	1.50	8.78	5.82	2.65	1.47	1.11	1.02	1.00	1.00	$m\bar{X}S^2$
		4.65	3.71	2.17	1.36	1.08	1.01	1.00	1.00	$r\bar{X}S^2$
		9.26	6.02	2.69	1.48	1.11	1.02	1.00	1.00	$m\bar{X}R$
		6.42	4.67	2.39	1.40	1.10	1.01	1.00	1.00	$r\bar{X}R$
	1.75	3.72	3.09	2.05	1.41	1.12	1.03	1.00	1.00	$m\bar{X}S^2$
		2.16	1.98	1.58	1.25	1.08	1.02	1.00	1.00	$r\bar{X}S^2$
		3.93	3.22	2.09	1.42	1.13	1.03	1.00	1.00	$m\bar{X}R$
		2.79	2.44	1.79	1.32	1.10	1.02	1.00	1.00	$r\bar{X}R$
	2.00	2.22	2.03	1.64	1.31	1.12	1.04	1.01	1.00	$m\bar{X}S^2$
		1.48	1.42	1.29	1.15	1.06	1.02	1.00	1.00	$r\bar{X}S^2$
		2.33	2.12	1.68	1.32	1.12	1.04	1.01	1.00	$m\bar{X}R$
		1.76	1.66	1.43	1.22	1.08	1.02	1.00	1.00	$r\bar{X}R$

References

1. Alwan, L.C. & Roberts, H.V. (1988) Time series modeling for statistical process control. Journal of Business and Economic Statistics **6**, 87-95
2. Amin, R., Frank, O. & Schmid, W. (1997) The effects of autocorrelation on the R–chart and the S^2–chart. Sankhya, Ser B **59**, Part 2, 229-255
3. Domangue, R. & Patch, S.C. (1991) Some omnibus exponentially weighted moving average statistical process monitoring schemes. Technometrics **33**, 299-314
4. Gan, F.F. (1995) Joint monitoring of process mean and variance using exponentially weighted moving average control charts. Technometrics **37**, 446-453
5. Harris, T.J. & Ross, W.H. (1991) Statistical process control procedures for correlated observations. Canadian Journal of Chemical Engineering **69**, 48-57
6. Kanagawa, A. & Arizono, I. (1997) Design of the $(\bar{x}, s)$ control chart based on Kullback–Leibler information. In: Frontiers in Statistical Quality Control **5**, Lenz, H.-J. & Wilrich, P.-Th. (Eds.), Physica-Verlag, Heidelberg, Germany, 183-192
7. Knoth, S., Schmid, W. & Schöne, A. (1998) Simultaneous Shewhart–Type Charts for the Mean and the Variance of a Time Series. Technical Report 107, Europe–University Frankfurt (Oder), Germany
8. Kramer, H. & Schmid, W. (1997) EWMA charts for multivariate time series. Sequential Analysis **16**, 131-154
9. Kramer, H. & Schmid, W. (1998) On the average delay of control schemes. In: E. von Collani et al. (Eds.), Advances in Stochastic Models for Reliability, Quality, and Safety, Birkhäuser, Boston
10. Lu, C.W. & Reynolds, Jr., M.R. (1998) Control charts for monitoring the mean and variance of autocorrelated processes. To appear in Journal of Quality Technology
11. MacGregor, J.F. & Harris, T.J. (1994) The exponentially weighted moving variance. Journal of Quality Technology **25**, 106-118
12. Mathai, A. & Provost, S.B. (1992) Quadratic Forms in Random Variables: Theory and Applications. Marcel Dekker, Inc., New York
13. Mittag, J.–H. & Stemann, D. (1997) Gage imprecision effect on the performance of the $\bar{X}$–S control chart. Technical Report, Fernuniversität Hagen, Germany
14. Montgomery, D. (1991) Introduction to Statistical Quality Control. Wiley, New York
15. Schöne, A. & Schmid, W. (1998) On the joint distribution of a quadratic and a linear form in normal variables. To appear in Journal of Multivariate Analysis
16. Schöne, A. (1997) Neue Entwicklungen der Statistischen Prozeßkontrolle bei korrelierten Daten. PhD thesis, University of Ulm, Germany
17. Schmid, W. (1995) On the run length of a Shewhart chart for correlated data. Statistical Papers **36**, 111-130
18. Schmid, W. (1997a) On EWMA charts for time series. In: Frontiers of Statistical Quality Control **5**, Lenz, H.-J. & Wilrich, P.-Th. (Eds.), Physica-Verlag, Heidelberg, Germany, 115-137
19. Schmid, W. (1997b) CUSUM control schemes for Gaussian processes. Statistical Papers **38**, 191-217
20. Stemann, D. (1997) Qualitätsregelkarten des Shewhart– und EWMA–Typs. PhD thesis, University of Dortmund, Germany

21. Vasilopoulos, A.V. & Stamboulis, A.P. (1978) Modification of control chart limits in the presence of data correlation. Journal of Quality Technology **10**, 20-30
22. Vander Wiel, S.A. (1996) Monitoring processes that wander using integrated moving average models. Technometrics **38**, 139-151
23. Wardell, D.G., Moskowitz, H. & Plante, R.D. (1994) Run–length distributions of special–cause control charts for correlated processes. Technometrics **36**, 3-27
24. Yashchin, E. (1993) Performance of CUSUM control schemes for serially correlated observations. Technometrics **35**, 35-52

Application of ISO 3951 Acceptance Sampling Plans to the Inspection by Variables in Statistical Process Control (SPC)

Olgierd Hryniewicz

Systems Research Institute, Warsaw, Poland

Abstract. A new approach to SPC which utilises acceptance sampling plans for inspection by variables from the International Standard ISO 3951 is proposed. It may be used as an alternative to Shewhart control charts or to Freund's acceptance control charts. The proposed new SPC procedure enables the process fraction nonconforming to be kept under control when the process standard deviation is unknown. A simple method for a cost-optimal design of this procedure is also proposed. It allows an appropriate sampling plan and an approximately optimal sampling interval to be found.

1 Introduction

Statistical procedures for Statistical Process Control (SPC) are the most frequently used procedures of Statistical Quality Control (SQC). The most popular among SPC procedures are Shewhart control charts introduced by W.A. Shewhart in the 1920's [12]. Shewhart's theory recognises two kinds of variability: random variability due to the wide variety of causes that are consistently present and not readily identifiable (random causes), and variability that can be attributed to some identifiable causes (assignable causes). Shewhart control charts are used to verify if a process is under statistical control, i.e. when its variability results only from random causes. A sample of n elements is drawn from a process, and a certain sample characteristic (mean value, standard deviation, range) is calculated. Then, the values of this characteristic for consecutive samples are plotted against the sample number. When an observed result falls beyond certain control limits an action should be taken as the process probably went out of control due to a certain assignable cause. The aim of this special action is to detect and to eliminate the reason for this disorder.

When we are interested in controlling the average level of a process an $\overline{X}$ –chart is used. When the spread of a process is of special interest a chart based on a within–sample range (R–chart) or a chart based on a within–sample standard deviation (s–chart) are used. It is recommended that both the average level and the spread of the process should be controlled simultaneously.

The usage of control charts is very simple, and the only problem is connected with their design. W.A. Shewhart proposed a very simple method for the construction of a control chart based on a principle of so–called 3–sigma control limits. Due to their simplicity the Shewhart control charts won undisputable support from practitioners, and are commonly used in practice despite some objections raised by statisticians. For a classical Shewhart $\overline{X}$ –chart a sample of n items is drawn from a process and the sample mean $\overline{x}$ is calculated. Then, the value of the sample mean is compared with two values $M - K$ and $M + K$. The value of M is either the target value for the quality characteristic of interest or the mean level of the controlled process estimated from at least 20 samples taken from this process. The critical value K is calculated as $K = A\sigma^*$, where σ^* is either the known value of the standard deviation σ of the measured quality characteristics or its estimator based on the results of measurements of a sufficiently large (min. 20) number of samples drawn from the controlled process, and A is an appropriately chosen constant. The values of A are given in many textbooks, and in the International Standard ISO 8258 [11]. When the observed value of $\overline{x}$ lies between $M - K$ and $M + K$ the process is considered to be under control, otherwise an alarm signal is generated. W.A. Shewhart proposed that K should be set to three standard deviations of $\overline{X}$, thus his control charts are sometimes called *"three-sigma control charts"*. From astatistical point of view the main weak points of the Shewhart control charts are the following:

- only probabilities of false alarms (type 1 errors) are controlled and kept sufficiently small;
- the expected magnitude of process disorders are not taken into account in the design of the chart;
- the Shewhart charts are sensitive only to large shifts of the process average or the process variation;
- the actual required quality level of the process is not taken into account;
- the problem of the length of a sampling interval (frequency of sampling) is insufficiently addressed;
- economic consequences (costs of sampling, costs of alarms, etc.) are not taken into account, etc.

Many authors have proposed remedies of a different type in order to remove these weaknesses. For example, cumulative sum control charts (so–called $CUSUM$–charts) have been proposed to detect shifts of small or medium magnitude. Many statistical methods for the control chart design have also been proposed with the aim of keeping decision errors under control. All these proposals are described in various textbooks like the book by Wetherill and Brown [14]. However, the last two of the objections listed above are seldom addressed in textbooks, and are discussed rather in papers of a theoretical character devoted to the problem of the "optimal design of control charts". The work in this area began with the pioneering work of Duncan [3]. Many papers which extend and refine the ideas of Duncan have been published during the last 40 years. A comprehensive review of relatively new results can be found in the paper of Ho and Case [4]. Despite the large number of different proposals as to how to design the control charts optimally these ideas have not found their way into practice. There are many possible reasons for this. Firstly, the mathematical models are usually very complicated and difficult to understand even by well trained practitioners. Secondly, the number of input parameters whose values have to be known is too large. Thirdly, the optimal solutions are found using numerical computational methods, and such results do not give any insight as to the structure of the optimal solutions. For all these reasons the problem of the optimal design of control chart becomes too complicated even for an experienced and well-trained user. There is also an additional, and only seldom recognised, reason for the failure in the implementation of the optimal control charts in practice. The users are looking for familiar "labels" that are well known to them. Procedures which do not have such labels are treated with suspicion as those which are not verified in practice, and thus not worthy of using.

In this paper we try to overcome the main problems mentioned above by proposing a new optimal, or sub-optimal, SPC procedure which utilises acceptance sampling plans for variables from the International Standard ISO 3951 [9]. This procedure is similar to that of acceptance control charts proposed by Freund, and described in the International Standard ISO 7966 [10]. In order to find approximately optimal solutions we apply some results of Hryniewicz [5] who generalised the results of von Collani [1],[2]. This approach allows the user to decrease the number of necessary input parameters to a level which may be acceptable in practice. The results are easy to calculate without the need to implement any numerical computational procedures. The basic statistical information which is needed for the calculations is contained in the international standards for SQC, and the additional information about the process does not need to be very precise. All of this makes the proposed procedure very simple and flexible, which is a prerequisite for its possible implementation into practice.

2 Acceptance sampling plans by variables as the SPC tools

Acceptance sampling plans by variables were introduced in the 1940's for the acceptance of lots consisting of items described by a quality characteristic which is measured on a continuous scale. The plans were designed for the case of normally distributed quality characteristic and have been widely used in practice. The construction of these plans have been discussed in many textbooks, for example in Schilling [13], and the most frequently used set of sampling plans is available in the International Standard ISO 3951 [9]. A sample of n items is taken from a lot of N items and the average value $\overline{X}$ of the quality characteristic of interest is calculated. Then, the procedure depends upon the knowledge of the standard deviation σ of the measured quality characteristic. In the case of a single upper specification limit U when the value of σ is known the observed value of the quality statistic

$$Q_U = \frac{U - \overline{X}}{\sigma} \tag{1}$$

is calculated and compared to an acceptability constant k. The result is negative (rejection) if $Q_U < k$, and positive (acceptance) otherwise. For a single lower specification limit L an equivalent quality statistic is given by

$$Q_L = \frac{\overline{X} - L}{\sigma} \tag{2}$$

and the decision criterion is similar. The decision criterion (1) has the same form as the decision criterion (control limit) in the acceptance control charts proposed in [10]. Thus, in the case of known σ the acceptance sampling plans by variables can be used as the alternative to the acceptance control charts.

In practice however, the value of σ is usually unknown, and is estimated either by a sample standard deviation s or by a sample range R. In the case of a single upper specification limit U a quality statistic

$$Q_U = \frac{U - \overline{X}}{s} \tag{3}$$

is calculated, and compared to an acceptability constant k. If $Q_U \geq k$ the decision is positive (acceptance), otherwise the decision is negative (rejection). A similar procedure is used in the case of a single lower specification limit. In the case of double specification limits the procedure is more complicated and usually the acceptance (or rejection) is determined graphically.

In [9] acceptance sampling plans by variables are indexed by two labels. First, the *Code Letter* determines the sample size n and is connected with the lot size interval $[N_{\min}, N_{\max}]$ (see: Table 1)

Table 1. Code Letters for different lot sizes

Code Letter	N_{min}	N_{max}
E	51	90
F	91	150
G	151	280
H	281	500
J	501	1200
K	1201	3200
L	3201	10000
M	10001	35000
N	35001	150000
P	150001	500000

Second, the *Acceptance Quality Limit (AQL)* which is the worst tolerable quality level of the production process, and together with the *Code Letter* determines the acceptance limit k. Typical acceptance sampling plans by variables (n, k) are presented in Table 2.

Table 2. Selected sampling plans from ISO/DIS 3951

Code	n			*AQL*	[%]		
Letter		0,10	0,15	0,25	0,40	0,65	1,0
E	9						1.615
F	13					1.830	1.712
G	18				2.025	1.910	1.770
H	25			2.215	2.102	1.969	1.829
J	35		2.399	2.289	2.160	2.028	1.862
K	50	2.569	2.461	2.336	2.209	2.052	1.885
L	70	2.631	2.510	2.389	2.239	2.082	1.904
M	95	2.670	2.553	2.410	2.261	2.093	1.965
N	125	2.711	2.574	2.432	2.274	2.154	2.021
P	160	2.733	2.597	2.447	2.334	2.209	2.083

Acceptance sampling by variables is not appropriate for judging the quality of individual lots. It may by used only for the inspection of continuing series of lots all supplied by one producer using one production process (see [9]). Therefore, sampling inspection by variables can be considered as the inspection of the process from which the inspected lots have been taken. The lots may be considered as portions of the inspected process, and the samples which are drawn from these lots as the samples from the process. Thus, the acceptance sampling of a continuing series of lots taken from the same production process may be used as the tool for process control. The only difference is in the type of action when the result of sampling is negative. We propose that in such a case the search for the assignable cause of the quality deterioration should be instigated rather than the rejection of the last portion of the process (inspected lot) .

The process control procedure using acceptance sampling by variables should be the following. First, determine the value of AQL for the controlled process. Second, determine the sampling interval, and treat it as the lot size. Then, using the rules of the ISO 3951 [9] determine the sample size n and the acceptance constant k. Now, at the end of the predetermined sampling interval draw a sample of n elements, and calculate for this sample the value of the appropriate quality statistic (using, for example, (3)). If this value is smaller than the acceptability constant k, the quality of the process is considered unacceptable, and the causes of this should be identified and removed.

By applying the procedure described above we can control the current quality level of a process. We can plot the observed values of the quality statistic against the sample number and, thus, observe the stability of the process, just as in the case of a classical control chart. However, by setting the control limit equal to the acceptance constant k we are able to verify additionally whether the current quality level of the process is still acceptable. This feature is not, unfortunately, present in the Shewhart control charts.

Acceptance sampling plans by variables proposed in the ISO 3951 [9] form a whole system with different inspection levels, three types of inspection (normal, tightened, reduced), switching rules, and the rules for the discontinuation of inspection. When a continuous series of lots is inspected the application of the whole sampling system assures a sufficient protection of a consumer against lots of a bad quality. The concept of normal, tightened and reduced inspection may also be proposed for the purposes of SPC. However, the procedure described in the ISO 3951 [9] does not seem to be useful in the context of SPC, and new rules typical for SPC should be proposed (for example, for changing the length of sampling intervals). Therefore, in order to simplify the procedure, we apply only these sampling plans from the ISO 3951 [9] which are proposed for a normal inspection.

3 Optimal choice of the control procedure

The procedure described in the previous section leaves the problem of the choice of the sampling interval open. However, when economic consequences of SPC are considered it is, as has been shown in [6], of critical importance. If the sampling interval is short the process deterioration may be revealed nearly immediately, but the inspection costs (with the costs of false alarms included) are high. On the other hand, if it is too long, the costs of inspection are low, but deteriorations are not revealed as quickly as necessary.

Many authors, after the pioneering paper of Duncan [3], tried to find an optimal solution. One of the most effective methods was proposed by von Collani [1],[2], and was further generalised in [5]. We use a certain simplification of this result for the optimal choice of the sampling interval. Let us introduce the following notation:

α - probability of a false alarm when the process operates at AQL level;
β - probability of not detecting a given shift in a quality level;
A_2 - average run length in the out–of–control state of the process, $A_2 = 1/(1-\beta)$;
τ - average time between consecutive disorders of the process;
a^* - unit inspection cost;
e^* - cost of a false alarm;
a - standardised unit inspection cost, $a = a^*/e^*$;
n - sample size;
S - standardised inspection cost, $S = a \cdot n$;
b^* - profit from one renewal of the process;
b - standardised profit from one renewal of the process, $b = b^*/e^*$;
h - sampling interval.

Detailed discussion of these quantities can be found in [1],[2], and [5]. It has been shown in [7] that, approximately, the problem of the optimal choice of a process inspection can be presented as the problem of the minimisation of the following function

$$G = \frac{b \cdot h \cdot (A_2 - 0.5)}{\tau} + (\alpha + S)\frac{\tau}{h} \tag{4}$$

It is easy to find that the optimal value of h which minimises (4is given by

$$h^* = \tau \cdot \sqrt{\frac{2}{2A_2 - 1} \cdot \frac{\alpha + S}{b}} \tag{5}$$

This equation gives nearly the same values of the approximately optimal sampling interval, as a similar but more precise result obtained in [5]. If we know the statistical properties of the inspection procedure (i.e. α, and A_2), two basic costs b and S, and the expected time between process failures τ, we can easily find the required optimal sampling interval.

The main obstacle in the design of optimal control procedures stems from a fact that the costs involved in the necessary calculations are usually either unknown or known only imprecisely. Therefore, the optimal procedures are difficult to find. However, if we need to find an appropriate acceptance sampling plan from the ISO 3951, and to use it as the SPC procedure we only need to indicate a range of possible values of the sampling interval. As these ranges are given in advance (see Table 1) we can formulate the following problem: *what are the values of the other parameters of the optimisation model for which a given SPC procedure (i.e. a given sampling plan, and a sampling interval which belongs to a given range of values) is optimal*?

Let $L(p)$ be the OC–function of an acceptance sampling plan, where p is a fraction nonconforming generated by the process under consideration. Suppose that $\alpha = 1 - L(AQL)$, and $\beta = L(\gamma \cdot \mathrm{AQL})$, where γ is the required discrimination ratio of the procedure. For the sampling plans given in ISO 3951 [9] the values of α and β can be easily found from the graphs of OC-curves for all necessary values of γ. Some selected results for $\gamma = 10$ are presented in Table 3.

Table 3. Values of $\alpha = 1 - L(AQL)$ and $\beta = L(\gamma \dot{A}QL)$ for selected sampling plans from ISO/DIS 3951 (for $\gamma = 10$)

Code Letter		AQL					
n		0,10	0,15	0,25	0,40	0,65	1,0
E	α						0.0742
9	β						0.2973
F	α					0.0701	0.0740
13	β					0.2779	0.1874
G	α				0.0607	0.0695	0.0664
18	β				0.2829	0.1805	0.1120
H	α			0.0534	0.0595	0.0633	0.0610
25	β			0.2717	0.1822	0.1034	0.0524
J	α		0.0426	0.0517	0.0533	0.0589	0.0465
35	β		0.2672	0.1680	0.1015	0.0447	0.0204
K	α	0.0370	0.0364	0.0412	0.0442	0.0404	0.0299
50	β	0.2197	0.1649	0.0910	0.0431	0.0161	0.0053
L	α	0.0339	0.0297	0.0366	0.0319	0.0290	0.0175
70	β	0.1215	0.0867	0.0355	0.0150	0.0037	0.0009
M	α	0.0274	0.0249	0.0249	0.0215	0.0165	0.0197
95	β	0.0615	0.0368	0.0134	0.0041	0.0007	0.0000
N	α	0.0248	0.0171	0.0176	0.0129	0.0213	0.0248
125	β	0.0240	0.0151	0.0039	0.0009	0.0000	0.0000
P	α	0.0189	0.0126	0.0115	0.0185	0.0300	0.0410
160	β	0.0090	0.0048	0.0009	0.0001	0.0000	0.0000

Moreover, assume that the standardised unit inspection cost a is known imprecisely and its value belongs to a certain interval $a \in [a_{\min}, a_{\max}]$.

Now, we can transform (5) as follows:

$$h^* = \sqrt{\frac{2(a+S)}{2A_2 - 1}} \cdot \sqrt{\frac{\tau^2}{b}} \tag{6}$$

We can see that the optimal value of the sampling interval h can be computed as the product of two factors. The first factor depends upon the statistical properties of the sampling plan, and upon the costs of sampling which are relatively easy to estimate. The second factor depends entirely upon the properties of the process itself, and its value is rather difficult to estimate even for experienced practitioners. However, some imprecise knowledge about the value of this factor may exist, even in a form of the interval of its possible values. Thus, we are usually able to calculate the interval for possible values of the optimal sampling interval given by (6), and then to take any value from this interval as the chosen sampling interval.

We can also propose an alternative method for the choice of the approximately optimal procedure. Suppose now that the ranges of the lot sizes presented in Table 1 are in fact the ranges of admissible sampling intervals. The question arises then about circumstances in which these admissible sampling intervals might be optimal. We will try to answer this question by finding for which values of the ranges given in Table 1 are optimal.

Let us denote the lower limit of the admissible sampling interval by $N_{\min}$, and the upper limit by $N_{\max}$. Now, we can find limits for f given by

$$f = \frac{\tau^2}{b} \tag{7}$$

for which a sampling interval $h \in [N_{\min}, N_{\max}]$ might be optimal. By simple transformation of (6) we find

$$f_m = f_{\min} = \frac{N_{\min}^2 (A_2 - 0.5)}{\alpha + n \cdot a_{\max}} \tag{8}$$

and

$$f_M = f_{\max} = \frac{N_{\max}^2 (A_2 - 0.5)}{\alpha + n \cdot a_{\min}} \tag{9}$$

where $\alpha = 1 - L(AQL)$, $A_2 = A_2(\gamma \cdot AQL) = 1/(1 - L(\gamma \cdot AQL))$, n is a sample size related to the sampling interval $[N_{\min}, N_{\max}]$, and $(a_{\min}, a_{\max})$ is an interval of the possible values of the standardised sampling unit cost a. An example of intervals $[f_m, f_M]$, for $\gamma = 10$, $a_{\min} = 0.01$, and $a_{\max} = 0.05$ are given in Table 4 (The entries in Table 4 which are of the form "$xxxEyy$" should be read as $xxx \cdot 10^{yy}$).

Table 4. Values of f_m and f_M for $\gamma = 10$

CL		*AQL*					
n		0,10	0,15	0,25	0,40	0,65	1,0
E	f_m						.4580E04
9	f_M						.4553E04
F	f_m					.1017E05	.8357E04
13	f_M					.9945E05	.8060E05
G	f_m				.2123E05	.1694E05	.1477E05
18	f_M				.2914E06	.2263E06	.1993E06
H	f_m			.5289E05	.4358E05	.3699E05	.3345E05
25	f_M			.7194E06	.5838E06	.4910E06	.4465E06
J	f_m		.1211E06	.9779E05	.8533E05	.7588E05	.7274E05
35	f_M		.3172E07	.2517E07	.2189E07	.1926E07	.1891E07
K	f_m	.4443E06	.3967E06	.3407E06	.3090E06	.2932E06	.2881E06
50	f_M	.1490E08	.1332E08	.1136E08	.1026E07	.9785E07	.9764E07
L	f_m	.1851E07	.1727E07	.1555E07	.1495E07	.1463E07	.1459E07
70	f_M	.8697E08	.8154E08	.7287E08	.7039E08	.6910E08	.6981E08
M	f_m	.1184E08	.1127E08	.1076E08	.1057E08	.1051E08	.1049E08
95	f_M	.7088E09	.6763E09	.6454E09	.6356E09	.6347E09	.6317E09
N	f_m	.1024E09	.1007E09	.9849E08	.9798E08	.9768E08	.9762E08
125	f_M	.9258E10	.9150E10	.8944E10	.8924E10	.8849E10	.8825E10
P	f_m	.1428E10	.1418E10	.1407E10	.1403E10	.1401E10	.1399E10
160	f_M	.7861E11	.7826E11	.7771E11	.7724E11	.7669E11	.7617E11

4 Example

Let us illustrate the procedure described above with a simple example. Suppose that a production process with a production rate 100 items per hour is to be controlled using an acceptance sampling plan from ISO 3951. From capability studies of this process we know that in an in–control state the fraction nonconforming should not exceed 0.4%, and the tenfold increase of this fraction should be detected as quickly as possible in order to determine and remove its causes. Suppose, that the unit cost of sampling is estimated as $a^* = 1$ money unit, and the cost of a false alarm is between 20 and 100 units, i.e. $e^* \in [20, 100]$. The average time between failures of the process is estimated as $\tau \in [180, 220]$ hours. The profit from keeping the process in–control during this period is estimated as $b^* \in [1000, 2000]$ money units.

From the data about the process we can find that $AQL = 0.4\%$, $\gamma = 10$, and $a \in [0.01, 0.05]$. Thus, we can find the appropriate sampling plan using the column labeled by the $AQL = 0.4\%$ in Table 4. From the data about τ, and b^* we find that the value of f, as defined in (7), belongs to the interval $[0.324 \cdot 10^7, 0.484 \cdot 10^8]$. This interval is entirely included in the interval which can be found in the cell of Table 4 which is labelled by the *Code Letter L* and $AQL = 0.4\%$. From Table 2 we see that the necessary sampling plan is $(n = 70, k = 2.239)$, and in Table 1 we can find that the range of the admissible values of the sampling interval is $[3201, 10000]$ items, i.e. $[32, 100]$ hours. Thus, setting h to any value from this interval is admissible. In our example, the interval of $h = 80$ hours may be chosen,i.e. the inspection of the process at the end of every 10-th production shift. In practice it means that we may apply a classical Shewhart control chart (with a much smaller sample size, and a shorter sampling interval – every 8 hours, for example) to control stability of the process under cosideration, and at every 10-th production shift we may use the already designed procedure to verify the quality requirements.

The example given above reveals some important differences between the procedure proposed in this paper and Shewhart control charts. In our case we need samples of at least 35 items to detect the increase of the process fraction nonconforming from 0.4% to 4%. Probability of a false alarm ($\alpha = 0.0319$) is substantially larger than in the case of a Shewhart control chart, but the average time to the detection of a disorder is quite small ($A_2 = 1.015$). For a Shewhart control chart the suggested sample size is much smaller ($n = 5$). It results in a very small probability of a false alarm, but the delay in the detection of process disorders might be quite substantial. If we additionally follow some suggestions that the process should be sampled once a day it may significantly increase the time in which the process remains out–of–control.

5 Conclusions

The proposed SPC procedure which uses acceptance sampling plans for inspection by variables from the International Standard ISO 3951 may be used as an alternative to Shewhart control charts or to Freund's acceptance control charts. The main advantage of the proposed procedure stems from the fact that it enables the process fraction nonconforming to be kept under control. The choice of an appropriate procedure is very easy, and it does not require any precise knowledge about the process. The numerical example given in this paper shows that even in the case of very imprecise information about some cost parameters an approximately optimal procedure might be found relatively easily. The proposed procedure utilises very well known sampling plans from the International Standard ISO 3951, and this fact may favourably influence those potential users who prefer to apply well known and widely approved statistical procedures.

The aim of this paper is to propose a simple and ready to use procedure. Mathematical models related to the problem have been simplified as far as possible. More thorough analysis of the properties of the proposed new SPC procedure requires, however, further investigations.

References

1. von Collani E. (1986) A Simple Procedure to Determine the Economic Design of an $\overline{X}$ -Control Chart. Journal of Quality Technology **18**, 145-151
2. von Collani E. (1989) The Economic Design of Control Charts. B.G. Teubner, Stuttgart
3. Duncan A.J. (1956) The Economic Design of $\overline{X}$–Charts Used to Maintain Current Control of a Process. Journal of the American Statistical Association **51**, 228–242
4. Ho, C., Case, K.E. (1994) Economic Design of Control Charts: A Literature Review for 1981 - 1991. Journal of Quality Technology **26**, 39–53
5. Hryniewicz O. (1992) Approximately Optimal Economic Process Control for a General Class of Control Procedures. In: Lenz H.J., Wetherill B., Wilrich P.T. (Eds.) Frontiers in Statistical Quality Control IV, Physica Verlag
6. Hryniewicz O. (1989) The Performance of Differently Designed p–Control Charts in the Presence of Shifts of Unexpected Size. Economic Quality Control **4**, 10–21
7. Hryniewicz O. (1999) Simple Method for the Design of Optimal Inspection Procedures. (submitted for publication)
8. International Standard ISO 2859-1, Sampling Procedures for Inspection by Attributes - Part 1: Sampling Schemes Indexed by Acceptable Quality Level (AQL) for Lot-by-lot Inspection
9. Draft International Standard ISO/DIS 3951-1: 1998 (E), Sampling Procedures for Inspection by Variables for Percent Nonconforming - Part 1: Introductory Specification for Single Sampling Plans Indexed by Acceptance Quality Limit (AQL) for Lot-by-lot Inspection - Single Quality Characteristic and Single AQL
10. International Standard ISO 7966 (1993) (E), Acceptance Control Charts
11. International Standard ISO 8258 (1991) (E), Shewhart Control Charts
12. Shewhart W.A. (1931) Economic Control of Quality of Manufactured Product. D. Van Nostrand, New York
13. Schilling E.G. (1982) Acceptance Sampling in Quality Control. Marcel Dekker Inc., New York
14. Wetherill G.B., Brown D.W. (1991) Statistical process Control. Theory and Practice. Chapman and Hall, London

Optimal Set-Up of a Manufacturing Process with Unequal Revenue from Oversized and Undersized Items

Shaul P. Ladany

Department of Industrial Engineering and Management
Ben-Gurion University of the Negev, Israel

Abstract. The optimal target-value to which the mean of a critical variable should be set-up, when the revenue per unit of oversized and undersized items are unequal, has been determined for the case where the variable is distributed normally. It has been proven analytically that the optimal target value should not be in the middle in-between the specification limits, but at a certain distance from it in the direction of the type of non-conforming item which generates a higher economic value per unit. That optimal distance was determined to depend on the spread of the specification limits, and on the relationship between the unit revenues obtained from oversized and undersized items. The use of the optimal target value is facilitated by the derivation of a simple closed-form formula. The increase in profit with optimal target value was found to reach up to about ten percent when the spread of the specification limits is narrow. Sensitivity of the profit to use of non-optimal target values was investigated.

1 Introduction

In many manufacturing processes in which a single variable of the product (for example, the diameter) is critical in determining the suitability of the manufactured product, the specification limits in respect to that variable define whether an item is conforming or not. In situations where the process is stable (i.e., there is practically no gradual deterioration), or where a stable process might suddenly change and such a change is immediately detected, and then the process is reset (for example when the tip of a turning tool suddenly breaks), it is common practice to set-up the mean of the variable at the middle distance in-between the specification limits. Such a traditional approach is suggested by Duncan [2], Montgomery [7], Ladany and Ben-Arieh [6], and is the underlying assumption for the initial set-up level of a process controlled by control charts, when Duncan [1], Ladany [3], Ladany and Bedi [4], Ladany and Raz [5] and others investigated the conditions for optimal use of control charts.

The underlying assumptions for this traditional approach are **(a)** that the distribution of the critical variable generated by the manufacturing process is symmetrical, and **(b)** that the economic value derived from each non-conforming item is either zero or equal for oversized and undersized items.

While in most manufacturing processes the first of these assumptions is indeed fulfilled, and in many cases the distribution of the process-generated variable is normal, the second assumption in many cases is contradicted. Quite often, the economic value of a non-conforming oversized item (i.e. item with the critical variable having a value above the upper specification limit) is quite different from the economic value of a non-conforming undersized item (i.e., item with the critical variable having a value below the lower specification limit). If, for example, the variable under consideration is the outside diameter of a bar, oversized items can be returned or reground at an additional cost, so that their economic value could be much higher, per unit, than the economic value of undersized items that could be sold only for scrap. Similarly, if the variable under consideration is the inner diameter of a hole, undersized holes can be rebored at an extra cost, but the reduction of an oversized hole with a sleeve or by any other means would require a much higher extra repair cost.

Thus, the optimal set-up is sought for processes in which oversized and undersized items have different economic values. It is determined in Section 2 for the case in which the critical variable is distributed normally. The benefits of using optimal set-up levels versus traditional mid-specification-limit levels are investigated in Section 3. Conclusions are provided in Section 4.

2 Analytical Analysis

Let the value of a critical dimension X of a part manufactured on a machine be distributed with a mean μ and a standard deviation σ. Let the spread of the specification limits be $2M\sigma$, where M is any real positive number.

If a manufactured item with a value x below the Lower Specification Limit of X, LSL_x, has an economic value of $r_L r_s$, and with a value x above the Upper Specification Limit of X, USL_x, has an economic value of $r_u r_s$ (where r_s is the economic value of an item which has a value x in-between the specification limits, and r_L and r_u are fractions valued between zero and one), then the expected economic value for one manufactured item, Π, is:

$$\begin{aligned} \Pi = {} & r_L r_s P(X < \mathrm{LSL}_x) + r_s P(\mathrm{LSL}_x \le X \le \mathrm{USL}_x) \\ & + r_u r_s P(X > \mathrm{USL}_x) \ . \end{aligned} \tag{1}$$

Dividing both sides of the equation by r_s, we obtain the relative expected economic value for one manufactured item (relatively to the economic value of one conforming unit), R:

$$R = r_L P(X < \mathrm{LSL}_x) + P(\mathrm{LSL}_x \le X \le \mathrm{USL}_x) + r_u P(X > \mathrm{USL}_x) \ . \tag{2}$$

When the mean μ of X is set-up at a distance $K\sigma$ from the mid-point in-between the specification limits toward the direction of the specification limit which processes the higher r_i $(i = L, u)$, and where K is a positive real number, as illustrated in Figure 1 (in the figure it is assumed that $r_u > r_L$),

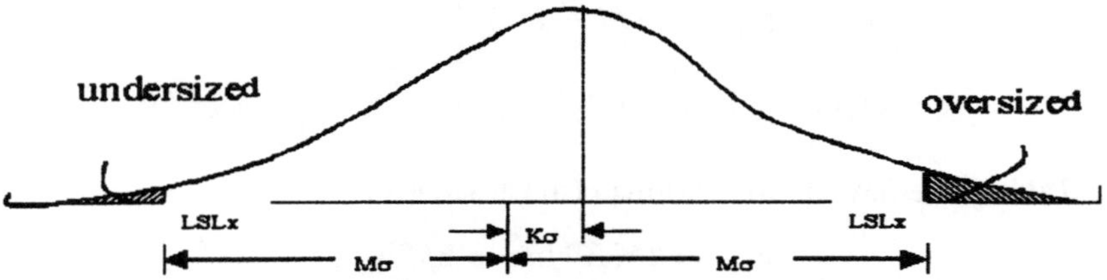

Fig. 1. Mean of the Distribution of X, set-up at a Distance of $K\sigma$ from the Mid-point Between Specification Limits

then the relative expected value for one manufacturing cycle will be

$$R = r_L F(-M - K) + [F(M - K) - F(-M - K)] + r_u [1 - F(M - K)] \ , \tag{3}$$

where $F(\bullet)$ is the distribution function of X at the value $x = (\bullet)\sigma$.

Equation (3) can be rewritten as

$$R = (r_L - 1) F(-M - K) + (1 - r_u) F(M - K) + r_u \ ,$$

or as

$$R = (1 - r_u) F(M - K) - (1 - r_L) F(-M - K) + r_u \ . \tag{4}$$

The maximal value of R can be attained when $\partial R / \partial K = 0$:

$$\frac{\partial R}{\partial K} = (-1)(1 - r_u) f(M - K) - (-1)(1 - r_L) f(-M - K) = 0 \ . \tag{5}$$

When X is a continuous variable with a probability density function $f(x)$, then Equation (5) provides the optimal value of K, K*, which satisfies

$$(1 - r_L) f(-M - K^*) = (1 - r_u) f(M - K^*) \ ,$$

or

$$\frac{f(M - K^*)}{f(-M - K^*)} = \frac{1 - r_L}{1 - r_u} = C \ . \tag{6}$$

It can be shown in many cases, including the normal case which will be dealt here, that the second derivative of R in respect to K at $K = K^*$ is negative, confirming that R is maximized.

When X is distributed normally, the standard normal probability density function, $f(z)$, prevails:

$$f(z) = \frac{1}{\sqrt{2\pi}} e \exp\left\{-\frac{1}{2} Z^2\right\} . \tag{7}$$

Substituting (7) into (6) we obtain

$$C = \frac{\frac{1}{\sqrt{2\pi}}\, e\exp\left\{-\frac{1}{2}(M-K^*)^2\right\}}{\frac{1}{\sqrt{2\pi}}\, e\exp\left\{-\frac{1}{2}(-M-K^*)^2\right\}}$$
$$= e\exp\left\{-\frac{1}{2}\left[(M-K^*)^2-(-M-K^*)^2\right]\right\} = e^{2MK^*} . \tag{8}$$

Taking the natural logarithms of both sides of (8)

$$\ln C = 2MK^* \ln e = 2MK^* ,$$

and

$$K^* = \frac{\ln C}{2M} . \tag{9}$$

The special case of this finding, when $r_L = r_u$ (i.e. when $C = 1$), provides as required $K^* = 0$.

The findings show that the optimal set-up level depends only on the spread in-between the specification limits, M, and on C. From (6) it can be seen that C has the meaning: the ratio of the fraction of loss from an undersized item, to the fraction of loss from an oversized item. $C \geq 1$.

The values of the optimal value of K, K^*, are illustrated in Figure 2 for different combinations of values of M and C.

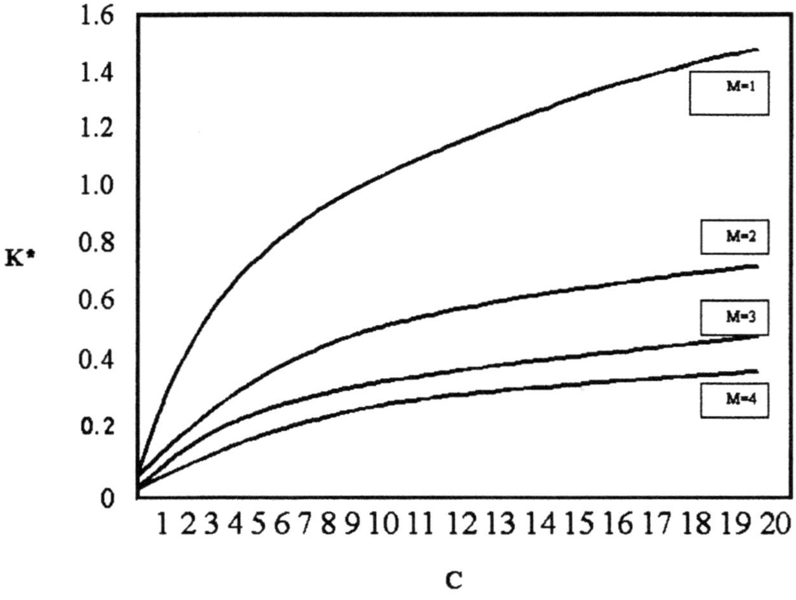

Fig. 2. Graphs of the optimal values of K, K^*, as a Function of M and C

3 The Benefits of Optimal Set-Up

In order to evaluate the economic benefit that optimal set-up would provide to the user, the relative *increase* in the expected economic value per one manufacturing cycle was calculated. This relative increase, E, was determined by relating the expected economic value per one manufacturing cycle when optimal set-up level is used, $\pi(K = K^*)$ (or $R(K = K^*)$), to the expected economic value per one cycle when traditional mid-specification limits set-up level is used, $\pi(K = 0)$ (or $R(K = 0)$). Thus

$$E = \frac{\pi(K = K^*)}{\pi(K = 0)} = \frac{R(K = K^*)}{R(K = 0)} . \tag{10}$$

Substituting (4) into (10) we obtain

$$E = \frac{(1 - r_u) F(M - K^*) - (1 - r_L) F(-M - K^*) + r_u}{(1 + r_u) F(M) - (1 - r_L) F(-M) + r_u} , \tag{11}$$

which can be further simplified by substituting the definition of C, (6), and the fact that $F(-M) = 1 - F(M)$, providing

$$E = \frac{F(M - K^*) - CF(-M - K^*) + \frac{r_u}{1 - r_u}}{[1 + C] F(M) - C + \frac{r_u}{1 - r_u}} \tag{12}$$

E was evaluated numerically for different combinations of M, C and r_u. The results are illustrated in Figure 3.

The graphs clearly demonstrate that for narrow spread of the specification limits of 2σ (i.e. $M = 1$), the use of the optimal set-up provides significant improvement in the economic value of the manufacturing process. When r_u (the fraction of the economic value of a conforming unit obtained for an oversized unit) is 0.9, and $C = 9$ (indicating that the fraction of the economic value of a conforming unit obtained for an undersized unit, r_L, is 0.1), the economic value of the production can be increased by about 10%. However, for the same $r_u = 0.9$, but for $C = 2.5$, the economic value of the production can be increased only by about 1%.

As r_u gets smaller, the percentage of increase in the economic value of the production increases for *identical* values of C, but it has to be remembered that since $r_L \geq 0$, for low values of r_u the upper bound of C decreases too, to a maximum of $1/(1 - r_u)$. Thus, for lower values of r_u the percentage of increase in the economic value of production cannot reach the same maximum levels as for higher values of r_u. Hence, with $r_u = 0.4$, at maximum only about 1.5% increase can be attained in the economic value of the production under optimal set-up.

As the spread in-between the specification limits gets wider, the percentage increase in the economic value of the production diminishes. Already with a spread of 4σ (i.e. $M = 2$), the maximum increase with $r_u = 0.9$ is about 1.5%, and with lower r_u's it is even less.

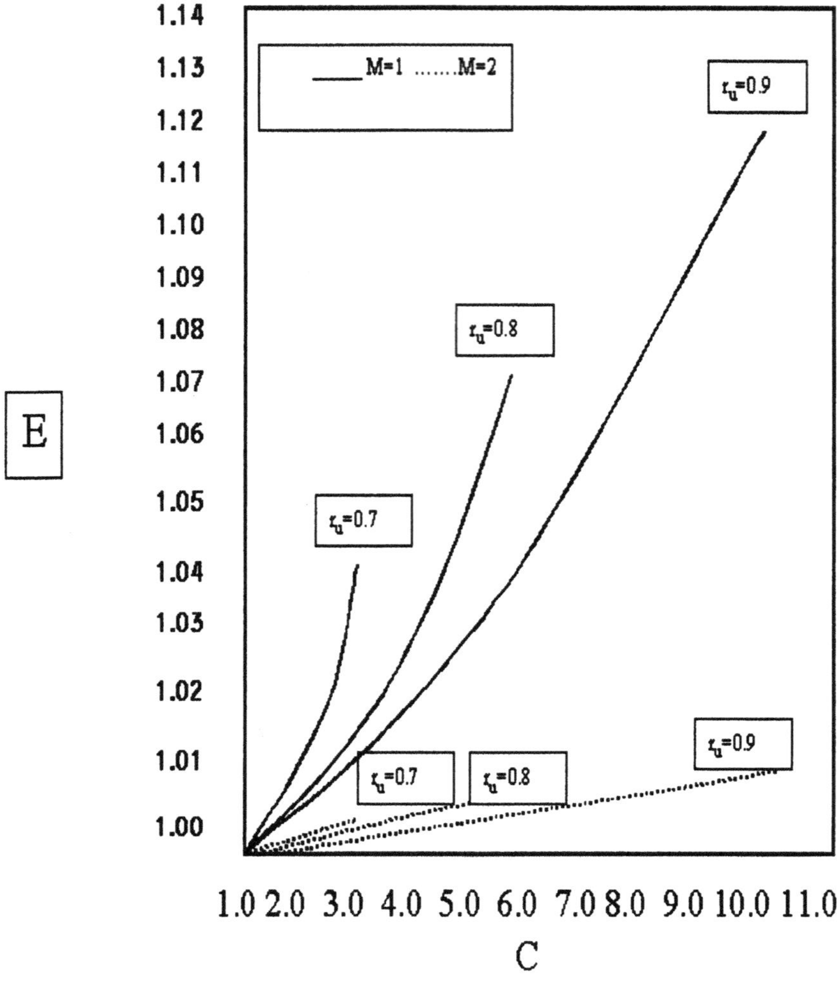

Fig. 3. The Relative Increase in the Economic Value per One Manufacturing Cycle when Optimal Set-up is Used (Instead of Mid-Specification Limit Set-up), for Different Combinations of M, C and r_u

When the spread is 6σ (i.e. $M = 3$) or more, the percentage increase in the economic value of the production with optimal set-up is low, and in most cases it is about 0.1% or less. Though this percentage increase in the economic value of the production, i.e. in the gross contribution to profit, seems to be unimpressive at first glance, the matter of fact is, that since cost is constant, the percentage increase in the *net* profit is considerably higher. Moreover, based just on the fraction of improvement in the economic value of the output, it is comparable, for example, to highly desired and praised improvements where the number of non-conforming units per million parts is reduced from 1,005 to 5, i.e. the fraction of conforming is increased from 0.998995 to 0.999995.

The effect of using optimal non-centered set-up with high M value can be revealed by the equivalent effect on the Process Capability Ratio, PCR, where PCR is defined as one sixth of the spread in-between the specification limits. For example, for the case of external turning where items with oversized diameter can be repaired at a very low additional cost, while items with undersized diameter have virtually no economic value, let say $r_u = 0.99$ and $r_L = 0.01$, $C = 0.99/0.01 = 99$. If $M = 3$, i.e. the conventional centered set-up has a PCR $= 1$, generating a fraction of 0.0027 of non-conforming units. The corresponding $K^* = 0.77$, and with such an optimal non-centered set-up the economic value of the output increases by 0.1144%. It is equivalent to the decrease in the per unit cost of also about 0.1144%. This decrease in the unit cost is the same one that results when a process with a centered set-up increases its fraction of conforming units from 0.9973 to $0.9973(1.001144) = 0.998405$, i.e. when M changes from 3 to 3.158. Hence, in the present example, the effect of using optimal non-centered set-up, can be viewed as a centered set-up process changing its PCR from 1 to 1.053.

4 Concluding Remarks

For normally distributed variables the derived closed solution determines the optimal level to which the mean of a stable process (which does not deteriorate gradually) should be set up. This level in terms of standard deviations, away from the mid-point in-between the specification limits, (in the direction of that specification limit for which non-conforming items with critical values beyond it have the higher economic value), depends only on the spread in-between the specification limits, and the ratio of the fraction of loss from an undersized item, to the fraction of loss from an oversized item.

The economic value of the output can be increased at best by an order of magnitude of 10% when the optimal set-up level is used, provided that the spread in-between the specification limits is narrow. Intuitively it seems that this value can be succeeded when r_u approaches 1, since then C increases. However, as illustrated in Fig. 3, increase in r_u reduces E for a given C. Thus the net effect of increase in r_u is only marginal on E. With wide spreads of 6 sigma or more in-between specification limits, the use of optimal set-up level will not provide any drastic economic benefits.

Sensitivity analysis of the relative expected economic value for one manufacturing cycle, R, to changes in the value of K, has revealed that if the value of K used is $0.5K^*$, the value of R would be about 2% below its maximal obtainable value with $K = K^*$ when $M = 1$ and $C = 10$, and $r_u = 0.9$, but only at most 0.2% below the maximal value with $M = 2$ (and any C and r_u). It was also evidenced that it is always preferable to upround K to $1.5K^*$ than to downround it to $0.5K^*$.

Optimal non-centered set-up decreases the number of undersized units (with virtually no economic value), and increases the number of oversized units (which can be repaired at a minimal additional cost). In many production processes automatic screening is implemented to prevent the forwarding of the non-conforming items to the next production stage (or to the final customer) and to prevent the creation of production "problems", losses and/or penalties, irrespective of the specific number of non-conforming units that might have been passed over. In such cases the screening is a fixed cost. When the cost of repair of oversized items is very low (and it already has been considered in the derivation of the optimal set-up position), while the economic value of undersized items is negligible, the combined net effect of the optimal non-central set-up together with the repair of oversized units is to reduce the total number of non-conforming units. Thus, it can also be viewed as the increase in the Process Capability Ratio from $M/3$ to $(M + K^*)/3$, where it is the one sided PCR for the Lower Specification Limit.

However, it is not advisable to set-up the mean offset by more than $K^*\sigma$ from the midpoint in the positive direction. Though it would have further decreased the number of undersized non-repairable non-conforming units, while simultaneously the screening and repair of oversized units would have increased the combined number of conforming units, the economic value of the production would decline.

When undersized items have a higher economic value per unit then oversized items, i.e., $r_L > r_u$, the derived results – formulae and graphs – can be used, with 2 changes: **(a)** The value used for r_u should be the highest between the *actual* r_u and r_L, and the value used for r_L should be the lowest between the *actual* r_u and r_L; and **(b)** The derived optimal K^* means that the optimal set-up level is at a distance K^* *below* the mid-point in-between the specification limits.

The optimal behavior for processes with non-normally distributed variables is not clear, and it is the subject of another investigation. Likewise, optimal set-up levels for unstable processes where the mean gradually deteriorates, and thus recalibration is periodically required, are presently the focus of another project.

Acknowledgement

The partial support of the Paul Ivanir Center of Robotics at Ben-Gurion University of the Negev is acknowledged.

References

1. Duncan, Acheson J. (1956) The Economic Design of Charts Used to Maintain Current Control of a Process. Journal of the American Statistical Association **51**, 228–2
2. Duncan, Acheson J. (1989) Quality Control and Industrial Statistics. 5th edition, Irwin, Homewood, Ill., 500–506
3. Ladany, Shaul P. (1973) Optimal Use of Control Charts for Controlling Current Production. Management Science, Theory, **19**(7), 763–772
4. Ladany, Shaul, P., Bedi, Dina N. (1976) Selection of the Optimal Set-up Policy. Naval Logistics Quarterly, **23**(2), 219–233
5. Ladany, Shaul P., Raz, Tzvi (1992) Optimal Dynamic Application of a Fraction-Defective Control-Chart to Control a Production Process. In Frontiers in Statistical Quality Control 4, edited by H.-J. Lenz, G.B. Wetherill and P.-Th. Wilrich, Physica-Verlag, Heidelberg, 169–179
6. Ladany, Shaul P., Ben-Arieh, David (1990) Optimal Industrial Robot-Calibration Policy. The International Journal of Advanced Manufacturing Technology **5**, 345–357
7. Montgomery, Douglas C. (1991) Introduction to Statistical Quality Control. 2nd edition, John Wiley & Sons, New York, 373–380

Frequency Distribution Supporting Recognition of Unnatural Patterns on Shewhart X-bar Chart

Jacek Nieckula

Systems Research Institute, Polish Academy of Sciences, Warszawa, Poland

Abstract. This paper describes a method supporting recognition of unnatural patterns on a Shewhart $\overline{X}$ chart by analysis of its frequency distribution. It has been implemented as a program that simulates data from a statistically controlled process as well as from a process affected by patterns. The following patterns are simulated: sudden shift, trend, stratification, systematic variable, cycle and mixture.

A user has the possibility to set up both the parameters of the process under statistical control and the parameters of a pattern. Then, he/she can observe static $\overline{X}$ and R charts or run dynamical simulation. During the simulation, it is possible to choose the moment when the pattern begins and when the pattern ends.

The further development of the method is the implementation of a trained neural network for recognition of the pattern.

1 Introduction

According to the Shewhart theory, the condition for process signals to be located between UCL and LCL is not sufficient to say that the process is under statistical control. The other requirement is the lack of variability that represents the real change in the process. This variability can appear in the form of patterns in the production data. The presence of patterns indicates that there are special causes affecting the process. These causes should be found and eliminated.

In many situation a pattern may only introduce certain fluctuations that are difficult to be recognized by the human controlling the process. This is the reason of the Type-II error. Such a process is called to be in statistical control and no alarm is generated. Also, no efforts are made to eliminate special causes. As the result, the process can suddenly collapse increasing losses.

Artificial intelligence methods seem to be very promising in recognition of special causes based on patterns in the process data. A few works refering to this field appeared in the eighties and in the nineties. They discussed two major problems: control chart selection and recognition of patterns. They were reviewed by Nieckula and Hryniewicz in [7].

Two dissertation has dealt with the problem of pattern recognition on Shewhart control charts. The work of Swift [5], describes an expert system that supports analysis of data on the average $\overline{X}$ chart. The author makes several introductory assumptions to make the study feasible: the relevant R chart was in control, control limits were known in advance, the characteristics of data generated for testing reflected well designed and thorough sampling plan, the overall flow of the patterns would be from "in-control" to "out-of-control" and back to "in-control".

Hwarng [6] applied neural network approach to pattern recognition on average $\overline{X}$ chart together with either range R chart or standard deviation s chart. Similarly to Swift [5], Hwarng assumed that the charts R or s were free of unnatural patterns, so that the investigation could be focussed on $\overline{X}$ chart only.

In both works [5] and [6] the patterns were recognized based directly on data performing average $\overline{X}$ chart. The only transformation used on the data was their normalization.

The method described below is not based on direct data from control charts but on data from frequency distributions of the average $\overline{X}$ chart or range R chart.

Moreover, it is oriented on the data taken from a real production process. The program is written in object oriented language C++. It has also classed responsible for input of raw process data instead of simulated data. The classes responsible for visualization of control charts and pattern recognition can work both on simulated data and real process data.

2 Process Model

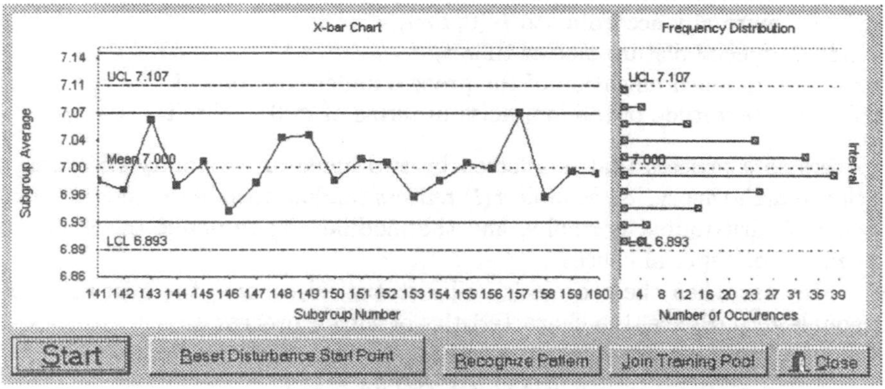

Fig. 1. Shewhart $\overline{X}$ chart with frequency distribution

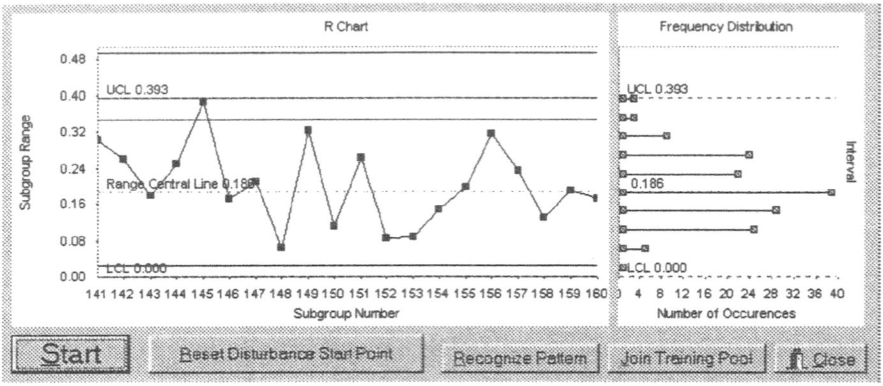

Fig. 2. Shewhart R chart with frequency distribution

The generator of the process data is based on modified equations used in [5] and [6]. The general equation

$$y(t) = \mu + x(t) + d(t), \tag{1}$$

where:

$y(t)$ — quality measurement at time t, or the tth quality measurement,
μ — process mean,
$x(t)$ — random normal variate at time t, distributed normally according to $N(0, \Gamma\sigma)$,
$d(t)$ — special disturbance at time t,
σ — standard deviation of the process under control and
Γ — magnitude of random noise in terms of σ, $0 < \Gamma \leq 1$

refers to a process that is affected by two kinds of variability according to Shewhart's theory, [8]. Module $x(t)$ reflects random variability due to "chance causes" distributed normally, and the module $d(t)$ responds to "assignable causes" or "special causes".

In case when the process is under statistical control, the expression $d(t)$ equals zero. Hence, the characteristics of such a process is

$$y(t) = \mu + x(t) \text{ for } \Gamma = 1. \tag{2}$$

Average $\overline{X}$ and range R charts for simulated process described by equation (2) are shown in Figures 1 and 2.

3 Generation of Patterns

Process out of statistical control is attributed to special causes. The main advantage of the application of control charts is the aid in detection of unnatural variability. Once such a state is detected, its causes should be identified and eliminated. The detection of unnatural variability means that the process is out of statistical control. If it is possible to recognize a pattern formed by values plotted on control chart, the cause can be easier assigned because certain patterns origin from known process bevavior.

Quality Control Handbook [1] mentions 14 patterns that are caused by assignable causes. They are: cycles, freaks, gradual change in level, grouping or bunching, instability, interaction, mixtures, stable form of mixture, stratification, sudden shift in level, systematic variables, tendency of one chart to follow another, trend and unstable form of mixture.

From the list mentioned above the following patterns can be simulated by the process data generator described in this paper: sudden shift, trend, stratification, mixture, systematic variable and cycle. Pattern equations, the generator is based on, are listed in Table 1.

Table 1. Equations used by pattern generator

Pattern	Equation
Sudden shift	$d(t,t_0) = (-1)^s u\sigma$, where
	s : 0 if $t \geq t_0$, and shift upward, 1 if $t \geq t_0$, and shift downward,
	u : the magnitude of the process shift in terms of σ,
	$u = 0$ if $t < t_0, u > 0$ if $t \geq t_s$,
	t_0 : the time point when the sudden shift occurs;
Trend	$d(t,t_0) = (t - t_0)\alpha\sigma$, where
	α: slope of the trend in terms of σ,
	t_0 : time point when the trend occurs;
Stratification	$y(t,t_0) = \mu + x(t,t_0) + d$, where
	$x(t,t_0)$: $N(0, \Gamma\sigma)$, for $t \geq t_0\Gamma$ is the random noise in terms of σ,
	$0 < \Gamma \leq 1$,
	d : an offset from the process mean,
	t_0 : time point when the stratification occurs;
Systematic variable	$d(t,t_0) = (-1)^t \Phi\sigma$, where
	Φ : the magnitude of the systematic pattern in terms of σ,
	above or below the process mean, $\Phi > 0$,
	t_0 : time point when the systematic occurs;
Cycle	$d(t,t_0) = \sin[2\pi(t - t_0)/\Omega]k\sigma$, where
	k : amplitude of the cycle in terms of $\sigma, k > 0$,
	Ω : period of the cycle,
	t_0 : time point when the cycle occurs;
Narrow mixture	$d(t,t_0,p) = (-1)^w m\sigma$, where
	p : a random number, $0 < p < 1$,
	m : the magnitude of the mixture pattern in terms of $\sigma, m > 0$,
	w : 0 if $p < p_s$ and $x(t) \in < 0, x_l >$, 1 if $p < p_s$ and $x(t) \in < -x_l, 0)$,
	p_s : a prespecified probability value which determines
	the shifting between two distributions,
	t_0 : time point when the mixture occurs,
	x_l : prespecified value of $x(t)$;

4 Simulation

To begin simulation, it is necessary to input two groups of data. The first one is required to simulate process under statistical control. This group of data is used to generate the values of the process described by the equation (2). The user should input the number of subgroups, subgroup size, process mean μ and process standard deviation σ. The second group of input data depends on the pattern. For every pattern, time point t_0 when the pattern occurs and pattern's magnitude in terms of σ should be given . Later, the start point can be changed during simulation.

All figures below refer to a process that consists of two componnents. $x(t)$ and $d(t)$. Before the simulation begins, user can change t_0 pressing the button "Reset Disturbance Start Point". As the result, the start of the pattern is suspended. Pressing "Reset Disturbance Start Point" triggers the pattern. Clicking on this button for the third time will switch the pattern off and the simulation will come back to the process under statistical control. Two other buttons, "Recognize Pattern" and "Join Training Pool", will serve to switch on the pattern recognizer and to append the pattern to the set used for training of the neural network.

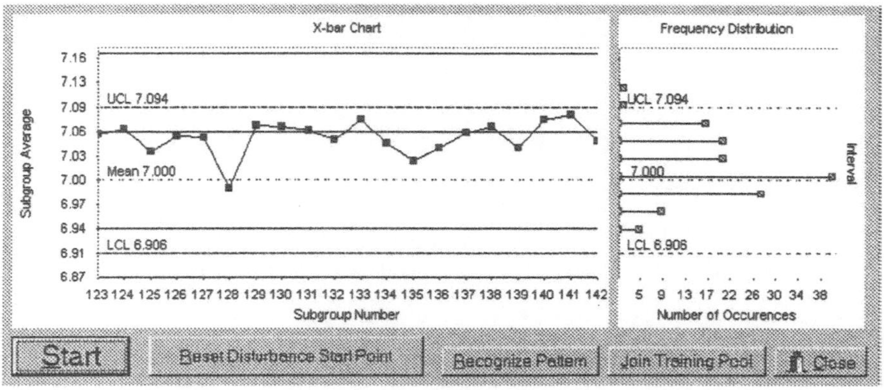

Fig. 3. Sudden shift—first phase

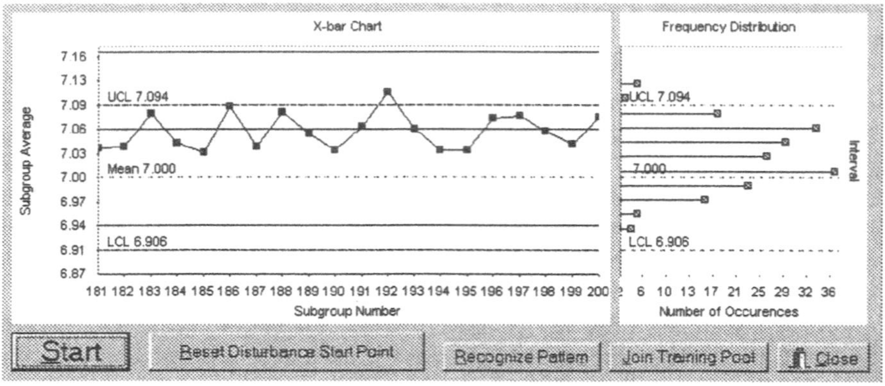

Fig. 4. Sudden shift—second phase

Figures 3 and 4 depict two phases of sudden shift. The upper part of the frequency distribution plot in Fig. 3 starts to grow. In the Fig. 4, the plotting has already two peaks. This pattern should be recognized between phase 1 and phase 2 based on growing bimodality of the frequency distribution.

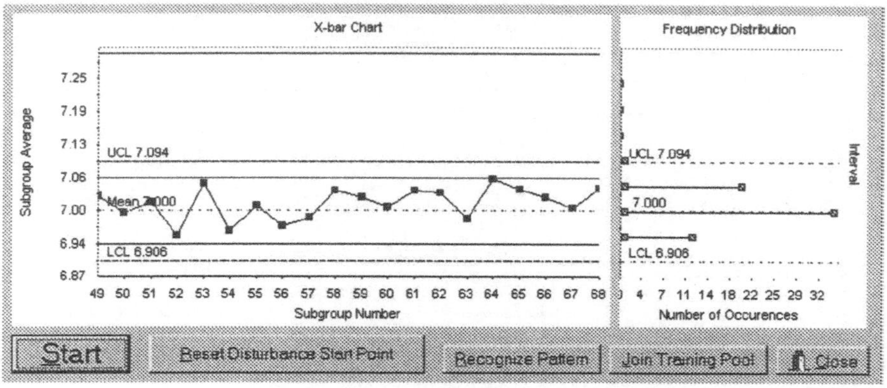

Fig. 5. Trend—first phase

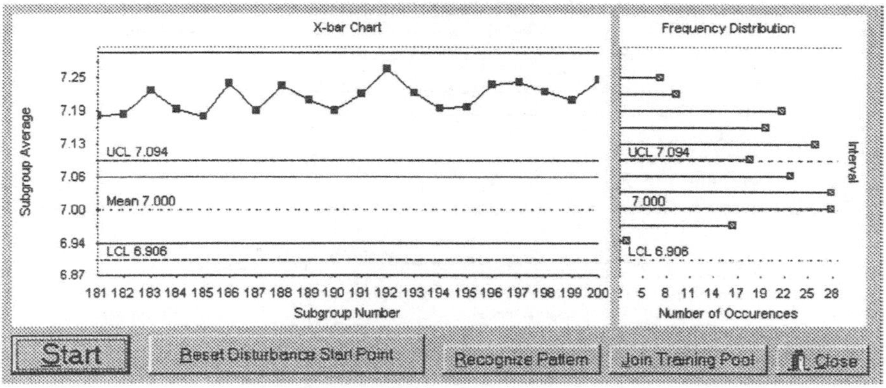

Fig. 6. Trend—second phase

Two aspects of trend in Figures 5 and 6 show increasing flat-topping of the frequency distribution. Based on this shape, trend can be recognized. It is worth emphasizing that, according to [1], up-and-down fluctuations may appear on a chart of a process under control. They can be misinterpreted as trends. Such temporary fluctuations will not be shown by the frequency distribution. Thus, the recognition of trends based on frequency distribution seems to be safer also in terms of the Type-I error.

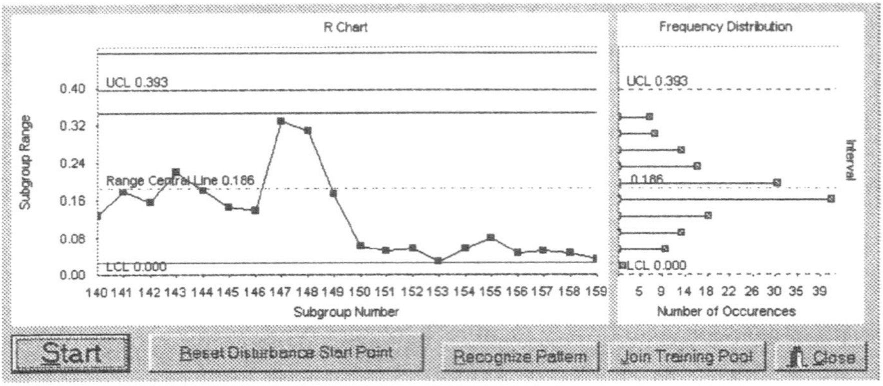

Fig. 7. Stratification—first phase

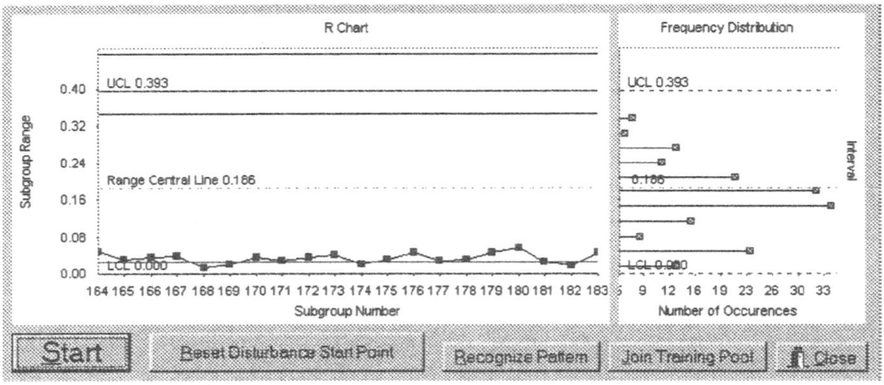

Fig. 8. Stratification—second phase

Stratification features unnaturally small fluctuation of data. It occurs if samples are taken from two different distributions. Usually, the result of such faulty sampling appears on range R chart. Figures 7 and 8 show simulated stratification. The distinguished feature is the peak or peaks that grows on the frequency distribution chart in the place where the central line of the stratification is located.

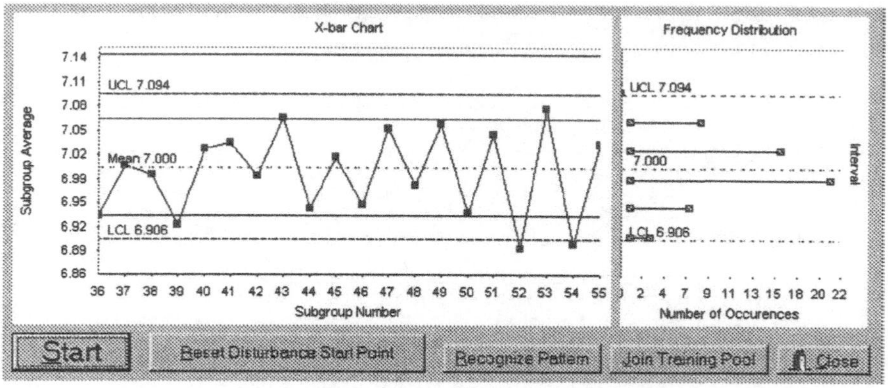

Fig. 9. Systematic variables—first phase

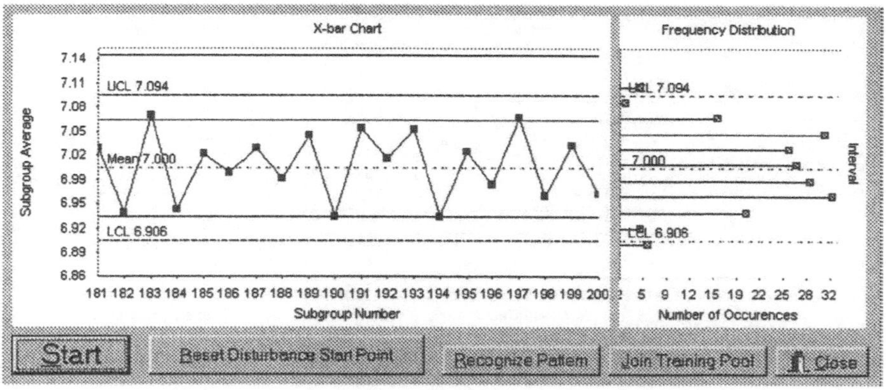

Fig. 10. Systematic variables—second phase

The definition of systematic variables, [1], states that they are predictable. The pattern demonstrated in Figures 9 and 10 has such a form that every second value is always greater than the preceding one and than the following one or vice versa. The plot of systematic variables resembles a sawtooth. When this pattern appears, frequency distribution becomes gradually wider. After some time it can become bimodal, as shown in Fig. 10.

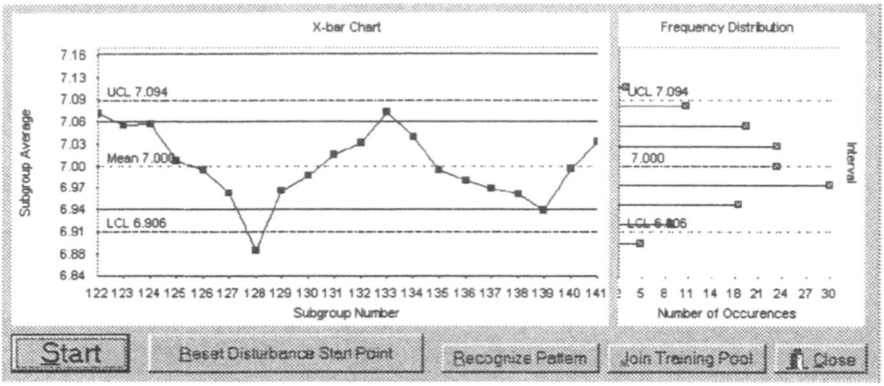

Fig. 11. Cycles—first phase

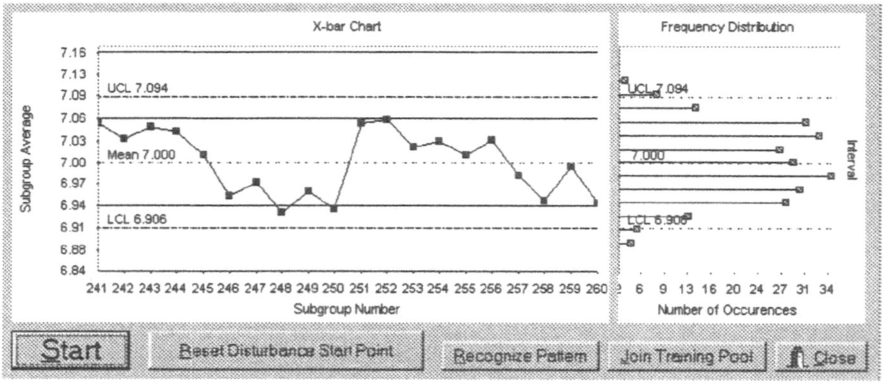

Fig. 12. Cycles—second phase

Cycles can be considered as a form of systematic variables. Regarding the frequency distribution, cycles produce longer phase of flat-topped plotting comparing with systematic variables. Such a shape is visualized in Fig. 11. Longer run of cyclic pattern can result in bimodal frequency distribution, Fig. 12.

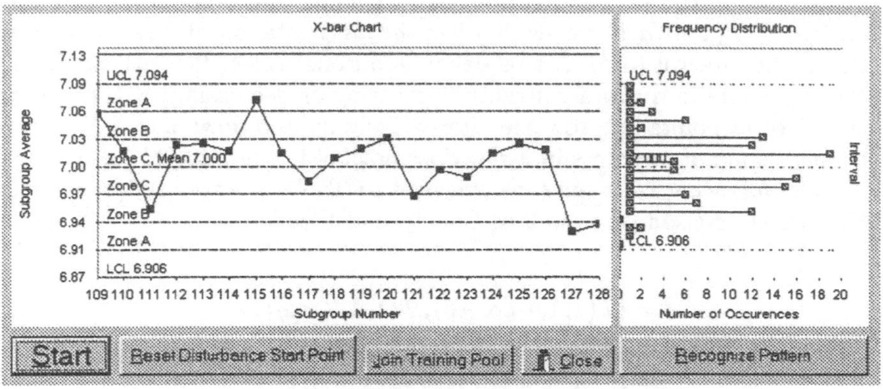

Fig. 13. Narrow Mixture

Narrow mixture shown in Figure Fig. 13 is difficult to be recognized on a traditional $\overline{X}$ chart because the simulated data falls between UCL and LCL. The frequency distribution shows that the pattern is bimodal. This happens because the disturbance acts in the interval $(-x_l, x_l)$, as described in Table 1. The Average Run Length (ARL) of the narrow mixture reflects ARL of the underlaying random normal variate $x(t)$. The narrow mixture is the example of a particularly danger disturbance that can be missed by the process supervisor.

5 Frequency Distribution vs. Supplementary Runs Rules

Shewhart $\overline{X}$ chart can be supported by supplementary rules that help to find the point when the process data start to deviate from the normal distribution and hence, the process is not in a state of statistical control. The tests may be particularly helpful before a point beyond the control limits occurs.

Table 2. Supplementary Runs Rules

	Test Number								
	1	2	3	4	5	6	7	8	9
Western Electric [1]	C_1	C_{10}	C_{11}	C_{12}					
Nelson [2]	C_1	C_{13}	C_{14}	C_{15}	C_{10}	C_{11}	C_{16}	C_{17}	
Champ and Woodal [3]	C_1	C_2	C_3	C_4	C_5	C_6	C_7	C_8	C_9

According to the notation applied by Champ and Woodal [3], a rule described by $T(k, m, a, b)$ means that k of the last m standardized sample means fall in the interval (a, b). Let us define additional rules: $R(k, m)$ means that k of m points in a row are steadily increasing or decreasing, $S(k, m)$ means that k of m points in a row are alternatively up and down and $U(k, m, a, b)$ means that k of m points in a row does not fall in the interval (a, b)

The following runs rules, as specified in Table 2, were used by Western Electric [1], Nelson [2] as well as Champ and Woodal [3]:

$$
\begin{aligned}
C_1 &= \{T(1,1,-\infty,-3),\, T(1,1,3,\infty)\}, \\
C_2 &= \{T(2,3,-3,-2),\, T(2,3,2,3)\}, \\
C_3 &= \{T(4,5,-3,-1),\, T(4,5,1,3)\}, \\
C_4 &= \{T(8,8,-3,0),\, T(8,8,0,3)\}, \\
C_5 &= \{T(2,2,-3,-2),\, T(2,3,2,3)\}, \\
C_6 &= \{T(5,5,-3,-1),\, T(5,5,1,3)\}, \\
C_7 &= \{T(1,1,-\infty,-3.09),\, T(1,1,3.09,\infty)\}, \\
C_8 &= \{T(2,3,-3.09,-1.96),\, T(2,3,1.96,3.09)\}, \\
C_9 &= \{T(8,8,-3.09,0),\, T(8,8,0,3.09)\}, \\
C_{10} &= \{T(2,3,-\infty,-2),\, T(2,3,2,\infty)\}, \\
C_{11} &= \{T(4,5,-\infty,-1),\, T(4,5,1,\infty)\}, \\
C_{12} &= \{T(8,8,-\infty,0),\, T(8,8,0,\infty)\}, \\
C_{13} &= \{T(9,9,-1,0),\, T(9,9,0,1)\}, \\
C_{14} &= \{R(6,6)\}, \\
C_{15} &= \{S(14,14)\}, \\
C_{16} &= \{T(15,15,-1,1)\}\ and \\
C_{17} &= \{U(8,8,-1,1)\}.
\end{aligned}
$$

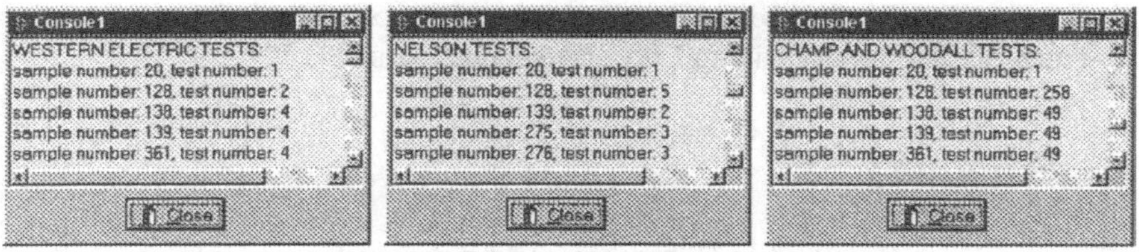

Fig. 14. Response to Additional Tests

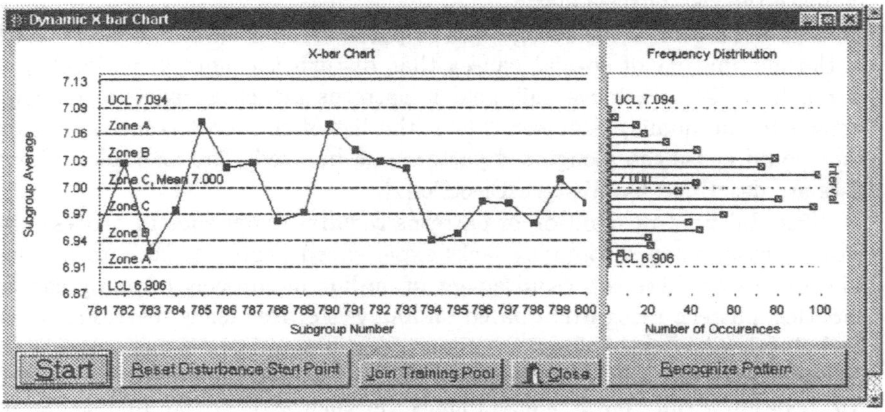

Fig. 15. Narrow Mixture

The run rules mentioned above were applied for the chart with narrow mixture shown in Fig. 13. The first alarm was generated for the sample number 20 by the rule C_1 as depicted in Fig. 14. The operator, except searching special causes, should consider here the natural variability as well. The second out-of-control signal was generated for the sample number 128 by the rules C_2, C_{10}, C_5 and C_8. The bimodality of the frequency distribution for sample 128 (see Fig. 13) could had been recognized before the run rules generated the alarm. The bimodality of the distribution of the narrow mixture is difficult to be recognized on the pure average $\overline{X}$ chart. Fig. 15 shows the final frequency distribution of 800 simulated samples. There is only one outlier. It can result from the natural variability of the process. The rest of the data fall between LCL and LCL. That is why the frequency distribution is helpful in case of such a pattern.

As described by Champ and Woodal [3], and Walker et al. [4], although the supplementary run rules increase the sensitivity of the Shewhart chart to small shifts in the mean, they reduce the ARL. Hence, the probability of false alarms increases. In case of Shewhart $\overline{X}$ chart, the analysis of the frequency distribution seems to be a valuable tool to find the time when a pattern occurs, identify it and find its special causes.

6 Summary

The approach described in this paper can be particularly helpful if a pattern occurs but the process data fall between control limits. The example of such a pattern is the narrow mixture. It would be difficult for an operator to recognize this disturbance observing the pure $\overline{X}$ chart. Application of run rules does not improve the recognition ability either. Moreover the run rules increase the risk of false alarm.

The main aim of the recognition of patterns on Shewhart control charts is the elimination of special causes that disturb the normal variability. It is much easier to achieve this goal if patterns are classified to a category known to the quality engineer. Then, the list of the most common reasons that could cause the recognized pattern can be used. The examples of such lists are described by Western Electric [1].

The ability of recognition of patterns requires experience in process control. In case, when a company lacks experienced people capable to monitor the process or there is a requirement of on-line monitoring of many characteristics, pattern recognition based on analysis of frequency distribution can be helpful. Although it loses the time order of the data, it can support the recognition of patterns when operator fails to identify the unnatural variability examining the "pure" control chart. In such a term the analysis of the frequency distribution can decrease the probability of the Type-II error.

The further development of the method would be the implementation of a neural network for recognition of patterns. The input layer of the network is to be fed by preprocessed values from the frequency distribution. The output layer would be responsible for the recognition.

References

1. Western Electric Company (1985) Statistical Quality Control Handbook. AT&T Technologies, Indianapolis, Indiana 46219
2. Nelson, L. S. (1984) The Shewhart Control Chart—Tests for Special Causes. Journal of Quality Technology **16**(4), 237–239
3. Champ, C. W. and Woodal W. H. (1987) Exact Results for Shewhart Control Charts With Supplementary Runs Rules. Technometrics **29**(4), 393–399
4. Walker E. and Philpot J. W and Clement J. (1991) False Signal Rates for the Shewhart Control Chart with Supplementary Runs Rules. Journal of Quality Technology **23**(3), 247–252
5. Swift, J. A. (1987) Development of a Knowledge Based Expert System for Control Chart Pattern Recognition and Analysis. PhD Thesis, Oklahoma State University, Stillwater, Oklahoma

6. Hwarng, H. B. (1991) Pattern Recognition on Shewhart Control Charts Using a Neural Approach". PhD Thesis, Arizona State University, Tempe, Arizona
7. Nieckula, J., Hryniewicz, O. (1997) Neural Network Support for Shewhart X-bar Control Chart. Systems Science **23**(1), 61–75
8. International Standard ISO 8258: (1991) Shewhart control charts, International Organization for Standardization. Casa Postale 56, CH-1211, Genève 20, Switzerland

Monitoring Processes with Data Available in Tabular Form Using Multiple Correspondence Analysis

Ross S. Sparks

CSIRO Mathematical and Information Sciences, Locked Bag 17, North Ryde, NSW 2113, Australia

Abstract. Monitoring service processes, when data are measured on an ordinal scale and recorded in tabular form, is considered. The paper addresses the following signals: changes in location, changes in the response profiles and changes in association. Multiple Correspondence Analysis (MCA) is used to visualise the nature of process changes and it provides a means of isolating possible causes of variation.

1 Introduction

In the manufacturing sector monitoring is common practice with the application of Shewhart [1], CUSUM and EWMA charts [2]. The benefits of applying these procedures are well documented in the literature for efficient management and control of processes in the manufacturing sector. Therefore, it is surprising that these monitoring tools are not generally applied in the service sector. On the few occasions the author has seen them applied, there has been inefficient use of the available information. For example, customer satisfaction questionaire data has been collected on an extended 5-point scale, but the information is reduced to a proportion of satisfied customers and the p−chart [2] is applied. In these situations, either the extended scale is wasteful or inefficient methods are used to display the information in the data.

In a competitive environment for the supply of services it may be necessary to monitor customer satisfaction. Monitoring the association between process performance measures and customer satisfaction to signal changes in customer perceptions is useful, particularly if it helps identify the most likely causes of these changes. It would also help if it provided insight into what factors or variables need to be controlled to maintain and possibly increase the number of satisfied customers.

This paper considers applications where customer perceptions of a fault repair process are assessed. Consider a measurement procedure where customers of a repair service are surveyed monthly. The responses to the survey are generally summarised by reporting the data in tabular form at the end of each month (e.g., see Example 1 below). This paper considers procedures for monitoring such data collected over a range of non-overlapping consecutive time periods, e.g., calendar months.

Example 1: A survey questionnaire is used to monitor customer satisfaction on a monthly basis. Customers during a particular month are phoned and asked to rate the service they received in the previous month on the likert scale: terrible, very poor, poor, neither poor nor good, good, very good, terrific. For each month, responses are accumulated to give monthly totals for all $q = 7$ possible responses. This is repeated over t months to give the matrix of counts. For example, with n_{ij} denoting the number of responses in category j for month i, this matrix is given by:

Month	*Terrible*	*Very Poor*	*Poor*	*Neither Poor nor Good*	*Good*	*Very Good*	*Terrific*
1	n_{11}	n_{12}	n_{13}	n_{14}	n_{15}	n_{16}	n_{17}
2	n_{21}	n_{22}	n_{23}	n_{24}	n_{25}	n_{26}	n_{27}
3	n_{31}	n_{32}	n_{33}	n_{34}	n_{35}	n_{36}	n_{37}
	$\vdots$						$\vdots$
t	n_{t1}	n_{t2}	n_{t3}	n_{t4}	n_{t5}	n_{t6}	n_{t7}

The total number of customers surveyed over the t months is $n_{..} = \sum_{i=1}^{t}\sum_{j=1}^{q} n_{ij}$. The proportion of people surveyed who fall into month i and category j is denoted as p_{ij}, the proportion of people surveyed who were customers in month i as r_i, and the average proportion of people surveyed that responded with category j for a month as c_j. As such,

$$p_{ij} = n_{ij}/n_{..}, \quad r_i = \sum_{j=1}^{q} n_{ij}/n_{..} \quad \text{and} \quad c_j = \sum_{i=1}^{t} n_{ij}/n_{..}$$

If, for all i and j,

$$p_{ij} = r_i c_j$$

then customer satisfaction responses are independent of month. In other words, the profile of category responses $\mathbf{p}_i = (p_{i1}\ p_{i2}\ p_{i3}\ p_{i4}\ p_{i5}\ p_{i6}\ p_{i7})$, does not differ across months. This would be the situation if there was no change in customer satisfaction over the t months considered. If, however, p_{ij} differed significantly from $r_i c_j$ for fixed i and all $j = 1, 2, \ldots, 7$, then customer satisfaction has changed for the ith month. How it has changed depends on how p_{ij} differs from $r_i c_j$ for each j. For example, if $p_{ij} > r_i c_j$ then there is a larger than expected number of customers that responded with satisfaction level j during month i.

Correspondence Analysis(CA), generally a two dimensional display of the matrix
$\{(p_{ij} - r_i c_j)/\sqrt{r_i c_j}\}$, is often used to display dependencies between row and

column categories of a data matrix [3] [4]. Therefore, CA not only offers a means of signalling changes in customer satisfaction, it also provides a graphical description of how it has changed. This paper will further develop CA as a useful tool for visualising process changes and we also use the CA biplot to signal several levels of unusual process behaviour. In other words, CA biplot is used to visualise process behaviour in a reduced dimensional space, but traditional hypothesis testing is used to signal unusual events from all dimensions of the data.

Section 2 considers several one dimensional displays or charts that are the equivalent of the Shewhart or conventional CUSUM charts used in manufacturing. Section 3 introduces CA as a useful monitoring tool. Section 4 examines applications of Multiple Correspondence Analysis (MCA) for monitoring several measures simultaneously. Examples of applications are considered in Section 5. In this paper, 2-dimensional CA or MCA biplots are considered, but the results can easily be extended to higher dimensional biplots. Finally some concluding remarks are made in Section 6.

2 Statistical Process Control Charts for signalling process changes for tabular data

Initially, we consider monitoring procedures retrospectively when no prior process information is available and expected values or target values for customer satisfaction profiles are unknown. This would be the situation if no charts have been applied in the past and we wish to establish whether the process is stable. In general, past data are used to define the expected profiles (long-term average) or to set a target value for monitoring.

The second part of this section considers the signalling of profiles which differ from a long-term average profile, or the signalling of deviations from a target profile.

2.1 Assessing whether a process is stable retrospectively

A stable process is defined as a process with profiles that do not differ significantly over time. A chi-squared statistic is used to assess stability. This statistic has the advantage of not requiring an estimate of the covariance matrix of the response profile. An out-of-control signal for the ith time period is obtained whenever

$$I_i = n_{i.} \sum_{j=1}^{q} (p_{ij}/r_i - c_j)^2/c_j > \chi_{df}^{2\,(\alpha)}$$

where $df = q - 1 - q/t + 1/t$ and $\chi_{df}^{2\,(\alpha)}$ is the upper tailed α of a chi-squared distribution with degrees of freedom df. (Note: Under the assumption that profiles are independent across time, the random variable $\sum_{i=1}^{t} I_i$ is approximately Chi-squared distributed with $(t-1)(q-1)$ degrees of freedom.)

A CUSUM chart is useful for detecting trends or small step changes in profiles and a time-series plot of

$$C_i = [\sum_{k=1}^{i} I_k]/\chi^{2\,(\alpha)}_{i\times df}$$

signals instability whenever $C_i > 1$.

Charts outlined in this section are useful for deciding retrospectively whether the process is stable. In the next section we consider monitoring data as it arrives to signal trends away from target profiles. This has the advantage of providing an early signal of changed process behaviour, for example, changes in customer satisfaction.

2.2 Monitoring whether a profile has shifted from its long-term average or from a target profile

Let $\mathbf{T}$ be the target vector or long-term mean profile for the proportion of each response for a particular question to customers and $n_{i.}$ be the number of people surveyed during the ith time period.

Monitoring to signal general changes in profiles

The Shewhart χ^2**-Chart:** A time series plot of a χ^2-statistic is considered. Define statistic d_{ij} as follows

$$d_{ij} = \sqrt{n_{i.}}(n_{ij}/n_{i.} - T_j)/\sqrt{T_j}$$

An equivalent to a Shewhart chart is the time series plot of $D_i = \sum_{j=1}^{q} d_{ij}^2$ with an out-of-control signal given whenever $D_i > \chi_q^{2(\alpha)}$. This chart monitors whether profiles have shifted generally from the target. It can be further improved by plotting the profiles $\left(d_{i1}\ d_{i2} \ldots d_{iq}\right)$ at each coordinate (i, D_i) on the chart of D_i (eg, see [5]). An application of a Shewhart chart using D_i as a control statistic is now presented.

Example 2: Approximately 2500 customers were surveyed each month for 24 months to assess customer satisfaction for a particular service. This involved monthly counts of satisfaction ratings on the six point Likert scale: extremely poor, very poor, poor, good, very good, excellent.

Figure 1 presents a chart for signalling profile shifts from their long-term average for the 24 months of data from Example 2. In Figure 1, the Shewhart χ^2-chart signals on the 7, 14 and 17 month.

A CUSUM chart: A chart similar to Page conventional CUSUM [6] is now considered. Let $C_0 = 0$ and for a suitable fixed value k let

$$C_i = max(0, C_{i-1} + (D_i - q)/\sqrt{2q} - k).$$

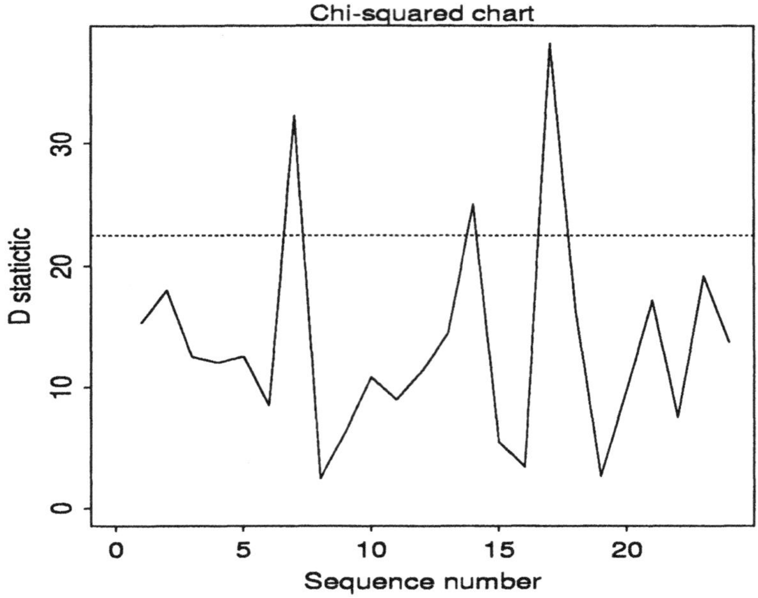

Fig. 1. An application of the Shewhart chart

An out-of-control signal is given when $C_i > h_C$ where h_C is chosen to give a specified in-control average run length (ARL). This CUSUM chart applied to the data for Example 2 is presented in Figure 2. The CUSUM statistic C_i was reset to zero after each signal. Notice that it signals on months 3, 7, 13, 17 and 24.

The Shewhart chart of D_i and Page's CUSUM of D_i are useful for signalling general changes in response profiles, but they fail to distinguish directional movements in customer satisfaction (such as customers becoming less satisfied) from other profile changes. Also, directional changes in customer satisfaction may not be efficiently signalled by functions of the D_i statistic. Therefore, statistics designed for signalling directional shifts in profile responses are now considered.

Monitoring directional shifts in location

Changes in profiles, such as a general increase in satisfaction ratings, are now considered. It is useful to have a metric that scales the categories to a one dimensional measure of directional change.

$Z-$**chart** Here we develop a control statistic that has an approximate standard normal distribution. In this paper we consider scaling the categories as ordinal ranks. For example categories terrible, very bad, bad, good, very good and excellent are scaled with ranks 1,2,3,4,5 and 6, respectively. If $c_{ij} = n_{ij}/n_{i.} - T_j$ then the control variable is taken as $d_i = \sum_{j=1}^{6} jc_{ij} = \mathbf{j}^t\mathbf{c}_i$ where

$$\mathbf{j}^t = \begin{pmatrix} 1 \; 2 \; 3 \; 4 \; 5 \; 6 \end{pmatrix} \qquad (1)$$

To standardise this variable, to produce a $Z-$score, we need to find the mean and variance of d_i.

If $p_j = \mathrm{E}[n_{ij}/n_{i.}]$ for all j, then the covariance matrix of random vector $\mathbf{c}_i = \begin{pmatrix} c_{i1} \; c_{i2} \ldots c_{iq} \end{pmatrix}$ is

$$\mathbf{\Sigma}_i = \begin{pmatrix} p_1(1-p_1) & -p_2p_1 & \ldots & -p_qp_1 \\ -p_2p_1 & p_2(1-p_2) & \ldots & -p_2p_q \\ \vdots & & \ddots & \vdots \\ -p_qp_1 & -p_qp_2 & \ldots & p_q(1-p_q) \end{pmatrix} / n_{i.}$$

Thus, random variable d_i has mean $\mu = \sum_{j=1}^{q} jp_j$ and variance equal to

$$\sigma_i^2 = \mathbf{j}^t\mathbf{\Sigma}_i\mathbf{j}$$

Therefore, a time sequence plot of $Z_i = (d_i - \mu)/\sigma_i$ versus i is useful for displaying trends in satisfaction. Directional changes are signalled by either $Z_i > 3$ (signalling increased satisfaction) or $Z_i < 3$ (signalling reduced satisfaction).

A CUSUM chart: Page conventional CUSUM [6] is now applied to random variable d_i. This often involves the simultaneous plot of lower and upper control statistics, S_i^L and S_i^U, on the same chart. For a fixed value $k_s > 0$, these control statistics are as follows: $S_0^L = S_0^U = 0$, with low-side values

$$S_i^L = \min(0, S_{i-1}^L + (d_i - \mu)/\sigma_i + k_s),$$

and high-side values

$$S_i^U = \max(0, S_{i-1}^U + (d_i - \mu)/\sigma_i - k_s).$$

In practice, $k_s = 0.5$ is often used. For a suitable h_s value, a signal is given whenever $S_i^U > h_s$ or $S_i^L < -h_s$.

The $p-$chart is well documented in the literature and is often used in situations where information is collected on an extended scale. Although this is an inefficient use of the information in the data, the $p-$chart is covered in this paper to illustrate its inefficiency in certain circumstances.

$p-$chart: Here the proportion of favourable responses (eg., customer responses are either good, very good or terrific) or the proportion of unfavourable responses (eg., customer responses are either bad, very bad or terrible) are charted on a $p-$chart. By categorising certain responses as unfavourable it therefore gives equal weight to categories bad, very bad and terrible and equal weight to good, very good and excellent. In situations where information is available on an extended scale this weighting discards information. Therefore, the application of the $p-$chart is generally not recommended in these instances.

A comparison of charts for signalling directional shifts in customer satisfaction

The charts described above are compared in Table 1 for simulated data. Average runlengths were calculated using 10 000 similations in both Tables 1 and 2 .

Data were generated as follows: Let the mean proportion falling into the $q = 6$ classes be $p_1 = T_1 = 0.125$, $p_2 = T_2 = 0.275$, $p_3 = T_3 = 0.275$, $p_4 = T_4 = 0.175$, $p_5 = T_5 = 0.1$ and $p_6 = T_6 = 0.05$.

1. A multivariate normal random vector of length q (denoted $\mathbf{x}$) is generated with mean equal to the $q \times 1$ zero vector and covariance equal to $\mathbf{D}_q$ with the jth diagonal element given by $p_j\,(1-p_j)/400$.
2. A random vector of responses falling into the jth class at a random time period is taken as the integer nearest to $400\,(p_j + x_j)/\sum_{i=1}^{6}(p_i + x_i)$ for responses that are on average on-target.
3. Out-of-control responses are generated as follows: Let $\rho_1 = -0.02146\delta$, $\rho_2 = -0.017718\delta$, $\rho_3 = 0.02146\delta$, $\rho_4 = 0.008859\delta$, $\rho_5 = 0.008859\delta$ and $\rho_6 = 0$. A directional shift in class frequencies is generated by taking the random vector of responses falling into the jth class as the natural number nearest to $400\,(p_j + x_j + \rho_j)/\sum_{i=1}^{6}(p_i + x_i + \rho_i)$.

A summary of the results for various shifts caused by increasing δ is given in Table 1. The value $\delta = 0$ corresponds to no shift in profiles.

In Table 1, it is clear that small persistent directional changes are signalled quicker using the CUSUM for directional shifts than any of the Shewhart charts. The CUSUM chart for general profile changes signals small location shifts ($0 < \delta \leq 2$) on average later than CUSUM for directional shifts, but significantly sooner when $\delta \geq 2.5$.

The $Z-$chart has a smaller ARL for all $\delta > 0$ in Table 1 than all other charts besides the CUSUM for directional changes. This chart is better than the CUSUM for directional changes for large location shifts, i.e. $\delta \geq 2.5$, but is worse for location shifts $0 < \delta \leq 2$.

The $p-$chart does not make use of all the information and therefore performs badly for the directional changes considered in Table 1.

In Table 2, data are generated as follows: (The mean proportion were generated as in Table 1.)

1. A random vector of length q (denoted $\mathbf{x}$) has x_j generated as a beta-distribution with parameters $400\,p_i$ and $400\,(1-p_i)$.
2. An on-target random vector of responses for the jth class is generated as the integer nearest to $400\,x_j/\sum_{i=1}^{6} x_i$.
3. Let ρ_j be defined as in Table 1, then a location shift in class frequencies is generated by taking the random vector of responses for the jth class as the integer nearest to $400\,(x_j + \rho_j)/\sum_{i=1}^{6}(x_i + \rho_j)$.

Table 1. Comparison of Shewhart and CUSUM charts using average runlength (ARL) and q=6 when errors are assumed normal with mean zero and variance $p_i(1-p_i)/400$.

	Shewhart	CUSUM	CUSUM	One dimensional	
	χ^2–Chart	general	directional	p-chart	Z–chart
h	14.575	7.4600	3.7180	2.3000	2.6900
k		0.5	0.5		
Profile shift δ	ARL(sde)	ARL(sde)	ARL(sde)	ARL(sde)	ARL(sde)
0.00	201.85(2.04)	201.11(2.05)	200.71(1.98)	199.99(2.00)	199.89(2.00)
0.25	70.18(0.70)	58.53(0.58)	11.54(0.07)	67.27(0.68)	36.80(0.36)
0.50	34.72(0.34)	25.74(0.24)	6.65(0.03)	45.20(0.44)	17.80(0.18)
0.75	20.34(0.19)	14.06(0.13)	5.02(0.02)	33.81(0.34)	10.87(0.10)
1.00	12.98(0.12)	8.88(0.08)	4.15(0.02)	26.65(0.27)	7.51(0.07)
1.50	6.61(0.06)	4.61(0.03)	3.25(0.01)	18.13(0.18)	4.38(0.04)
2.00	4.05(0.04)	3.06(0.02)	2.77(0.01)	13.47(0.13)	2.98(0.02)
2.50	2.83(0.02)	2.34(0.01)	2.46(0.01)	10.46(0.10)	2.26(0.02)
3.00	2.15(0.02)	1.90(0.01)	2.25(0.01)	8.40(0.08)	1.86(0.01)
4.00	1.51(0.01)	1.45(0.00)	1.99(0.01)	5.98(0.05)	1.42(0.01)
5.00	1.23(0.01)	1.23(0.00)	1.80(0.01)	4.52(0.04)	1.21(0.01)

This paper deals with data collected on monthly intervals where one false alarm every 5 years was considered an adequate error rate. In term of ARL this corresponds to an in-control ARL of 60, the value used in Table 2.

In Table 2, it is clear that small persistent profile changes ($\delta \leq 0.75$) of the form considered are signalled quicker using the CUSUM for directional changes than the Shewhart charts. CUSUM for directional changes signals small location shifts ($0 < \delta \leq 3$) on average sooner than CUSUM for general changes, but significantly later when $\delta \geq 4$. The Shewhart p–chart performs consistently worse than the Z–chart, illustrating that the p–chart is inappropriate in certain instances.

The procedures described in this section are useful for signalling special causes but they fail to provide detailed information on the nature of this special cause. A procedure that explores the nature of signals is now considered.

Table 2. Comparison of Shewhart and CUSUM charts using ARL and q=6 when the beta distribution is used to generate the data.

	Shewhart χ^2−Chart	CUSUM general	CUSUM directional	One dimensional p-chart	One dimensional Z−chart
k		0.5	0.5		
h	29.3758	7.4650	12.7742	2.6890	3.1762
Profile shift δ	ARL(sde)	ARL(sde)	ARL(sde)	ARL(sde)	ARL(sde)
0.00	60.21(0.59)	60.85(0.44)	60.05(0.41)	61.60(0.61)	61.20(0.61)
0.25	40.95(0.41)	32.41(0.19)	16.74(0.06)	29.67(0.29)	24.40(0.24)
0.50	28.32(0.28)	21.90(0.11)	12.19(0.04)	22.55(0.22)	14.76(0.14)
0.75	20.33(0.20)	16.54(0.08)	10.08(0.03)	18.58(0.18)	10.23(0.10)
1.00	15.11(0.15)	13.35(0.06)	8.81(0.02)	15.83(0.15)	7.79(0.07)
1.50	9.35(0.09)	9.65(0.04)	7.27(0.02)	12.24(0.12)	5.23(0.05)
2.00	6.43(0.06)	7.59(0.03)	6.35(0.01)	9.83(0.09)	3.86(0.03)
2.50	4.75(0.04)	6.28(0.02)	5.75(0.01)	8.32(0.08)	3.05(0.02)
3.00	3.71(0.03)	5.38(0.02)	5.28(0.01)	7.24(0.07)	2.52(0.02)
4.00	2.52(0.02)	4.22(0.01)	4.63(0.01)	5.64(0.05)	1.93(0.01)
5.00	1.91(0.00)	3.50(0.01)	4.18(0.01)	4.64(0.04)	1.60(0.01)

3 Introduction to CA as a monitoring tool for visualising tabular data

Let matrix $\mathbf{N}$ be a $t \times q$ matrix of counts n_{ij} for t time periods and q possible responses. Note that $n_{..} = \sum_{i=1}^{t} \sum_{j=1}^{q} n_{ij}$, $\mathbf{r} = (r_1\ r_2 \ldots r_t)$ and $\mathbf{c} = (c_1\ c_2 \ldots c_q)$ where r_i and c_j are defined as in Section 1. Also $\mathbf{P} = \mathbf{N}/n_{..}$. Let $\mathbf{D_c}$ and $\mathbf{D_r}$ be diagonal matrices with diagonal elements equal to the elements of vector $\mathbf{c}$ and $\mathbf{r}$, respectively.

Consider the singular value decomposition(SVD) of matrix $\mathbf{Q} = \mathbf{D_r}^{-1/2}(\mathbf{P}-\mathbf{rc}^t)\mathbf{D_c}^{-1/2}$, i.e.,

$$\mathbf{Q} = \mathbf{U}\mathbf{D}_\lambda\mathbf{V}^t$$

where: $\mathbf{U}^t\mathbf{U} = \mathbf{I}_q$, $\mathbf{V}^t\mathbf{V} = \mathbf{I}_q$ and $\mathbf{D}_\lambda$ is a diagonal matrix with diagonal elements $\lambda_1 \geq \lambda_2 \geq \ldots \geq \lambda_q$.

Let $\mathbf{V}_p$ and $\mathbf{U}_p$ be the first $p(< q)$ rows of $\mathbf{V}$ and $\mathbf{U}$, respectively, and $\mathbf{D}_{\lambda_p}$ be the matrix consisting of the first p rows and p columns of $\mathbf{D}_\lambda$. The

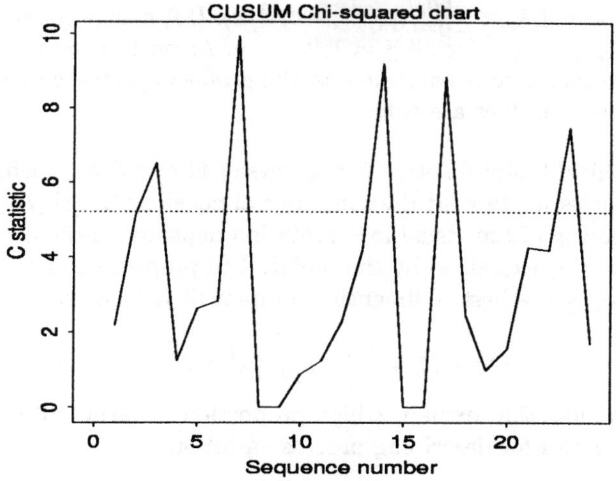

Fig. 2. An application of the CUSUM chart

Correspondence Analysis (CA) biplots [3] in p-dimensions is the display of co-ordinates taken from the rows of

$$\mathbf{G} = \mathbf{D}_{\mathbf{c}}^{-1/2}\mathbf{V}_p\mathbf{D}_{\lambda_p}$$

and

$$\mathbf{F} = \mathbf{D}_{\mathbf{r}}^{-1/2}\mathbf{P}\mathbf{G}\mathbf{D}_{\lambda_p}^{-1} = \mathbf{D}_{\mathbf{r}}^{-1/2}\mathbf{U}_p\mathbf{D}_{\lambda_p}.$$

Let $\mathbf{g}_i^t$ be the ith row of $\mathbf{G}$ and $\mathbf{f}_j^t$ be the jth row of $\mathbf{F}$. When $p = q$, note the results that follow from this scaling:

- $\mathbf{F}\mathbf{D}_{\lambda_p}^{-1}\mathbf{G}^t \approx \mathbf{D}_{\mathbf{r}}^{-1}(\mathbf{P} - \mathbf{r}\mathbf{c}^t)\mathbf{D}_{\mathbf{c}}^{-1}$ and thus $\mathbf{f}_i^t\mathbf{D}_{\lambda_p}^{-1}\mathbf{g}_j \approx p_{ij}/(r_ic_j) - 1$. Therefore, if the corresponding elements of vectors $\mathbf{f}_i$ and $\mathbf{g}_j$ are all of the same sign, then $p_{ij} > r_ic_j$. Similarly, if the corresponding elements of $\mathbf{f}_i$ and $\mathbf{g}_j$ are all of the opposite sign, then $p_{ij} < r_ic_j$. For example, if the two-dimensional space describes the data adequately, then $p_{ij} > r_ic_j$ whenever $\mathbf{f}_i$ and $\mathbf{g}_j$ are in the same quadrant, and $p_{ij} < r_ic_j$ whenever $\mathbf{f}_i$ and $\mathbf{g}_j$ are in the opposite quadrant.
- $\mathbf{g}_j^t\mathbf{g}_j \approx \sum_{i=1}^{t}(p_{ij}/c_j - r_i)^2/r_i$ and thus if $\mathbf{g}_j$ is nearly equal to the zero vector then $p_{ij} \approx r_ic_j$ for all $i = 1, 2, \ldots, t$. As such, column category j is independent of time. In Example 1 of Section 1, if the survey data produce the same proportion of "Terrific" ratings for each survey period then this rating category is independent of time.

- Similarly, $\mathbf{f}_i^t\mathbf{f}_i \approx \sum_{j=1}^{q}(p_{ij}/r_i - c_j)^2/c_j$. If $\mathbf{f}_i$ nearly equals the zero vector then $p_{ij} \approx r_i c_j$ for all $j = 1, 2, \ldots, q$. As such, time period i displays a row profile that is equivalent to the profile expected under the hypothesis that row profiles are equal.

Graphical biplot displays of the rows of **G** and **F** are useful for describing dependencies for tabular data in lower dimensions [2] [3] [7].

Reliability of the visual interpretation depends on the proportion of variation (inertia) explained by the biplot. The proportion of the total variation explained by the best p-dimensional biplot fit is equal to:

$$(\lambda_1^2 + \lambda_2^2 + \ldots + \lambda_p^2)/(\lambda_1^2 + \lambda_2^2 + \ldots + \lambda_q^2).$$

CA biplots that explain a high proportion of variation in two dimensions are very useful for describing process variation.

3.1 Describing an unstable process

Chi-squared charts are useful for indicating differences in row(time period) profiles. For periods of instability, CA biplots can be used to describe how these profiles differ. Figure 3, for example presents the CA biplot of the survey data. The two-dimensional presentation in Figure 3 explains approximately 85% of the variation (inertia). As discussed previously Figure 1 signals in months 7, 14 and 17. The CA biplot indicates the following:

The 7th month has a higher than expected number of very good (VG) ratings and good ratings (G) and a lower than expected number of excellent (X) and extremely poor (EP) ratings.

The 14th month has a higher than expected number of excellent ratings and a lower than expected number of good (G), poor (P) and very poor (VP) ratings.

The 17th month has a higher than expected number of excellent and extremely poor ratings and a lower than expected number of good and very good ratings.

The first three months have higher than expected number of very good ratings and good ratings but a lower than expected number of excellent and extremely poor ratings. This may explain why the CUSUM signals at month 3 in Figure 2.

Month 20, 21 and 23 indicate a general trend to the poorer ratings with months 22 and 23 indicating an increase in excellent ratings. These could be the reasons why the CUSUM signals at month 23 in Figure 2.

It is sometimes difficult to observe periods with small step changes in row profiles using the CA biplot, and therefore the biplot of cumulative sums of the $\mathbf{g}_i$'s, and $\mathbf{f}_j$'s is often useful for displaying such changes.

CUSUM of the $\mathbf{g}_i$'s: A plot of the CUSUM row co-ordinates $\mathbf{G}$, i.e., the plot of co-ordinates $\mathbf{g}_1, \mathbf{g}_1 + \mathbf{g}_2, \ldots, \sum_{i=1}^{T} \mathbf{g}_i$, and the rows of $\mathbf{F}$ is often useful for describing drifting trends in the profiles. For example, the chart in Figure 4 contains the CUSUM of row profile information for the customer satisfaction example considered in Section 2. It is clear from this CUSUM (see Figure 4) that for the first 8 months there was a trend towards the region with higher than expected proportion of responses of very good and good ratings. This is followed by a trend away from the poorer ratings for months 11 to 16, then from months 17 to 23 there is a trend towards an increase in poorer ratings.

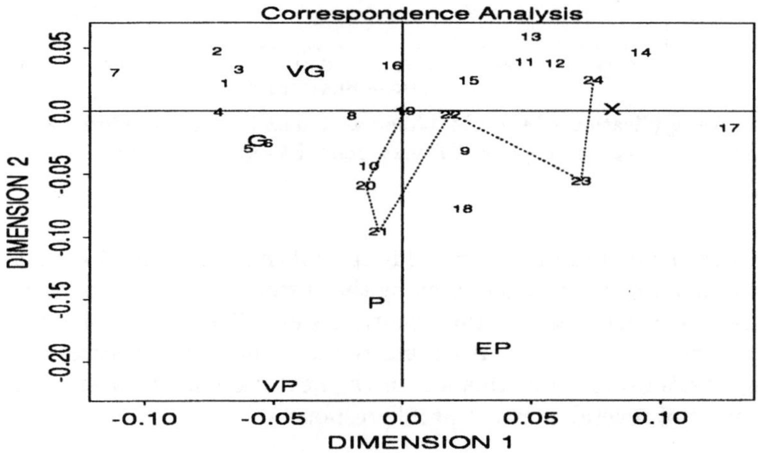

Fig. 3. An application of the CA biplot as a monitoring procedure: X-excellent, VG-very good, G-good, P-poor, VP-very poor, EP-extremely poor

3.2 Visualising how profiles shift from their long-term average or their target profile

Let vector $\mathbf{T}$ be the long-term average or target profile for relative frequencies of responses, e.g., for a question with a yes-no response if it is expected that 50% of responses are yes and 50% of responses are no, then the mean profile for the relative frequencies is $\mathbf{T}^t = \left(1/2\ 1/2\right)$.

The CA procedure with $\mathbf{c}$ replaced by the known $q \times 1$ target vector $\mathbf{T}$, where

$$\mathbf{Q} = \mathbf{D}_r^{-1/2}\left(\mathbf{P} - \mathbf{r}\mathbf{T}^t\right)\mathbf{D}_{\mathbf{T}}^{-1/2}, \tag{2}$$

is useful for visualising signalled deviations from target. Interpretations from the biplot display are very similar to those outlined in Section 2.

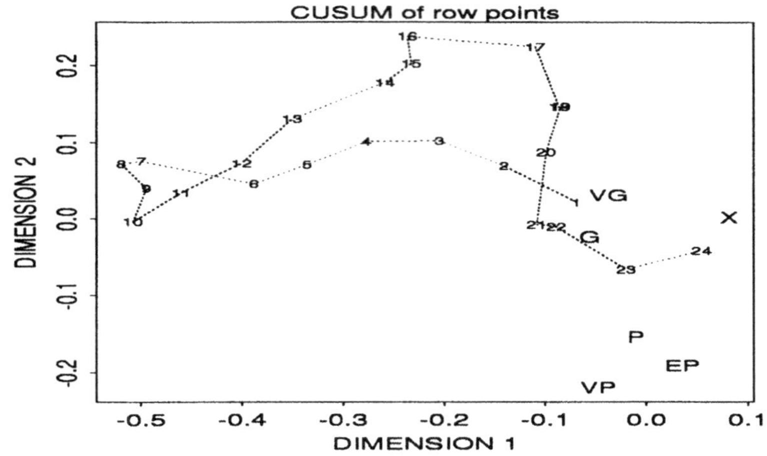

Fig. 4. An application of the CA biplot as a monitoring procedure: X-excellent, VG-very good, G-good, P-poor, VP-very poor, EP-extremely poor

Processes often involve monitoring several characteristics, for example, in monitoring a repair process as seen by the customer, we may be interested in the customer's perception of the repairer's knowledge, the speed of the repair and the repairer's manner. All these relate to how the customer perceives the repairer's overall performance. In the next section we consider the joint monitoring of several question-profile responses.

4 Monitoring responses to multiple questions

One approach to jointly monitoring several related characteristic of a process is to use Multiple Correspondence Analysis (MCA) [3] [4]. Other approaches are considered in this paper.

Example 3: Monitoring a fault repair process may involve the following separate units:

The fault reporting process;
The repair process.

All customers are asked to rate their satisfaction with: the process overall, the reporting process and the repair process. Assume that a six point Likert scale in Example 2 is used. If a particular customer rated overall satisfaction as very good, the reporting process as good, and the repair process as excellent, then the indicator vector for this customer is (vertical lines used to indicate separate questions)

$$(0\,0\,0\,0\,1\,0|0\,0\,0\,1\,0\,0|0\,0\,0\,0\,0\,1)\,.$$

The first six elements relate to the response for overall satisfaction, the second set of six relates to the reporting process and the last six relates to the repair process. Each set of six elements contains a single 1 and five zeros, e.g., if the customer responds very good then the fifth element of the six is set equal to 1 with all others zero. The indicator matrix consists of all customer response vectors. If customers' responses are bound together in rows then the indicator matrix takes the form

$$\mathbf{Z} = \left(\mathbf{Z}_1 \; \mathbf{Z}_2 \; \mathbf{Z}_3 \right)$$

where $\mathbf{Z}_j$ is the indicator matrix for the jth question.

If $\mathbf{Z}$ is split into grouped customer responses for t time periods, then $\mathbf{Z}_{ij}$ is the indicator matrix for the jth question for a service during the ith time period. If n customers are surveyed for each time period then $\mathbf{Z}_{ij}$ is an $n \times 18$ matrix. The ith row of the counts matrix $\mathbf{N}_j$ is equal to $\sum_{k=1}^{n} \mathbf{z}_{k\,ij}^t$ where $\mathbf{z}_{k\,ij}^t$ is the kth row of $\mathbf{Z}_{ij}$.

Two MCA procedures are discussed. The first is a slight variation to applying CA for a single question and the second is a variation on MCA [3] [3] [4].

For the applications considered in this paper there were legal and confidential reasons why the data could not be supplied in incidence matrix form and, therefore, the data were provided in summary tabular form.

4.1 Monitoring responses to multiple questions for data in tabular form

Let the count matrices for t consecutive time periods measure overall satisfaction, satisfaction with reporting the fault, and satisfaction with the repair. These are measured on a q response Likert scale and recorded as matrices of counts

$$\mathbf{N} = \left(\mathbf{N}_1 \; \mathbf{N}_2 \; \mathbf{N}_3 \right).$$

The $\mathbf{N}_j$ matrix is found by summing over the appropriate subsets of rows of $\mathbf{Z}_j$.

The total customers surveyed over all t time periods is $n_{..} = \mathbf{1}_t^t \mathbf{N}_j \mathbf{1}_q$ for all j, assuming surveyed customers respond to all three questions. If, in each survey period, the same number of customers are sampled, then $\mathbf{r}_j = \mathbf{N}_j \mathbf{1}_q / n_{..} = \mathbf{1}_t / t$ for all j. The proportion of people who responded for each category for the jth question is given by vector $\mathbf{c}_j = \mathbf{1}_t^t \mathbf{N}_j / n_{..}$ and the proportion of customers who were surveyed in time period i and responded category j are recorded in matrix $\mathbf{P}_j = \mathbf{N}_j / n_{..}$. Then the usual matrix used in CA as defined in (2) is

$$\mathbf{Q}_j = t^{-1/2}(t\,\mathbf{P}_j - \mathbf{1}_t \mathbf{c}_j^t)\mathbf{D}_{\mathbf{c}_j}^{-1/2}, \qquad (3)$$

where $\mathbf{D}_{\mathbf{c}_j}$ ia a diagonal matrix with diagonal elements equal to the elements of vector $\mathbf{c}_j$. When $t > 3q-3$, the singular value decomposition of matrix $\left(\mathbf{Q}_1\ \mathbf{Q}_2\ \mathbf{Q}_3\right)$ is given by

$$\left(\mathbf{Q}_1\ \mathbf{Q}_2\ \mathbf{Q}_3\right) = \mathbf{U}\mathbf{D}_\lambda \left(\mathbf{V}_1^t\ \mathbf{V}_2^t\ \mathbf{V}_3^t\right)$$

where $\mathbf{U}^t\mathbf{U} = \mathbf{I}_{3q-3}$, $\sum_{j=1}^{3} \mathbf{V}_j^t\mathbf{V}_j = \mathbf{I}_{3q-3}$, $\mathbf{V}_j\mathbf{V}_j^t = \mathbf{I}_{q-1}$, $\mathbf{V}_j\mathbf{V}_k^t = \mathbf{0}_{q-1}$ for $k \neq j$ and $\mathbf{D}_\lambda$ is a diagonal matrix with diagonal elements $\lambda_1 \geq \lambda_2 \geq \ldots \geq \lambda_{3q-3} \geq 0$. Let $\mathbf{V}_{j\,p}^t$ be the matrix consisting of the first p rows of $\mathbf{V}_j^t$, $\mathbf{U}_p$ be the matrix consisting of the first p rows of $\mathbf{U}$ and $\mathbf{D}_{\lambda\,p}$ be the diagonal matrix consisting of the first p diagonal elements of $\mathbf{D}_\lambda$.

Then the p-dimensional biplot has as coordinates the rows of matrices

$$\mathbf{G} = \begin{pmatrix} \mathbf{G}_1 \\ \mathbf{G}_2 \\ \mathbf{G}_3 \end{pmatrix} = \begin{pmatrix} \mathbf{D}_{\mathbf{c}_1}^{-1/2}\mathbf{V}_{1\,p} \\ \mathbf{D}_{\mathbf{c}_2}^{-1/2}\mathbf{V}_{2\,p} \\ \mathbf{D}_{\mathbf{c}_3}^{-1/2}\mathbf{V}_{3\,p} \end{pmatrix} \mathbf{D}_{\lambda\,p}$$

and

$$\mathbf{F} = \mathbf{U}_p\mathbf{D}_{\lambda\,p}/\sqrt{t}.$$

These are plotted to assess whether the process is stable. (An example using $p = 2$ is presented in Figure 5). An alternate approach is to use Gabriel CMP biplot [8] that examines the SVD of

$$\mathbf{N}_z = \mathbf{N}\mathbf{D}_s^{-1} = \mathbf{U}_z\mathbf{D}_{\lambda,z}\mathbf{V}_z^t$$

where $\mathbf{D}_s$ is a diagonal matrix of the standard deviations of the columns of $\mathbf{N}$. This biplot is then achieved by a $p-$dimensional plot of the rows of

$$\mathbf{G}_z = \mathbf{V}_{z,p}\mathbf{D}_{\lambda\,p}$$

and

$$\mathbf{F} = \mathbf{U}_{z,p}.$$

In Figure 5 we consider the monitoring of tabular data of customer satisfaction ratings for: the reporting process overall, the manner of the person taking details of the report and the information that this person provided to the customer on the nature of the fault and how it would be handled. It is used to illustrate a biplot approach to signalling and describing various process changes. Using asymptotic hypothesis tests, changes are displayed in and around the biplot using grey scales. Let training data be available in the form of incidence matrices $\mathbf{Z} = \left(\mathbf{Z}_1\ \mathbf{Z}_2\ \mathbf{Z}_3\right)$ to estimate the covariance matrix

$$\mathbf{\Sigma}_p = \begin{pmatrix} \mathbf{\Sigma}_{11}\ \mathbf{\Sigma}_{12}\ \mathbf{\Sigma}_{13} \\ \mathbf{\Sigma}_{21}\ \mathbf{\Sigma}_{22}\ \mathbf{\Sigma}_{23} \\ \mathbf{\Sigma}_{31}\ \mathbf{\Sigma}_{32}\ \mathbf{\Sigma}_{33} \end{pmatrix}.$$

That is, $\widehat{\mathbf{\Sigma}}_{ij} = \mathbf{Z}_i^t(\mathbf{I}_n - \mathbf{1}\,\mathbf{1}^t/n)\mathbf{Z}_j/n$ for all i and j. All tests considered below assume that the same number of customers are surveyed each month. The following sub-sections describe information on, around or linked to the biplot.

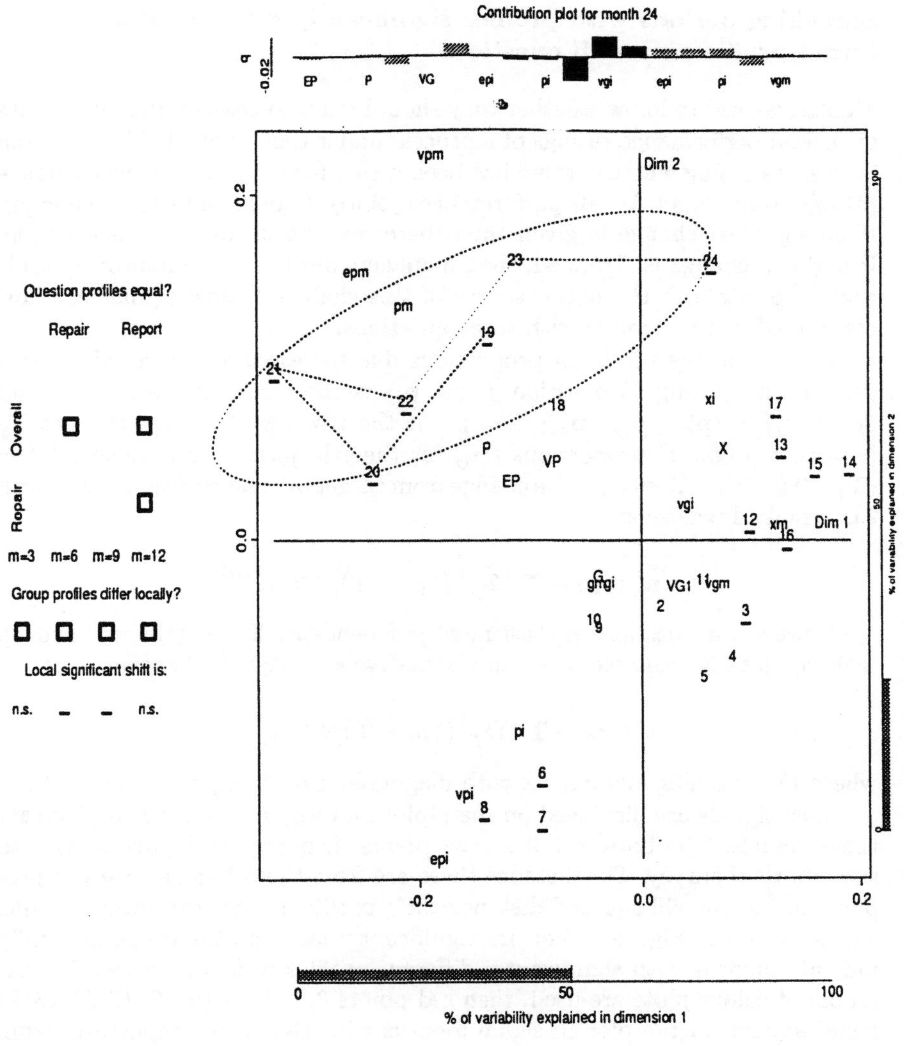

Fig. 5. Monitoring the Reporting Process for 24 months (p-poor, e-extremely, v-very, g-good, x-excellent, m-manner, i- information, capitals-overall)

Signalling periods with profiles significantly different from the target profile across all questions

Managers need to know whether they should react to changes in one or more of several performance ratings of a process under their control. The first step is often assessing whether there has been a significant change in performance ratings from target for all performances jointly (global tests for a change). If no signalled change is given then there may be no need to take action. If a global change is signalled, then managers need to find monitoring tools that help establish the major source of this change. Global signals are now considered in an example with three questions.

Let $\mathbf{p}_{ij}$ be the profile of proportions due to responses surveyed at time period ith relating to question j, i.e. $\mathbf{p}_{ij}$ is the i row of matrix $\mathbf{P}_j$, and denote $\mathbf{p}_i^t = (\mathbf{p}_{i1}^t \quad \mathbf{p}_{i2}^t \quad \mathbf{p}_{i3}^t)$, i.e., $\mathbf{p}_{ij}$ is the ith row of matrix $\mathbf{P}_j$. Let $\mathbf{T}_j$ be target profile of proportions $t\,\mathbf{p}_{ij}$ for the jth question and denote $\mathbf{T}^t = (\mathbf{T}_1^t \quad \mathbf{T}_2^t \quad \mathbf{T}_3^t)$. Then a profile change from target over all questions for month i is signalled whenever

$$n_{..}\,(t\,\mathbf{p}_i - \mathbf{T})^t \widehat{\boldsymbol{\Sigma}}_p^{-1} (t\,\mathbf{p}_i - \mathbf{T})/t > \chi^{2\,(\alpha)}_{3q-3}$$

However, if questions are assumed independent, $\boldsymbol{\Sigma}_p$ is unknown and no training data is available, then an alternative is to signal whenever

$$n_{..}\,(t\,\mathbf{p}_i - \mathbf{T})^t \mathbf{D}_T^{-1} (t\,\mathbf{p}_i - \mathbf{T})/t > \chi^{2\,(\alpha)}_{3q-3},$$

where $\mathbf{D}_T$ is a diagonal matrix with diagonal elements equal to vector $\mathbf{T}$.

Such signals are displayed on the biplot and Figure 5 is used to illustrate these signals. The integer $1, 2, \ldots, 24$ points displayed in Figure 5 refer to the month of survey. The distance these are from the origin generally is proportional to the chi-squared distance their profile is from the target profile. Those points in Figure 5 that are significantly far from the origin generally indicate months with significantly different profiles from the target. For example, if colour plots are used, then red points $3, 7, 8, 14, 19, 20, 21, 22, 23, 24$ could appear on the plot to signal months with significant departures from their long-term mean profile.

For efficient monitoring of processes using the CA biplot, managers need to be able interpret signals from the biplot and thus be informed of the nature of signalled changes.

Interpreting profile changes using the biplot

The coordinates on the biplot are given by $\mathbf{f}_i^t$ the ith row of $\mathbf{F}$ and $\mathbf{g}_j$ the jth row of $\mathbf{G}$. Note that, an indication of a stable process is

$$\mathbf{f}_i^t \mathbf{D}_{\lambda_p}^{-1} \mathbf{g}_j \approx t\,p_{ij}/c_j - 1 \quad \text{for all } i, j$$

or an indication of a process that is generally on-target is

$$\mathbf{f}_i^t \mathbf{D}_{\lambda_p}^{-1} \mathbf{g}_j \approx t\,p_{ij}/T_j - 1 \quad \text{for all } i, j.$$

Thus, for a particular question-response, if we find that $\mathbf{f}_i^t \mathbf{D}_{\lambda_p}^{-1} \mathbf{g}_j > 0$ for all poorer ratings and $\mathbf{f}_i^t \mathbf{D}_{\lambda_p}^{-1} \mathbf{g}_j < 0$ for all better ratings, then we can generally conclude that month i is rated worse than target. For example in the two-dimensional biplot, if $\mathbf{f}_i$ and $\mathbf{g}_j$ are in the same quadrant then $\mathbf{f}_i^t \mathbf{D}_{\lambda_p}^{-1} \mathbf{g}_j > 0$. If $\mathbf{f}_i$ and $\mathbf{g}_j$ are in the opposite quadrants then $\mathbf{f}_i^t \mathbf{D}_{\lambda_p}^{-1} \mathbf{g}_j < 0$. Therefore, months 6, 7 and 8 in Figure 5 was a period where information provided to customers was rated very badly relative to the target. Months 19 to 23 was a period where manner of the people receiving the customers' fault reports received relatively poorer ratings. Months 2 to 5 was a period where overall satisfaction and manner received relatively more very good performance ratings and relatively fewer poor, very poor and extremely poor ratings. Months 13 to 17 was a period where relatively more extremely good ratings were received for all questions.

The last 6 months are connected by a dashed line to indicate performance trends in the fault reporting process. For example, note that there appears to be a clear trend towards the poorer performance in manner of the person receiving the fault report. Less obvious trends are often difficult to detect with monthly signals, but the cumulative sums of deviations from target can often be useful.

Signalling whether the CUSUM of profile deviations from the target is significantly different from the zero vector

For monthly data it is convenient to CUSUM quarterly, that is, consider signals for cumulative sums of the last m months with $m = 3, 6, 9$ and 12. A significant change from the target profile for the latest m months is signalled whenever

$$n_{..} \sum_{i=t-m+1}^{t} (t\,\mathbf{p}_i - \mathbf{T})^t \;\; \widehat{\boldsymbol{\Sigma}}_p^{-1} \sum_{i=t-m+1}^{t} (t\,\mathbf{p}_i - \mathbf{T})/(mt) > \chi_{3q-3}^{2\,(\alpha)}.$$

For example, in Figure 5 the ellipse around the last 6 months signals a significant shift away from the target, i.e., $m = 6$ signals. The ellipse is approximately the 90 percentile point around the mean of the significant group of months using the chi-squared distribution. Since no ellipse appears for $m = 3, 9$ and 12 these groups do not differ significantly in profile from their target.

The two signals considered so far involve global tests of significance jointly over all questions. When these tests indicate a signal, then managers wish to establish which questions contributed to this gobal signal. Therefore, signalled changes in profiles for individual questions are useful for deciding where to direct future attention.

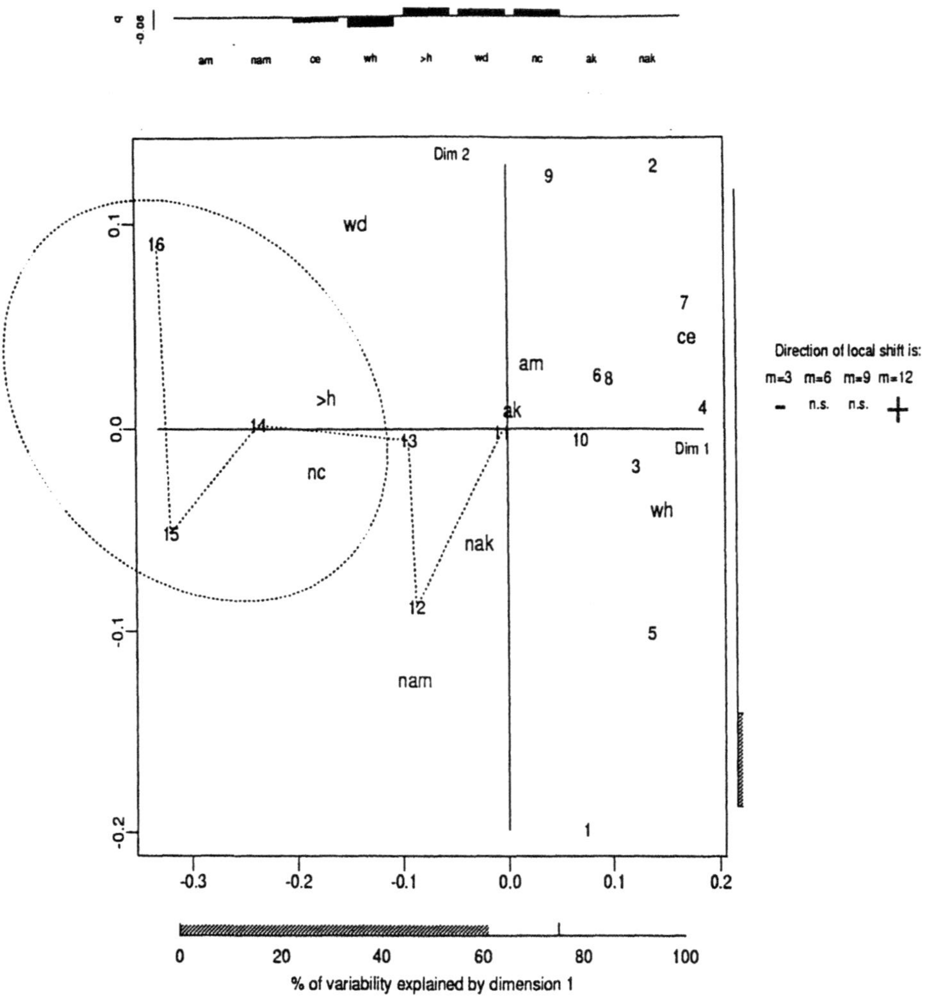

Fig. 6a. Chart of the Past 16 Months Data for the Appointment Process and 1st region (a-appointment, m-made, k-kept, n-no, ce-come early, wh-within 1 hour, >h-later than 1 hour, wd-within a day, nc-not come)

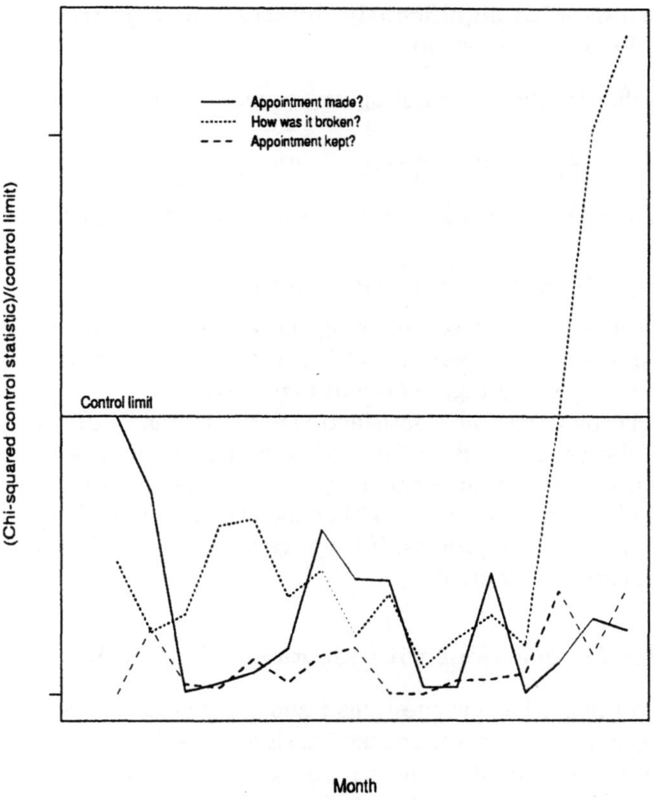

Fig. 6b. Chi-squared charts for customer responses to questions about the appointment process for the 1st region

Signalling periods with significantly different profiles from the target profile for each question

A change in profile for the jth question at the most recent observation t is signalled if

$$n_{..}\,(t\,\mathbf{p}_{tj}-\mathbf{T}_j)^t\boldsymbol{\Sigma}_{jj}^{-1}(t\,\mathbf{p}_{tj}-\mathbf{T}_j)/t>\chi_{q-1}^{2\,(\alpha)}$$

when $\boldsymbol{\Sigma}_{jj}$ is known. Otherwise, an alternative approach is to signal whenever

$$n_{..}\,(t\,\mathbf{p}_{tj}-\mathbf{T}_j)^t\mathbf{D}_{\mathbf{T}_j}^{-1}(t\,\mathbf{p}_{tj}-\mathbf{T}_j)/t>\chi_{q-1}^{2\,(\alpha)},$$

where $\mathbf{D}_{\mathbf{T}_j}$ is a diagonal matrix with diagonal elements equal to $\mathbf{T}_j$. This signal is not displayed on the biplots, but it is displayed on a chart linked to the biplot. For example, see Figure 6b and Figure 6d.

Questions relating to customer satisfaction with different parts of the process are nearly always collected on the same Likert scale. In these instances it is useful to know whether the response profiles to these questions are the same. This would inform managers whether the satisfaction ratings are similar for different parts of the process. If they are not, this may indicate areas where improvements are required.

Similar profiles for questions with identical Likert scales

If $\mathbf{G}_i \approx \mathbf{G}_j$ then the profiles for questions i and j are generally very similar. Profile means $\bar{\mathbf{p}}_t = \left(\bar{\mathbf{p}}_{t1}\ \bar{\mathbf{p}}_{t2}\ \bar{\mathbf{p}}_{t3}\right)$, are used to signal whether or not questions i and j differ in profiles. For the i and j pairwise comparison a signal is given whenever

$$n_{..}\,(\bar{\mathbf{p}}_{ti}-\bar{\mathbf{p}}_{tj})^t\boldsymbol{\Sigma}_d^{-1}(\bar{\mathbf{p}}_{ti}-\bar{\mathbf{p}}_{tj})>\chi_{q-1}^{2\,(\alpha)}$$

where $\boldsymbol{\Sigma}_d=\boldsymbol{\Sigma}_{ii}+\boldsymbol{\Sigma}_{jj}-2\,\boldsymbol{\Sigma}_{ij}$.

In Figure 5 all questions are recorded on the same six point Likert scale and therefore pairwise checks of similarity of their profiles are possible. The result of signals are recorded as a matrix of boxes on the left-hand-side of Figure 5 under "Question profiles equal?". A box signals significantly different profiles, while "n.s." indicates no significant difference. In Figure 5, all boxes appear, thus signalling that all pairwise profiles differ significantly.

In the biplot, identical profiles on the same Likert scale have respective ranked points coinciding on the biplot. Therefore, similar profiles have respective ranked points very close together in the biplot. In Figure 5 profiles differ most for the ratings exceptionally poor, very poor and poor, but are fairly similar for the better rating categories.

In addition to signalling whether performance ratings are on target, managers are often interested in whether or not customer satisfaction ratings are changing locally. For monthly data, managers may be interested in assessing whether satisfaction ratings of customers for the last quarter differed significantly from the previous 21 months satisfaction ratings. It would also be interesting to signal local changes once a process change has been made, monitoring whether this has caused a change in customer satisfaction.

Signalling a general shift in average profiles for a group of months from the average profiles for the remaining months

A change in the mean profile for the most recent m observations from the mean profile for the remaining $t-m$ is signalled whenever

$$n_{..}m(t-m)\,(\bar{\mathbf{p}}_m - \bar{\mathbf{p}}_{[-m]})^t \widehat{\boldsymbol{\Sigma}}_p^{-1} (\bar{\mathbf{p}}_{mj} - \bar{\mathbf{p}}_{[-m]}) > \chi_q^{2(\alpha)}$$

where $\bar{\mathbf{p}}_{[-m]}$ is the sample mean profile for the remaining $t-m$ observations.

Note that in Figure 5 $t = 24$ and local changes in group profiles are considered for group sizes $m = 3, 6, 9$ and 12. These changes are signalled on the left hand-side of Figure 5 below the heading "Group profiles differ locally?". A box indicates significant differences in profiles, while "n.s." indicates non-significant differences.

Figure 5 signals that the mean profiles for the most recent $m = 3, 6, 9$ and 12 months differ significantly from the mean profiles of the remaining $n-m = 21, 18, 15$ and 12 months, respectively. The biplot in Figure 5, indicates that months 24 to 19 are very different from the remaining months. It is also clear that most of the recent 12 months (24 to 13) have positive values for dimension 2, while the first 12 months mostly have negative dimension 2 values. Therefore, the biplot provides visual confirmation of these differences. Often managers want to know whether satisfaction ratings have improved. Signalling general changes in profiles can often miss detecting significant directional changes in customer satisfaction ratings.

Signalling a directional shift in average profiles for a group of months from the average profiles for the remaining months in the biplot

Let $\mathbf{J}^t = \left(\mathbf{j}^t\ \mathbf{j}^t\ \mathbf{j}^t\right)$, where $\mathbf{j}^t$ is defined by (1). Then a directional shift towards better performance for the most recent m months relative to the remaining months is signalled whenever

$$\sqrt{n_{..}m(t-m)}\,\mathbf{J}^t(\bar{\mathbf{p}}_m - \bar{\mathbf{p}}_{[-m]})/\sqrt{\mathbf{J}^t\widehat{\boldsymbol{\Sigma}}_p\mathbf{J}} > 3$$

and directional shift towards worse performance signals whenever

$$\sqrt{n_{..}m(t-m)}\,\mathbf{J}^t(\bar{\mathbf{p}}_m - \bar{\mathbf{p}}_{[-m]})/\sqrt{\mathbf{J}^t\widehat{\boldsymbol{\Sigma}}_p\mathbf{J}} < -3.$$

These are analogous to the 3-sigma control limit used in the control literature [2].

Directional signals are recorded on the right hand-side of Figure 5 just below "Local directional shift is:". A negative signs signal months with significantly poorer ratings, a positive sign signals significantly better ratings, while "n.s." indicate nonsignificant differences. In Figure 5, the most recent $m = 6$ and 9 months signal relatively poorer ratings than the ratings for remaining $n - m = 18$ and 15 months, respectively. The biplot in Figure 5 clearly indicates that months 24 has relatively more excellent ratings (see month 24's position in the biplot and the excellent cells in the barplot above Figure 5 are all higher than average), but months 23 to 18 indicate on average poorer ratings. Therefore, the biplot provides visual confirmation why group sizes $m = 6$ and 9 signal and reasons why group sizes $m = 3$ and 12 fail to signal.

Directional changes jointly over all questions are useful for signalling an overall change in customer satisfaction. Managers are generally more interested in controlling directional changes in customer satisfaction, for example maintaining an improved customer rating.

Signalling directional changes for individual months over all questions

For month t, we signal overall a better performance rating whenever

$$\sqrt{n_{..}}\,\mathbf{J}^t(t\,\mathbf{p}_t - \mathbf{T})/\sqrt{t\,\mathbf{J}^t\widehat{\boldsymbol{\Sigma}}_p\mathbf{J}} > 3$$

and overall a worse performance rating whenever

$$\sqrt{n_{..}}\,\mathbf{J}^t(t\,\mathbf{p}_t - \mathbf{T})/\sqrt{t\,\mathbf{J}^t\widehat{\boldsymbol{\Sigma}}_p\mathbf{J}} < -3.$$

In Figure 5, months with significantly better performances have the number of the month underlined, e.g., months 3, 13, 14, 15, 16 ,17 and 24. Months 13 to 17 signal a period when performances were rated significantly better than target. Underlining these points in different colours are useful for distinguishing significantly better performances from significantly worse performances. Months with significantly worse performances have the number of the month underlined (in red on a colour plot), e.g. months 6, 7, 8, 19, 20, 21, and 22. Note that there are months that signal directional changes but not general profile changes, i.e., months 6, 13, 15, 16, and 17. Month 23 signalled a profile change that is not a directional change.
Managers are interested in which questions have contributed to a globally signalled directional change.

Signalling directional changes for individual questions

This is carried out as in Section 2.2.2, that is, for the most recent month signal a better performance rating whenever

$$\sqrt{n_{..}}\,\mathbf{j}^t(t\,\mathbf{p}_{tj} - \mathbf{T}_j)/\sqrt{t\,\mathbf{j}^t\widehat{\mathbf{\Sigma}}_{jj}\mathbf{j}} > 3$$

and a worse performance rating whenever

$$\sqrt{n_{..}}\,\mathbf{j}^t(t\,\mathbf{p}_{tj} - \mathbf{T}_j)/\sqrt{t\,\mathbf{j}^t\widehat{\mathbf{\Sigma}}_{jj}\mathbf{j}} < -3$$

where the vector $\mathbf{T}$ is as defined in Section 4.1.1. Let $\mathbf{q}_t$ be the last row of $\mathbf{Q}$ given by (4). A *contribution plot* is a barplot of $\mathbf{q}_t$ (e.g., see above Figure 5) which is useful for displaying deviations from target for the most recent month t (month 24 in Figure 5). Therefore, the signalled shift for the 24th month in Figure 5 results from more excellent ratings for the reporting process overall than would have been achieved by a target profile performance, more very good and excellent ratings for the information provided and more extremely poor, very poor, poor and good ratings for manner. This indicates that month 24 is a period where information provided was rated significantly better than target (signalled by shading the respective bars dark grey (instead of light grey) in the barplot).
The *contribution plot* presents the current scaled deviation from the target profile and helps describe profile changes. Also, the size and direction of each bar in the plot of $\mathbf{q}_t$ describes the magnitude and direction of the shift for all category responses. The signalled directional changes in individual questions are indicated by shading the respective bars black (instead of light grey) for better rated performances (Figure 5 and Figure 6c) and darker grey for worse rated performances (Figure 6a).

If a different number of customers are surveyed at each time period and if all customers respond to all 3 questions within each time period, then the biplot procedure above is carried out by replacing $\mathbf{1}_t/t$ by $\mathbf{r} = \mathbf{N}_j\mathbf{1}_q/n_{..}$. The testing procedures for signals are notationally more messy but still straight forward.

When monitoring profile deviations from the target profile or the long term mean then replace vector $\mathbf{c}_j$ in (3) by their target profiles $\mathbf{T}_j$, and then proceed as outlined in this section.

4.2 An illustration of monitoring several nested questions

The following example of nested questions for an appointment process in a service industry is considered.

Example 4: On a monthly basis, customers of each month were asked:

- Was an appointment made? They were asked to respond either with appointment made (am) or with no appointment made (nam).
- If they responded "am" to the previous question, then they were asked whether the appointment was kept? The responses were appointment kept (ak) and appointment not kept (nak).

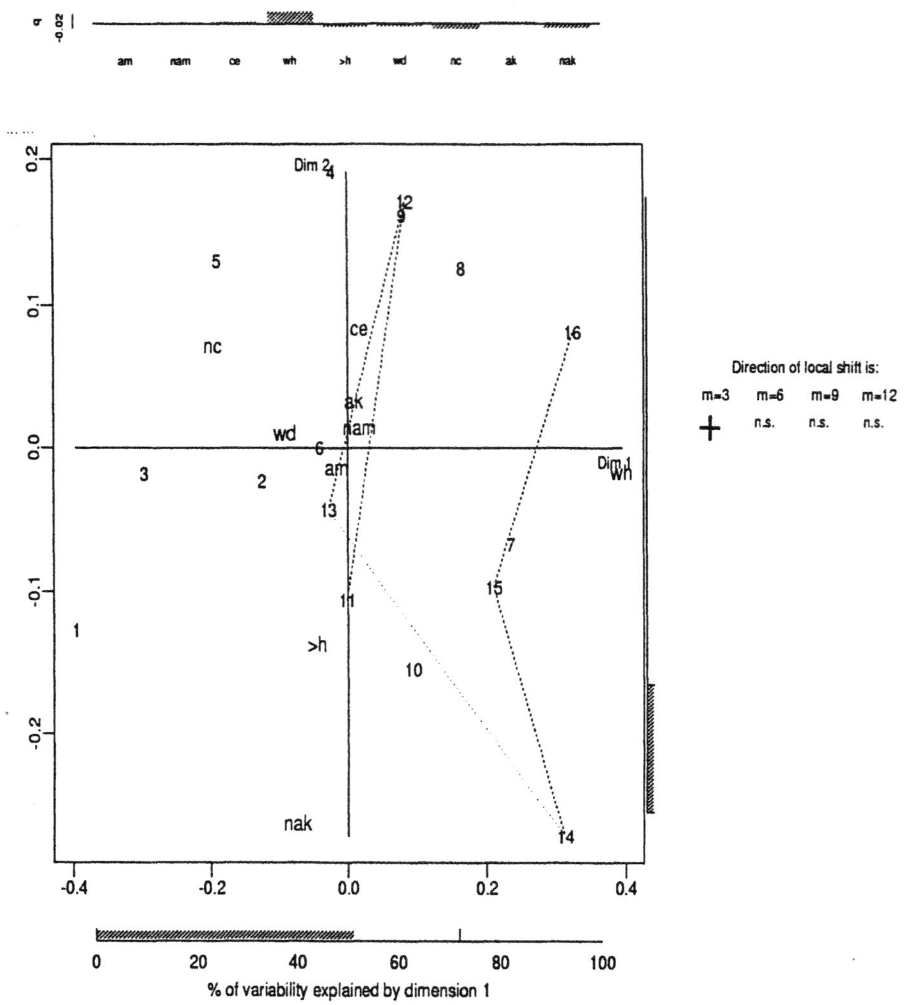

Fig. 6c. Chart of the Past 16 Months Data for the Appointment Process and 2nd region (a-appointment, m-made, k-kept, n-no, ce-come early, wh-within 1 hour, >h-later than 1 hour, wd-within a day, nc-not come)

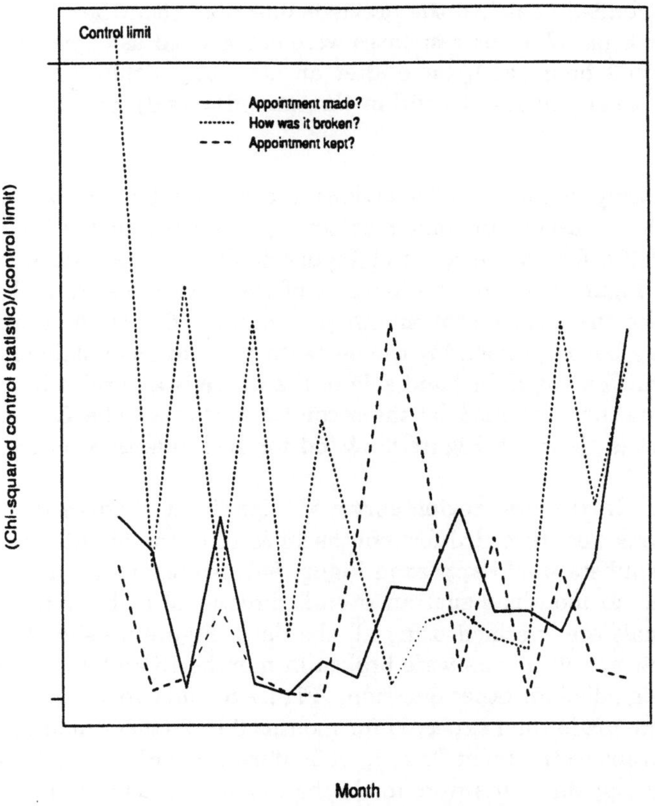

Fig. 6d. Chi-squared charts for customer responses to questions about the appointment process for the 2nd region

- If they responded "nak" to the previous question, then they were asked how was it broken? Their responses were categorised as came early(ce), came within 1 hour (wh), came after an hour but within 2 hours(>h), came later than 2 hours but within the same day (wd), or did not come (nc).

The monitoring of this set of questions is carried out as in Section 4.1. Displays which are useful for joint monitoring of such data are illustrated in Figure 6a & 6b for one region and Figure 6c & 6d for a second region. At the top of Figure 6a & 6c is a barplot of the deviations from the long term average for the most recent month (i.e., month 16). At the bottom of these figures is a barplot indicating the percentage variation explained by the major axis, while on the right-hand side of the figure is a barplot indicating percentage variation explained by the second major axis. The Chi-squared Shewhart chart is shown in Figure 6b & 6d for each question on the same axis.

Region 1: In the first region surveyed, significant changes in profiles over all questions were recorded for months 15 & 16 in the biplot (for colour graphics red numbers would appear in Figure 6a). No points are underlined, indicating that no months signal an overall directional shift in rated performance. Signals are specified using all the data. Figure 6b signals profile changes for how appointments were broken in months 15 and 16. No profile changes were signalled for other questions. Figure 6a indicates that more appointments were made than expected for months 6 to 9 (these months are in the same quadrant as the point "am"), while during months 3, 13 & 16 proportionally less appointments were made than expected. The barplot above Figure 6a signals a directional shift in profiles for how an appointment was broken for month 16. The shift is towards poorer performance ratings, i.e., less repairers missing appointments by coming early(*ce*) or within the hour (*wh*) and more repairers missing appointments by not coming(nc), coming later than 2 hours but within the same day(wd) and coming later than 1 hour but within 2 hours(>h). The ellipse around months 14, 15 and 16 signals a significant CUSUM change in profiles overall for these three months. Since there is no other ellipse in Figure 6a, the latest $m = 6, 9$ and 12 fail to signal CUSUM differences in profiles.

Significant directional shifts are signalled on the right of Figure 6a. Here cumulative quarters are tested for differences with the remaining months. The past quarter signals a significant poorer performance, while the past twelve months signalled a significantly better performance.

Region 2: For the second region surveyed, month 1 in Figure 6d signals a change in profiles for how an appointment was broken and the biplot (Figure 6c) indicates that this is likely to be coming late for appointments. The respective bars in the barplot above Figure 6c indicate no significant directional shift for how the appointment was broken, but higher than expected proportion of "ce"(came early) and "wh"(within the hour), indicating that a higher than expected proportion of appointments were broken by coming early or within the hour during month 16.

Significant directional shifts are signalled on the right of Figure 6c. Here cumulative quarters are tested for differences with the remaining months. The past quarter signals a significant better performance overall.

The CA biplot applied to tabular data for several questions indicates similar or correlated profiles. The direction of the profiles for ordinal data are useful for interpreting the nature of the association. For example, in Figure 6a points "am" and "fam" have similar orientation from the origin and points "nam" and "nfa" have similar orientation from the origin, indicating that the responses to these questions tend to be positively associated. This association indicates that months with a higher than expected proportion of appointments made are likely to also have a higher than expected proportion of follow-up appointments made. An interesting feature of the biplot for Region 2 is that the profiles of appointment made and appointment kept are negatively associated. This occurs, for example, when a region makes an appointment to meet customer need even when no repairer is capable of making the appointment.

In the situation where profiles are not positively associated (e.g., appointment made and appointment kept in Figure 6c) it is difficult to assess from the biplot whether, in recent months, there has been a general improvement in the appointment process. In these cases, the tests for direction shifts for the local m months is therefore very useful. On the right-hand-side of Figure 6c we see that the past three months signal a significantly improved performance rating, but this was not significant for the past 6,9 or 12 months.

Although, in this section it has been illustrated that the CA biplot is useful for displaying associations between profiles, this association does not necessarily imply a correlation between customer responses to questions leading to these profiles. The CA biplot as applied in this section, therefore fails to display the relationship between the individual's responses to questions. The next section considers monitoring the relationship between customers responses to particular questions and their overall satisfaction ratings.

4.3 Monitoring correlations between question-responses

It is useful to monitor the relationship between customer responses to questions relating to within process performance and overall process performance. For example, a major driver of customer satisfaction at a restaurant is the quality of the food. In these instances, it may be useful to monitor the relationship between customers' rating of the quality of food and their overall satisfaction with the restaurant. Procedures which monitor such data are now considered.

Greenacre [3] showed that CA of indicator matrices similar to (1) with two questions where $\mathbf{Z} = \begin{pmatrix} \mathbf{Z}_1 & \mathbf{Z}_2 \end{pmatrix}$, is equivalent to canonical correlation between matrices $\mathbf{Z}_1$ and $\mathbf{Z}_2$. The principal coordinates for this analysis were shown to be identical to CA applied to the counts matrix $\mathbf{N} = \mathbf{Z}_1^t\mathbf{Z}_2$ (see Greenacre [3] page 131). Therefore, a reasonable method for monitoring relationships is to display and compare the question-response profiles of the matrices $\mathbf{N}$ over time. The chi-squared test is used to signal significant changes in correlation. If training data produces a value of c for $\mathrm{tr}(\mathbf{Q}^t\mathbf{Q})$, then provided there have been no changes in associations for new data, $\mathrm{tr}(\mathbf{Q}^t\mathbf{Q})$ is expected to also equal c. Since questions are dependent, $m\mathrm{tr}(\mathbf{Q}^t\mathbf{Q})$ has approximately a non-central chi-squared distribution. The non-centrality parameter,when m customers are surveyed, is estimated by $\lambda = mc - (q_1 - 1)(q_2 - 1)$. For new data, a change in relationships is signalled using 3-sigma limits, given by either

$$m\,\mathrm{tr}(\mathbf{Q}^t\mathbf{Q}) > mc + 3\sqrt{2[(q_1 - 1)(q_2 - 1) + 2\lambda]}$$

for increased associations, or

$$m\,\mathrm{tr}(\mathbf{Q}^t\mathbf{Q}) < mc - 3\sqrt{2[(q_1 - 1)(q_2 - 1) + 2\lambda]}$$

for decreased associations. If the association between question-responses signals an increase (a signal on the high-side) then all category-response codes relating to this signal are coloured black, and if the association decreases significantly (a signal on the low-side) then all related codes are coloured dark grey, otherwise they are coloured light grey.

To aid the interpretation when a signal is given we isolate the categories that provide most of the inertia for the signal. Note that $\mathrm{tr}(\mathbf{Q}^t\mathbf{Q}) = \sum_{i=1}^{q_1}\sum_{j=1}^{q_2}(p_{ij}-r_ic_j)^2/(r_ic_j)$. Therefore, the contribution the ith row category-response makes to $\mathrm{tr}(\mathbf{Q}^t\mathbf{Q})$ is given by $IR_i = \sum_{j=1}^{q_2}(p_{ij} - r_ic_j)^2/(r_ic_j)$. Similarly, the contribution for jth column category-response is $IC_j = \sum_{i=1}^{q_1}(p_{ij} - r_ic_j)^2/(r_ic_j)$. The relative gain or loss of inertia for new data relative to the training data is useful for indicating changes in association. Let $RIR_i(RIC_j)$ be the ratio of $IR_i(IC_j)$ values for new data to training data. Barplots of $(RIR_1 - 1, RIR_2 - 1, \ldots, RIR_{q_1} - 1)$ are used to assess the relative contribution to changes of each row. Deviation of the bar heights from 0 indicates empirical changes in relative inertia, and thus indicates changes in dependence. Any bar above zero indicates increased dependence and any bar below zero indicates reduced dependence.

A similar plot for column contributions is applicable.

It is easy to extend this to consider more than two questions. Consider, for example, simultaneously examining the relationship between overall satisfaction (with incidence matrix $\mathbf{Z}_1$) and :

1. satisfaction with the reporting process (with incidence matrix $\mathbf{Z}_2$)
2. satisfaction with the repairer's process (with incidence matrix $\mathbf{Z}_3$)
3. satisfaction with the time it took to correct the fault. (with incidence matrix $\mathbf{Z}_4$)

If the relationship between satisfaction with the repairer and overall satisfaction with the organisation increases then this means the repair process may be becoming more important in detemining a satisfied customer overall. If simultaneously to this the relationship between satisfaction with the reporting process and the organisation overall decreases, this indicates that the reporting process appears to becoming less important to customers while the repairer's process is becoming more important.

The CA biplot can be applied to the following counts matrices

$$(\mathbf{N}_1\ \mathbf{N}_2\ \mathbf{N}_3) = (\mathbf{Z}_1^t\mathbf{Z}_2\ \mathbf{Z}_1^t\mathbf{Z}_3\ \mathbf{Z}_1^t\mathbf{Z}_4)$$

with defined $\mathbf{Q}_j$ as in (3). Singular value decomposition of the matrix $(\mathbf{Q}_1\ \mathbf{Q}_2\ \mathbf{Q}_3)$ is

$$(\mathbf{Q}_1\ \mathbf{Q}_2\ \mathbf{Q}_3) = \mathbf{U}\mathbf{D}_\lambda\,(\mathbf{V}_1^t\ \mathbf{V}_2^t\ \mathbf{V}_3^t)$$

where: $\mathbf{U}^t\mathbf{U} = I_{q-1}$, $\sum_{j=1}^{3}\mathbf{V}_j^t\mathbf{V}_j = I_{q-1}$ and $\mathbf{D}_\lambda$ is a diagonal matrix with diagonal elements $\lambda_1 \geq \lambda_2 \geq \ldots \geq \lambda_{q-1} \geq 0$. Let $\mathbf{V}_{j\,p}^t$ be the matrix consisting of the first p rows of $\mathbf{V}_j^t$ and $\mathbf{D}_{\lambda\,p}$ be the diagonal matrix consisting of the first p diagonal elements of $\mathbf{D}_\lambda$.

Let $\mathbf{c}_j = \mathbf{1}_q^t\mathbf{N}_j$ and $\mathbf{r}^t = \mathbf{N}_j\mathbf{1}_q$. The coordinates taken as the rows of

$$\mathbf{G} = \begin{pmatrix}\mathbf{G}_1\\ \mathbf{G}_2\\ \mathbf{G}_3\end{pmatrix} = \begin{pmatrix}\mathbf{D}_{\mathbf{c}_1}^{-1/2}\mathbf{V}_{1\,p}\\ \mathbf{D}_{\mathbf{c}_2}^{-1/2}\mathbf{V}_{2\,p}\\ \mathbf{D}_{\mathbf{c}_3}^{-1/2}\mathbf{V}_{3\,p}\end{pmatrix}\mathbf{D}_{\lambda\,p}$$

and

$$\mathbf{F} = \mathbf{D}_{\mathbf{r}}^{-1/2}\mathbf{U}_p\mathbf{D}_{\lambda\,p}$$

are plotted to assess whether associations are stable.

Let $\mathbf{f}_i^t$ be the ith row of $\mathbf{F}$ and $\mathbf{g}_j$ be the jth row of $\mathbf{G}$. Note that $\mathbf{f}_i^t\mathbf{D}_{\lambda_p}^{-1}\mathbf{g}_j \approx p_{ij}/(r_ic_j) - 1$. Therefore, if the corresponding elements of vectors $\mathbf{f}_i$ and $\mathbf{g}_j$ are all of the same sign, then $p_{ij} > r_ic_j$ and thus row i and column j are positively associated. Similarly, if the corresponding elements of $\mathbf{f}_i$ and $\mathbf{g}_j$ are all of the opposite sign, then $p_{ij} < r_ic_j$ and thus row i and column j are negatively associated. For example, Question 1 is positively associated with overall satisfaction if the respective elements of vectors $\mathbf{f}_i$ and $\mathbf{g}_j$ are mostly of the same sign, and negatively associated if the respective elements of vectors $\mathbf{f}_i$ and $\mathbf{g}_j$ are mostly of the opposite sign. Although determining associations is important, this section monitors whether these associations have changed with time.

Figure 7 displays the biplot for a particular region and the example used is the one mentioned above. Figure 7a displays the biplot for a chosen month, and Figure 7b presents the biplot for the training data. Figures 7a & 7b are both on the same scale, therefore a comparison of of these figures is used to assess changes in associations for this month relative to the training data. The display in Figure 7b explains roughly 80% of the variation in the training data and this is marked by a black vertical tick on the barplot below Figure 7a.

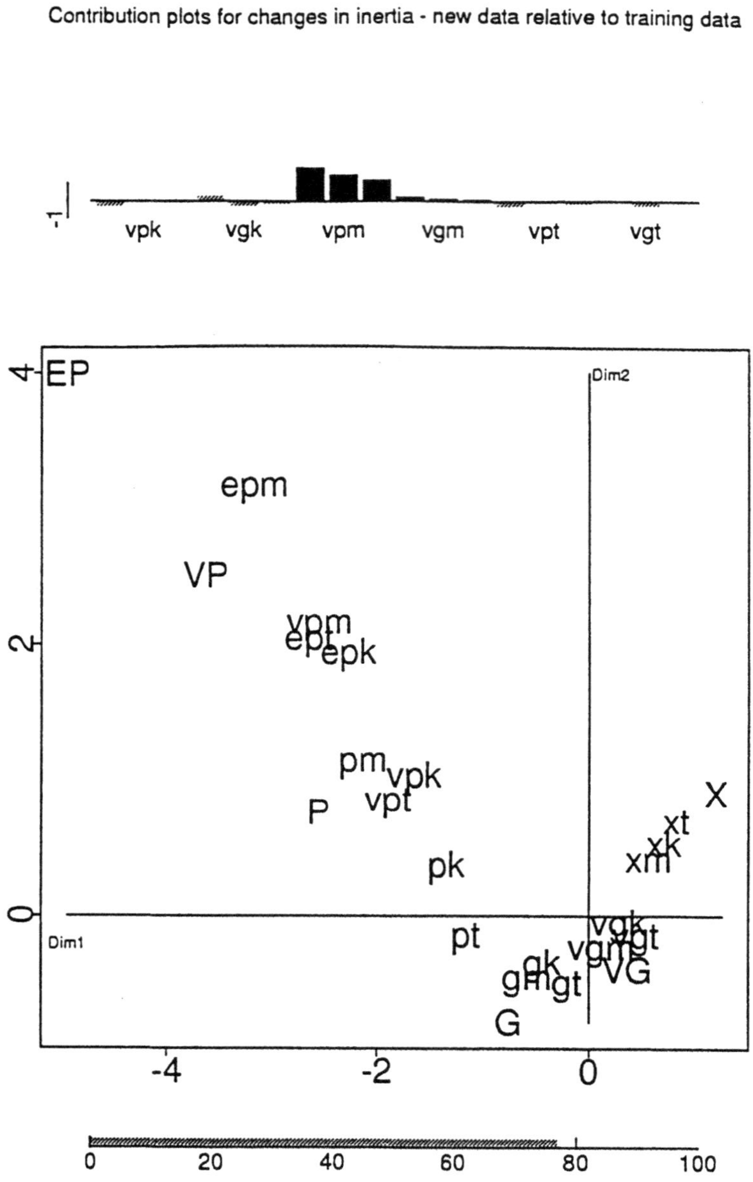

Fig. 7a. Correspondence Analysis for the current month's data (p-poor, g-good, e-extremely, v-very, p-poor, x-excellent)

Contribution plots for changes in inertia - new data relative to training data

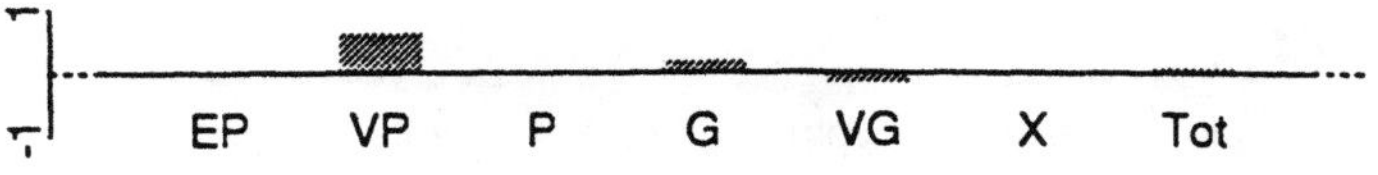

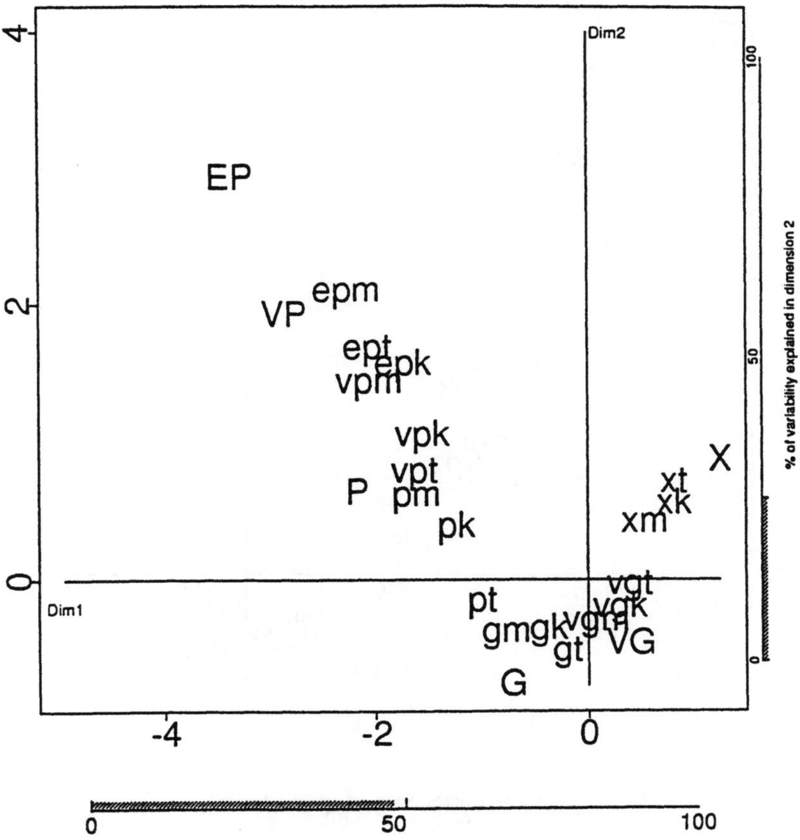

Fig. 7b. Correspondence Analysis for the training data (k-knowledge, m-manner, t-repair-time, capital letters-overall satisfaction))

Contribution plots for changes in inertia - new data relative to training data

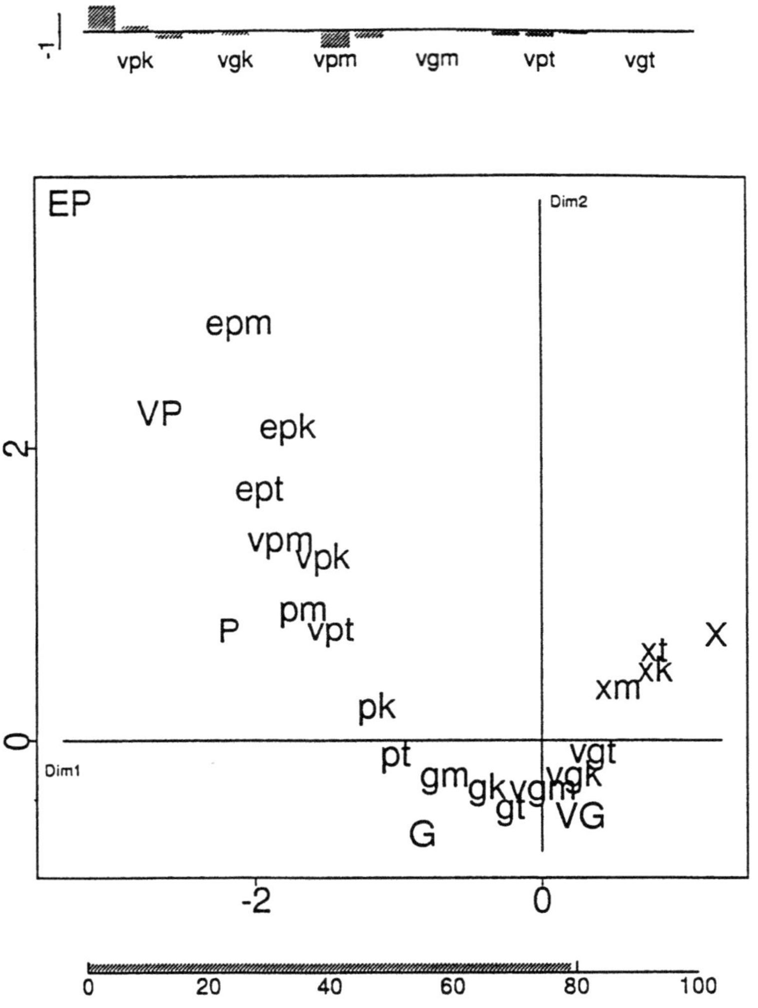

Fig. 7c. Correspondence Analysis for the current month's data (p-poor, g-good, e-extremely, v-very, p-poor, x-excellent)

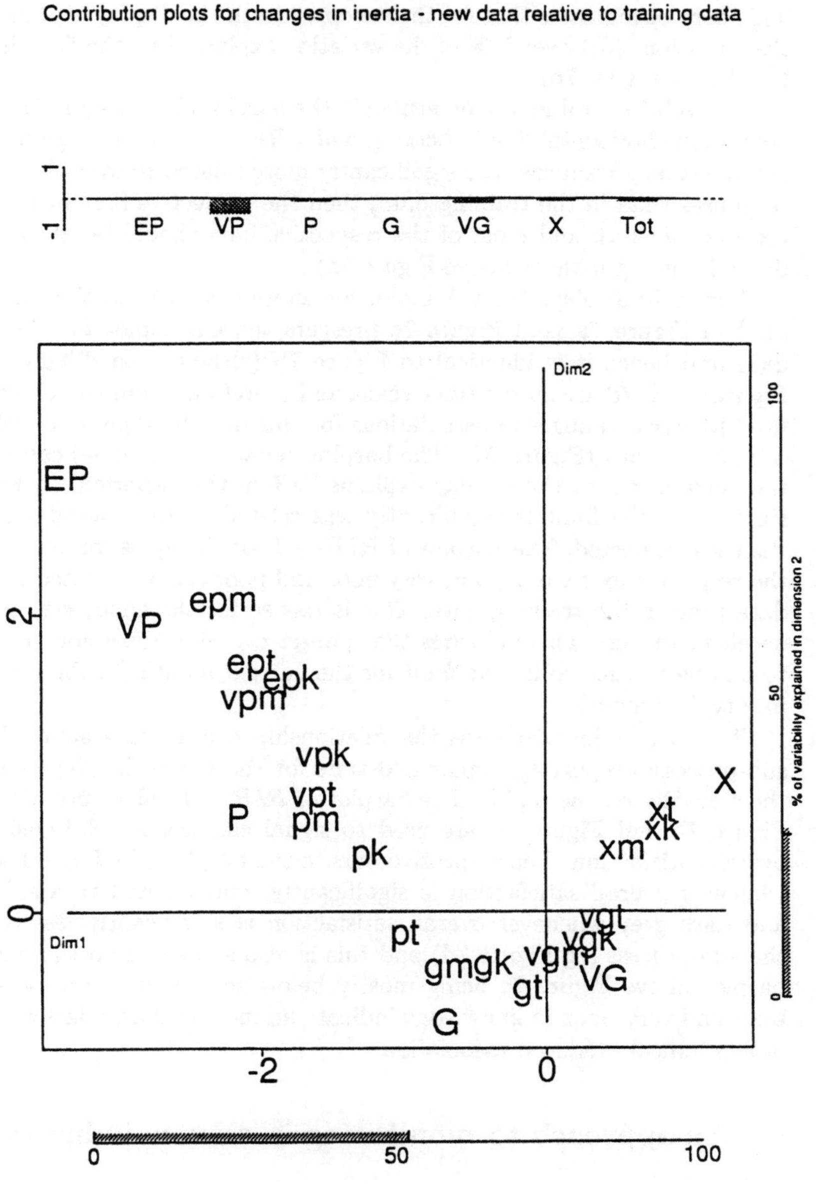

Fig. 7d. Correspondence Analysis for the training data (k-knowledge, m-manner, t-repair-time, capital letters-overall satisfaction))

This barplot indicates that for the current month the display explains 77% of the variation. Just over 50% of the variation explained by the first dimension (see below Figure 7b).

The $RIC_i - 1$ values are presented in the barplot above Figure 7a, with the continuous horizontal line indicating when $RIC_i = 1$ (no change in inertia). If a question's responses are significantly more related to overall satisfaction responses than in the training data, then the respective bars in the barplot are shaded black and most of the respective bars should be above the line dashed line (eg manner above Figure 7a).

Figure 7c displays the CA biplot for another month for the same region used in Figure 7a, and Figure 7c presents the CA biplot for the training data and hence it is identical to Figure 7b (although on different scales). Figures 7c & 7d are on the same scale, and therefore a comparison of these is used to assess changes in associations for this month (Figure 7c) relative to a typical month (Figure 7d). The barplot below Figure 7c indicates that for the current month the display explains 79% of the variation. In Figure 7c, time to fix the fault is significantly less related to the overall satisfaction than was expected. The barplot of $RIR_i - 1$ for this question indicates that the responses extremely poor, very poor and poor are less related in the new data than in the training data. This is less so for the good, very good and excellent ratings. This indicates that poorer overall satisfaction is less likely to be due to time to fix the fault for the current month (in Figure 7c) than in a typical month.

The CA biplot visualises the relationship overall satisfaction has with sub-processes reporting, repair and time for the month by the similarity of their profiles on the biplot. The barplot of $RIR_i - 1$ values presented above Figure 7b and Figure 7d are used to signal changes in relationships with overall satisfaction. The respective bars in the barplot of $RIR_i - 1$ are black whenever overall satisfaction is significantly more related to a sub-process and dark grey whenever overall satisfaction is significantly less related to the sub-process (eg Figure 7d) and this is also evident from the bars of the barplot above Figure 7c being mostly below zero. Only bars for extremly poor and very poor in knowledge indicate an increased association, while all others indicate reduced association.

5 An approach to monitoring in service industries

Monitoring procedures are more useful if they help identify causes of variable performances in outcome measures. Managers are often more focussed on outcomes to processes and when outcomes are poor they need to have a systematic way of looking for the most probable causes. In this section we design a system for monitoring a fault repair process discussed previously.

The flow chart in Figure 8 outlines a possible approach to managing customer satisfaction when dealing with a fault repair process. In this regard,

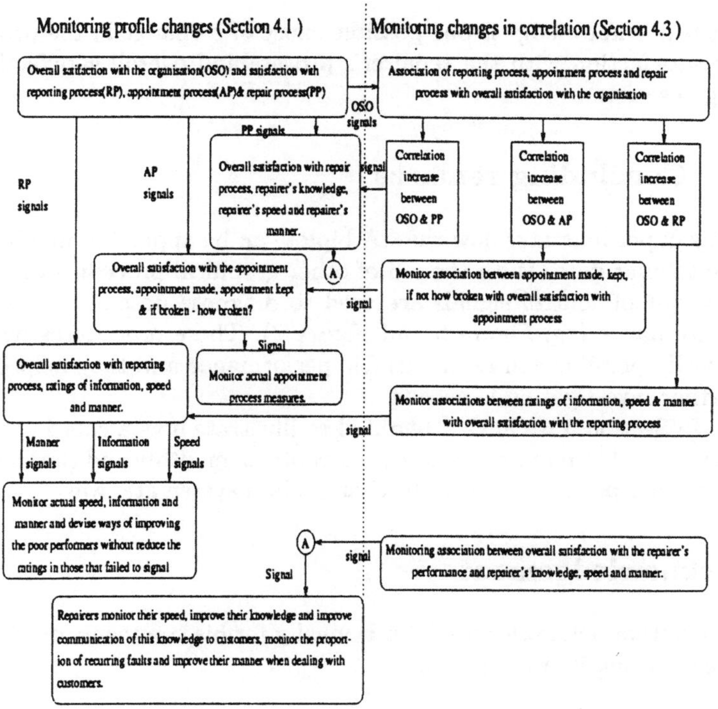

Fig. 8. Organisational chart for an example of monitoring a process with related parts

the outcome measure that most managers wish to control, besides costs, is customers' overall satisfaction with an organisation. This outcome is generally a consequence of all the organisation has done or has not done for the customer during the process of meeting a customer's repair needs. For example, a recent spate of dissatisfaction with an organisation could be due to events within the reporting process, within the repairer's process or to some other cause. Simultaneous monitoring of related sub-process outcome measures and overall organisation outcome measures is often useful for identifying what sub-processes have caused this spate. For example, if overall satisfaction with the organisation and the repairer's process both signal profile changes in recent times and an increased correlation is signalled between these variables, then it is very likely the change in overall satisfaction with the organisation is due to changes in the repairer's performance. The repairer's process could be investigated to find what has changed. If it is found that the repairer's knowledge has signalled a change in profile and there is a signalled increase in its correlation with overall satisfaction with the repairer's process, then repairer's knowledge is one possible cause of a change in the profiles of overall satisfaction with the repairer's process and overall satisfaction with the organisation.

6 Concluding remarks

This paper illustates how the CA biplot can be applied to monitor multivariate tabular data. With the aid of other graphical procedures it is illustrated how out of control signals are used to diagnose possible causes of variable performance (eg, Figure 6 and Figure 7). These procedures help managers identify possible causes of variable performance and they help decide where further investigation is needed.

In Section 5 we have attempted to illustrate how a coordinated approach can be used to monitor customer perceptions monthly and thus help managers to be in a position to react to changes in a systematic way.

Acknowledgement

The author acknowledges John Donnelly for his comments on this paper and the resulting improvements.

References

1. Shewhart, W. (1931) Economic control of quality of manufactured product. Macmillan, London
2. Montgomery, D.C. (1991) Introduction to Statistical Quality Control. 2nd ed. John Wiley & Sons, New York
3. Greenacre, M.J. (1984) Theory and Application of Correspondence Analysis. Academic Press, Inc., London
4. Gower, J.C. and Hand, D.J. (1996) Biplots. Chapman & Hall, London
5. Fuchs, C. and Benjamini, Y. (1994) Multivariate profile charts for statistical process control. Technometrics **36**, 182–195
6. Page, E.S. (1954) Continuous Inspection Schemes. Biometrika **41**,100-114
7. Benzecri, J.-P. (1992) Correspondence Analysis Handbook. Marcel Dekker, Inc, New York
8. Gabriel, K.R. (1995) Biplot display of multivariate categorical data. Chapter 9 of Recent Advances in Descriptive Multivaiate Analysis edited by W.J. Krzanowski

Monitoring a Proportion Using CUSUM and SPRT Control Charts

Marion R. Reynolds, Jr.[1] and Zachary G. Stoumbos[2]

[1] Virginia Polytechnic Institute and State University, Blacksburg, VA 24061-0439
[2] Rutgers, The State University of New Jersey, Newark, NJ 07102-1895

Abstract. In many process monitoring problems, items from the process are classified into one of two categories, usually called defective and nondefective, and the objective is to detect any change in the proportion p of defective items produced by the process. In some applications, there may be a continuous stream of inspection data from the process, as would occur with 100% inspection. In other applications, the inspection data may be in the form of samples taken from the process. The traditional control chart for monitoring p is the Shewhart p-chart. For this chart, the proportion defective for segments or samples of n items is determined and plotted. There are a number of disadvantages to using the Shewhart p-chart. In particular, the p-chart is not efficient for detecting small changes in p, and when the target value is close to 0, the discreteness and skewness of the binomial distribution frequently produces a p-chart with undesirable properties. This paper evaluates and compares several control charts which are alternatives to the Shewhart p-chart. A CUSUM chart for monitoring p can be based on the binomial distribution when items are grouped into segments or samples. A CUSUM chart can also be based on the Bernoulli distribution when inspection results are recorded and used individually as they are obtained. These two CUSUM charts are compared in terms of the time required to detect a shift in p for the case of 100% inspection and for the case of samples of items. When samples are taken from the process, a highly efficient control chart can be based on using sequential sampling at each sampling point. In particular, an SPRT can be applied at each sampling point and the decision of this test used to determine whether to signal or go to the next sampling point. It is shown that this SPRT chart will detect shifts in p much faster than CUSUM charts and the Shewhart p-chart.

1 Introduction

In many process monitoring applications, the items produced by the process will be classified into one of two categories, usually called defective and nondefective (or conforming and nonconforming), and the proportion of defective items produced by the process will be used as a measure of product quality. If p is used to represent the probability that an item is defective, then the objective of process monitoring with a control chart is to detect any special cause that has produced a change in p from an in-control value p_0. In most applications the primary objective would be to detect an increase in p, which would correspond to a deterioration in product quality.

In some applications the inspection process will produce a continuous stream of inspected items. This would occur, for example, with 100% inspection, which is becoming more common with the use of automatic inspection systems. A continuous stream of inspected items could also occur with less than 100% inspection. For example, if the process produces an item every 6 seconds, but the inspection process can only inspect one item per minute, then the inspection process running continuously will inspect 10% of the process output.

In other applications, samples of items may be inspected during certain inspection periods. For example, samples of n = 100 items might be taken for inspection twice per shift, at the start and halfway through the shift. The samples of 100 items might be actual samples of the production during the time periods that inspection is being done, or they might consist of 100 consecutive items selected at the start and at the middle of the work shift.

The objective of this paper is to review some control charts that can be used to monitor p in the case of a continuous stream of inspected items and in the case of samples of inspected items taken from the process.

2 The Shewhart *p*-Chart

The traditional control chart used to monitor p is the Shewhart p-chart (see Woodall (1997) for a general review of work on the problem of monitoring p). When samples of n items are taken from the process, the p-chart is based on plotting the proportion of defective items observed in these samples. In the case of a continuous stream of inspected items, the samples would correspond to segments of n consecutive items. If T is used to represent the total number of defectives in a sample of n, then the control statistic T/n is plotted on the p-chart.

There are a number of disadvantages to using the p-chart. The exact distribution of T/n is determined from the fact that T has a binomial distribution (assuming that p is constant and items are independent). The binomial distribution is discrete and this means that, for given values of p_0 and n, there are a limited number of possible values for the probability of a false alarm. In practice, binomial probabilities are usually not considered in determining the control limits for the p-chart. Instead, the control limits are usually set at three standard deviations from the target p_0 (i.e., at $p_0 \pm 3\sqrt{p_0(1-p_0)/n}$ (three-sigma limits)), as is usually done for a normally distributed variable. In most quality control applications, p_0

will be small, and n will not be large enough for the normal distribution to be a good approximation to the discrete binomial distribution. The result is that the false alarm probability can be much different than what would be expected for a normal distribution. In addition, the lower control limit may be negative, so there is no effective lower control limit. Ryan and Schwertman (1997) provide a detailed discussion of the problems associated with determining the control limits for the Shewhart p-chart.

In addition to problems caused by the discreteness of the binomial distribution, the p-chart has the disadvantage of being ineffective for detecting small changes in p. Faster detection of small changes in p can be achieved by using other types of charts instead of a p-chart. Several other control charts for monitoring p will be discussed in the remaining part of this paper.

3 The Bernoulli CUSUM Chart for a Continuous Stream of Inspected Items

Consider now the situation in which there is a continuous stream of inspected items. When inspected items are classified as defective or nondefective, the results for the ith item inspected can be represented as a Bernoulli observation X_i, which is 1 if the ith item is defective and 0 otherwise. The parameter p then corresponds to $P(X_i = 1)$. If a sample consists of a segment of n consecutive items, then let

$$T_k = \sum_{i=n(k-1)+1}^{nk} X_i \tag{1}$$

be the total number of defectives in the kth sample. The Shewhart p-chart would plot T_k/n.

The Bernoulli CUSUM chart is based directly on the individual observations X_1, X_2, ..., without using a summary statistic based on any grouping of the items into samples. For detecting an increase in p, the Bernoulli CUSUM control statistic is

$$B_i = \max(0, B_{i-1}) + (X_i - \gamma),\ i = 1, 2, ..., \tag{2}$$

where γ is the reference value. This CUSUM chart will signal that there has been an increase in p if $B_k \geq h$, where h is the control limit. The starting value B_0 would usually be 0 unless a headstart is used.

Reynolds and Stoumbos (1999a, b, c) investigated the properties of the Bernoulli CUSUM chart and compared it to the binomial CUSUM chart and to the p-chart. A geometric CUSUM chart based on the time between defectives was considered by Bourke (1991). Reynolds and Stoumbos (1999a) show that the geometric CUSUM chart is equivalent to the Bernoulli CUSUM chart, and also show how to apply the Bernoulli CUSUM chart when it is important to detect a decrease in p (improvement in quality). The problem of detecting a decrease in p will not be explicitly considered in the current paper.

The reference value γ for the Bernoulli CUSUM statistic can be chosen using the representation of a CUSUM chart as a sequence of tests, where each test is a *sequential probability ratio test* (SPRT) (see Reynolds and Stoumbos (1999a, b, c) and references therein). To determine the value of γ, it is necessary to specify a value $p_1 > p_0$ which represents an out-of-control value of p which should be detected quickly. For a given in-control value p_0 and out-of-control value p_1, define the constants r_1 and r_2 by

$$r_1 = -\log \frac{1-p_1}{1-p_0} \quad \text{and} \quad r_2 = \log \frac{p_1(1-p_0)}{p_0(1-p_1)}. \tag{3}$$

Then, starting from the basic definition of an SPRT, it can be shown that the appropriate choice for γ is

$$\gamma = r_1 / r_2. \tag{4}$$

For simplicity in plotting and for ease in developing statistical properties, it will usually be convenient if $\gamma = 1/m$, where m is an integer. For example, if $p_0 = 0.01$ and p_1 is chosen to be $p_1 = 0.03$, then this will give $\gamma = 0.01824 = 1/54.83$. In this case, γ can be decreased slightly down to 1/55 by a slight decrease in p_1. The value of p_1 required is approximately 0.02984. In general, if $\gamma = 1/m$, where m is an integer, then the two possible values for $X_i - \gamma$ will be $-1/m$ and $(m-1)/m$, and then the possible values of B_i will also be integer multiples of $1/m$. This will be convenient for plotting the chart. If p_0 and p_1 are small, then a slight change in p_1 will be sufficient to make $\gamma = 1/m$, where m is an integer. The precise specification of p_1 will not be critical in most applications, so this slight change in p_1 will be of no practical consequence.

4 Measures of Control Chart Performance

The performance of a control chart in detecting changes in a process is usually measured by the *average run length* (ARL), which is defined either as the expected number of samples required for a control chart to signal, or as the expected number of individual observations required to signal. In this paper it will be useful to have separate measures for the number of samples, the number of observations, and the time required by a control chart to signal.

Define the *average number of samples to signal* (ANSS) to be the expected number of samples of n observations required for the chart to signal. Define the *average number of observations to signal* (ANOS) to be the expected number of individual observations required for the chart to signal. Define the *average time to signal* (ATS) to be the expected time required for the chart to signal. In these definitions, the number of samples, the number of observations, and the time are measured relative to the start of the process when monitoring is initiated.

The ANSS, ANOS, and ATS computed for some $p \neq p_0$ measure the performance of the chart in detecting a shift to this p. The computation of these measures can be done assuming that the shift is present when the process starts. For example, suppose that the production process produces one item per minute, 100% inspection is used, and a p-chart is applied to segments of n = 100 consecutive items. If the ANSS for a shift to a particular p is 5 samples, then this corresponds to an ANOS of (5)(100) = 500 items and an ATS of approximately 500 minutes or 8.33 hours. Thus, the expected number of items produced at the shifted value of p is 500, and the expected time required to detect the shift is 8.33 hours. In evaluating a control chart for a particular application or in comparing different control charts, it would, of course, be desirable to have a small ATS at values of p which are a significant distance from p_0. Recall that in the definition of the Bernoulli CUSUM statistic, p_1 was defined as a value of p which should be detected quickly. In practice, it would usually be desirable to consider a range of values of p around p_1 and to have a chart with good performance for all of these values of p.

When $p \neq p_0$, there are several problems with using the ANSS, ANOS, and ATS when these measures are computed relative to the start of the process, as is usually done. One problem is that the shift in p may occur after the process has been running for some time, rather than at the start of the process. In this case, the control statistic of a CUSUM chart may not be at the starting value as assumed in the computation. Another problem is that a shift which occurs in the future may occur while a sample is being taken. If a shift occurs while a sample is being taken, then the distribution of the number of defectives in this sample will be the sum of two binomial random variables with different values of p. This fact is not considered in traditional evaluations of the performance of control charts.

One approach to a more realistic computation of performance measures for a CUSUM chart is based on the assumption that the control statistic has reached a steady-state or stationary distribution by the time a shift in p occurs. The ATS computed under this assumption is usually called the *steady-state average time to signal* (SSATS). The *steady-state* ANSS (SSANSS) and the *steady-state* ANOS (SSANOS) are defined similarly. The computation of the steady-state properties must account for the fact that the shift in p may occur while sampling or between samples, so two more assumptions will be used. The first is that the probability of the shift occurring in any particular interval is proportional to the length of the interval. The second is that when the shift occurs in an interval, the position of the shift within the interval is uniformly distributed over the interval. Methods for computing the steady-state properties for control charts for monitoring p are given in Reynolds and Stoumbos (1998, 1999c).

5 Evaluating the ANOS of the Bernoulli CUSUM Chart

If the Bernoulli CUSUM control statistic B_i given by (2) takes on values which are integer multiples of $1/m$, then it follows that the Bernoulli CUSUM chart can be modeled as a Markov chain. Standard methods can then be used to find the ANOS. In particular, if $\boldsymbol{Q}$ is the transition probability matrix for the Markov chain and $\boldsymbol{I}$ is the identity matrix which is of dimensions equal to $\boldsymbol{Q}$, then the ANOS is the sum of the elements of the row of $(\boldsymbol{I}-\boldsymbol{Q})^{-1}$, corresponding to the starting state. In the case of a continuous stream of inspected items, the ATS of the Bernoulli CUSUM chart is simply the product of the ANOS and the time between items.

If p_0 and p_1 are close to zero, as they would be in many applications, the dimension of $\boldsymbol{Q}$ may be quite large, and thus it may be inconvenient to work directly with $\boldsymbol{Q}$ in computing the ANOS. Reynolds and Stoumbos (1999a) obtained an expression for the ANOS that does not involve matrices. This expression was obtained by using the fact that the increment $X_i - \gamma$ added in B_i takes on only two values, and this implies that each row of $\boldsymbol{Q}$ has at most two non-zero elements. The fact that $\boldsymbol{Q}$ has a very simple structure allows for an explicit solution for the ANOS to be obtained.

Although an explicit expression for the ANOS can be obtained, as described above, this expression is relatively complicated. As an alternative to this expression for the ANOS, Reynolds and Stoumbos (1999a) derived a *corrected diffusion* (CD) theory approximation to the ANOS. This CD approximation is based on an extension of work done by Siegmund (1979, 1985) for the case of continuous distributions (see also Stoumbos and Reynolds (1996a)). The CD approximations are usually very accurate and provide a relatively simple method for designing the Bernoulli CUSUM chart.

The CD approximation to the ANOS of the Bernoulli CUSUM chart uses an adjusted value of h, which will be denoted by h^*, in a relatively simple formula. This adjusted value of h is

$$h^* = h + \varepsilon(p_0)\sqrt{p_0 q_0}, \qquad (5)$$

where $\varepsilon(p)$ can be approximated by

$$\varepsilon(p) \approx \begin{cases} 0.410 - 0.0842(\log(p)) - 0.0391(\log(p))^3 \\ -0.00376(\log(p))^4 - 0.000008(\log(p))^7, & \text{if } 0.01 \le p < 0.5 \\ \frac{1}{3}\left(\sqrt{\frac{1-p}{p}} - \sqrt{\frac{p}{1-\mathrm{p}}}\right), & \text{if } 0 < p < 0.01. \end{cases} \tag{6}$$

When $p = p_0$, the CD approximation to the ANOS is

$$\text{ANOS}(p_0) \approx \frac{e^{h^* r_2} - h^* r_2 - 1}{|r_2 p_0 - r_1|}. \tag{7}$$

If it is desirable to determine the control limit h to give a desired value for the in-control ANOS, equation (7) can be used to find the required value of h^*, and then (5) and (6) can be used to find the required value of h. Finding h^* using (7) can be accomplished by simple trial and error or using a numerical algorithm. Stoumbos and Reynolds (1999a) give a Newton-Raphson procedure that can be used to find h^*, if it is desirable to write a simple program to find h using (7).

In most applications, it will be desirable to determine how fast a shift from p_0 to p_1 will be detected. The CD approximation to the ANOS when $p = p_1$ is

$$\text{ANOS}(p_1) \approx \frac{e^{-h^* r_2} + h^* r_2 - 1}{|r_2 p_1 - r_1|}. \tag{8}$$

Note that h^* uses p_0 even though the ANOS is being approximated at p_1. Approximations to the ANOS for other values of p and a discussion of the accuracy of the CD approximation are given in Reynolds and Stoumbos (1999a).

6 The Binomial CUSUM Chart

Although the Bernoulli CUSUM chart would appear to be a natural chart to use in the case of a continuous stream of inspected items, the traditional chart used for any problem involving p is the Shewhart p-chart. The p-chart would be applied by designating segments of n consecutive items as a sample. Another chart that has sometimes been used in this situation is the CUSUM chart based on plotting a cumulative sum of T_1, T_2, ... (see, e.g., Gan (1993)). Each T_k has a binomial distribution, so this CUSUM chart will be called the binomial CUSUM chart. For detecting an increase in p, the binomial CUSUM control statistic is

$$S_k = \max(0, S_{k-1}) + (T_k - n\gamma), \quad k = 1, 2, ..., \tag{9}$$

where $n\gamma$ is the *reference value*, and the parameter γ is given by (4). This CUSUM chart will signal that there has been an increase in p if $S_k \geq h$, where h is the control limit.

7 Comparison of Control Charts for a Continuous Inspection Stream

The performance of the p-chart, the Bernoulli CUSUM chart, and the binomial CUSUM chart in detecting shifts in p will now be evaluated in terms of an example. The performance of these charts will be compared by setting up all charts so that ANOS(p_0) is approximately the same for the charts, and then comparing the SSANOS values for the charts for a range of values of p above p_0.

Consider a process in which all output is inspected automatically. Suppose that the current monitoring scheme for this process designates segments of $n = 100$ consecutive items as a sample and uses a Shewhart p-chart. The in-control value of p is $p_0 = 0.01$. If the p-chart for this problem is set up with three-sigma control limits, then the lower control limit is negative, and thus there is no effective lower control limit. The upper control limit is equivalent to giving a signal if $T_j \geq 4$. This control limit gives an in-control ANSS of only 54.42, which is much below the value of 740.8 that would be expected from three-sigma limits for a normal distribution. A more reasonable false alarm rate for the p-chart can be obtained by adjusting the upper control limit so that a signal is given if $T_j \geq 5$. This results in an in-control ANSS of 291.35. Each sample consists of 100 items, which corresponds to an in-control ANOS of 29135 items.

The method given in Reynolds and Stoumbos (1999c) was used to compute SSANOS values for the p-chart for various shifts in p. This method accounts for the fact that the shift in p could occur within a segment of n items. The resulting SSANOS values are given in column 2 of Table 1 (the ANOS is given in the row corresponding to $p = 0.01$). This p-chart is not very effective for detecting small shifts in p. For example, if p increases from 0.01 to 0.02, then the SSANOS will be 1976. If, for example, items are one minute apart, then this corresponds to 32.93 hours. The performance of the p-chart for detecting small shifts can be improved by using a larger value of n, but this would hurt performance for large shifts.

Table 1. ANOS (p = 0.01) and SSANOS (p > 0.01) values for the Shewhart p-chart, Bernoulli CUSUM chart, and binomial CUSUM chart

	Shewhart p-Chart	Bernoulli CUSUM Chart	Binomial CUSUM Chart
p	$n = 100$	$n = 1$	$n = 100$
0.01	29135	29234	28264
0.02	1977	977	999
0.03	564	331	366
0.04	284	193	234
0.05	190	136	180
0.07	125	86	131
0.10	96	55	102
0.20	72	25	74
0.50	59	9	59
1.00	55	5	54

To design a Bernoulli CUSUM chart for this problem, suppose that process engineers decide that it would be desirable to quickly detect any special cause which increases p from 0.01 to 0.03, and that the in-control ANOS should be roughly 29135 (the value corresponding to the p-chart in current use). From a previous discussion of the case of p_0 = 0.01 and p_1 = 0.03, it was shown that decreasing p_1 slightly from 0.03 to about 0.02984 would give γ= 1/55, and thus m = 55. Using trial and error to solve (7) to give ANOS(p_0) $\approx$ 29135, results in a value of h^* of 5.034. Then, using (5) and (6) to convert to h gives $\varepsilon(p) = 3.277$, $\varepsilon(p_0)\sqrt{p_0 q_0} = 0.326$, and h = 4.708. Using h = 4.708 is equivalent to using h = 259/55 = 4.709, because h can be taken to be an integer multiple of 1/55. The exact in-control ANOS using this value of h is 29249, so the design method based on the CD approximation gives a control limit with in-control ANOS very close to the desired value.

Values of the SSANOS for the Bernoulli CUSUM chart designed above are given in column 3 of Table 1. Comparing the SSANOS values of the Bernoulli CUSUM chart and the p-chart in Table 1 shows that the Bernoulli CUSUM chart will detect shifts in p much faster than the p-chart. For example, if p shifts to 0.02, then the expected number of items required to detect this shift is 977 items for the Bernoulli CUSUM chart, but 1977 items for the p-chart. If p shifts to 0.10, the p-chart requires an average of 96 items to detect this shift, while the Bernoulli CUSUM chart requires an average of 55 items.

Table 1 also gives SSANOS values for the binomial CUSUM chart for the case of $n = 100$, $p_1 \approx 0.02984$, and $\gamma = 1/55$. The value of h was adjusted to give an in-control ANOS as close as possible to the corresponding value of 29135 for the p-chart. This results in $h = 195/55$, which gives an in-control ANOS of 28264 items, slightly below the values for the p-chart and the Bernoulli CUSUM chart.

The binomial CUSUM chart in Table 1 is a little slower than the Bernoulli CUSUM chart for small shifts in p, and considerably slower for very large shifts. Grouping items into samples will always be a disadvantage in detecting very large shifts in p with the binomial CUSUM chart, because no matter how many defectives are suddenly observed, this CUSUM chart must wait until a sample is completed before a statistic can be computed and a signal given.

8 Plots of the Control Charts

To illustrate how the p-chart, binomial CUSUM chart, and Bernoulli CUSUM chart would look when plotted for a sequence of data, 500 Bernoulli observations were generated. The first 250 were generated with $p = 0.01$, and the last 250 with $p = 0.04$ to simulate a process shift half way through the sequence. The Bernoulli observations are plotted in Figure 1. Most of the observations are 0, and this shows up as a solid horizontal line at 0.

A p-chart based on segments of $n = 100$ is plotted in Figure 2. A binomial CUSUM chart also based on segments of $n = 100$ is plotted in Figure 3. A Bernoulli CUSUM chart is plotted in Figure 4. The design of these charts is the same as in Table 1.

Both the p-chart and the binomial CUSUM chart signal at the end of the 5th segment of 100 items. The Bernoulli CUSUM chart signals very early in the 5th segment, at observation 402. The p-chart and the binomial CUSUM chart are plotting summary statistics from segments of 100 items, so these charts do not give complete information about the data from the process. In contrast, the plot of the Bernoulli CUSUM chart indicates the location of each defective in the sequence of inspected items, and this extra information may be useful in identifying the location and cause of the shift in p.

9 Monitoring Based on Samples from the Process

In some applications, the inspection process may not run continuously because inspection is destructive or expensive. In this situation, samples of items would be selected and inspected at certain times, and there would be periods when no sampling or inspection is done. For example, samples of $n = 100$ items might be taken for inspection twice per shift, at the start and halfway through the shift.

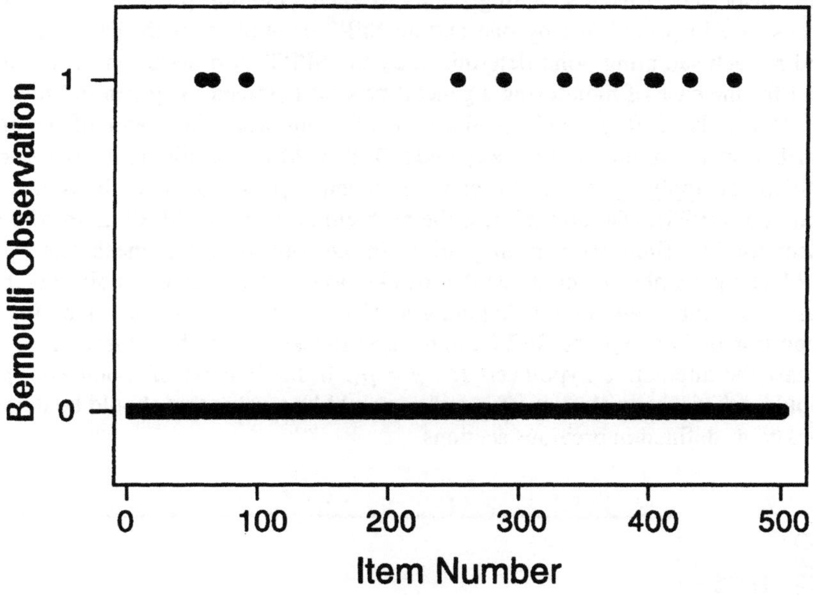

Fig. 1. The 500 Bernoulli observations.

In almost all published evaluations of the statistical properties of control charts it is assumed that the time required to take a sample is short enough that it can be neglected relative to the time between samples. This assumption will be used here. Let d be the time between samples. For the sample taken at sampling point k, let the Bernoulli random variable X_{ki} be defined by $X_{ki} = 1$ if the ith item at sampling point k is defective, and $X_{ki} = 0$ otherwise. Also let $T_{kj} = \sum_{i=1}^{j} X_{ki}$ be the total number of defectives after j items have been inspected in sample k.

The p-chart and the binomial CUSUM chart would be based on the sequence $T_{1n}, T_{2n}, \ldots$, where the sample size is always n at each sampling point. A chart, called the SPRT chart, which allows the sample size at each point to depend on the data will be considered next.

10 The SPRT Chart

The SPRT chart is based on applying a sequential test (an SPRT) to the individual items inspected at each sampling point. At each sampling point, items from the process are inspected one by one and an SPRT is applied, with the sample size used at each sampling point determined by the SPRT. A description of the SPRT chart for the case of monitoring a general process parameter is given in Stoumbos and Reynolds (1996b), and for the case of monitoring the mean of a normal distribution in Stoumbos and Reynolds (1997). More details about the current problem of applying the SPRT chart to monitor p are given in Reynolds and Stoumbos (1998), who also address the problem of using the SPRT chart to detect a decrease in p (improvement in quality). In the context of hypothesis testing, the SPRT is a general sequential test that can be used to test a simple null hypothesis against a simple alternative hypothesis. For the case of a test involving the proportion defective p, the SPRT can be used to test the null hypothesis H_0: $p = p_0$ against the alternative hypothesis H_1: $p = p_1$. In the context of monitoring p, p_0 would be the in-control value of p, and p_1 would be a value that should be detected quickly, as defined in previous sections.

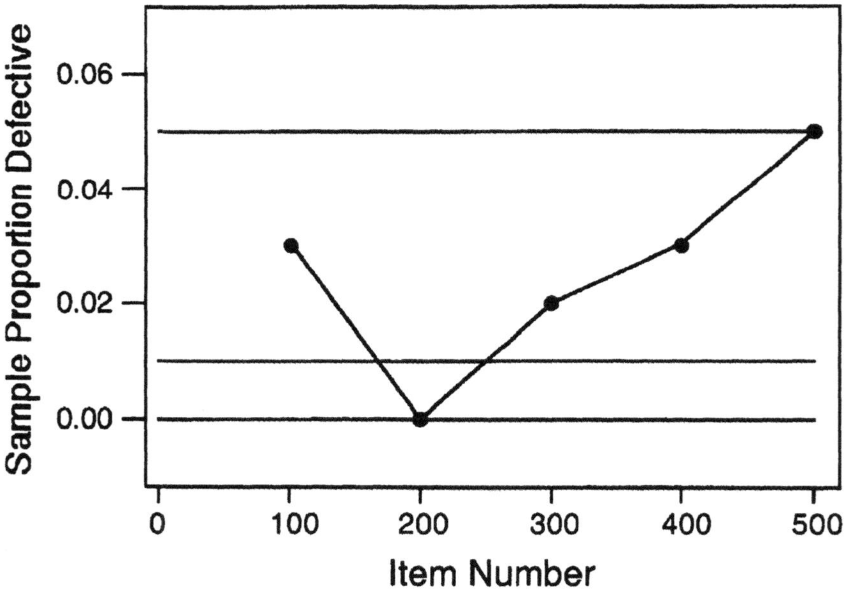

Fig. 2. The Shewhart p-chart for samples of $n = 100$.

The statistic used by the SPRT is defined in terms of the log of the likelihood ratio using the density, $f(x;p) = p^x(1-p)^{1-x}$, of X_{ki}. After the *j*th item is inspected at sampling point *k*, this statistic is

$$S_{kj} = \sum_{i=1}^{j} \log \frac{f(X_{ki};p_1)}{f(X_{ki};p_0)} = \sum_{i=1}^{j} (r_2 X_{ki} - r_1) = r_2 T_{kj} - r_1 j, \tag{10}$$

where the constants r_1 and r_2 are defined by (3). The SPRT chart requires the specification of two constants a and b, $b < a$, for the SPRT to be applied at each sampling point. The SPRT continues sampling at sampling point k, as long as $b < S_{kj} < a$, but stops sampling as soon as this inequality is violated. If $S_{kj} \le b$ before $S_{kj} \ge a$, then the SPRT applied at sampling point k accepts H_0: $p = p_0$, and the decision is that the process is in control. This means that sampling is stopped until sampling point $k+1$ is reached, and at sampling point $k+1$, another SPRT is applied. But if $S_{kj} \ge a$ before $S_{kj} \le b$, then the SPRT applied at sampling point k rejects H_0. This is taken as a signal that there has been a change in p, and action should then be taken to find and eliminate the cause of this change in p. Thus, the SPRT chart involves applying an SPRT at each sampling point and giving a signal whenever an SPRT rejects H_0.

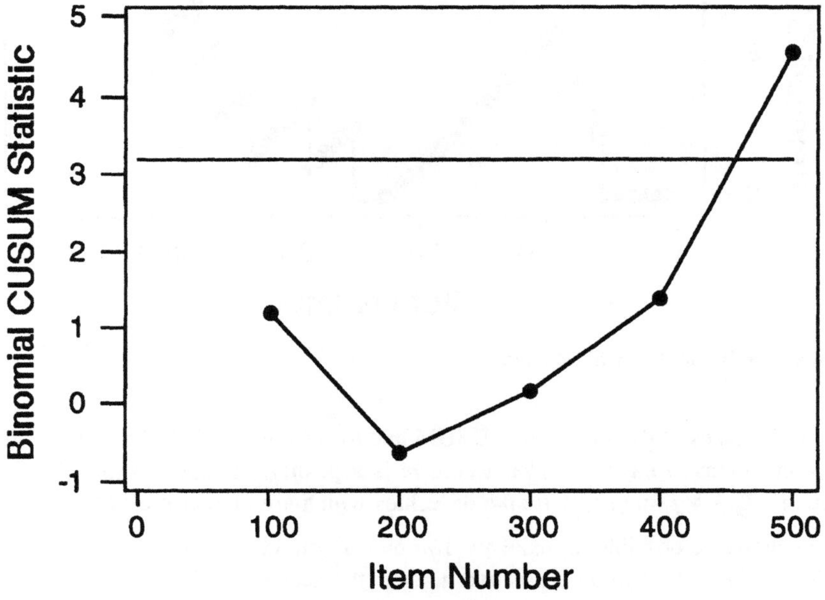

Fig. 3. The binomial CUSUM chart for samples of $n = 100$.

The inequality

$$b < S_{kj} < a \tag{11}$$

determines when the SPRT continues sampling and is usually called the *critical inequality* of the SPRT. In some applications, it may be more convenient to conduct the SPRT by dividing S_{kj} by r_2 to obtain an equivalent critical inequality. If $p_1 > p_0$, then this equivalent critical inequality is

$$g < T_{kj} - \gamma j < h, \tag{12}$$

where $g = b/r_2$, $h = a/r_2$, and γ is given by (4).

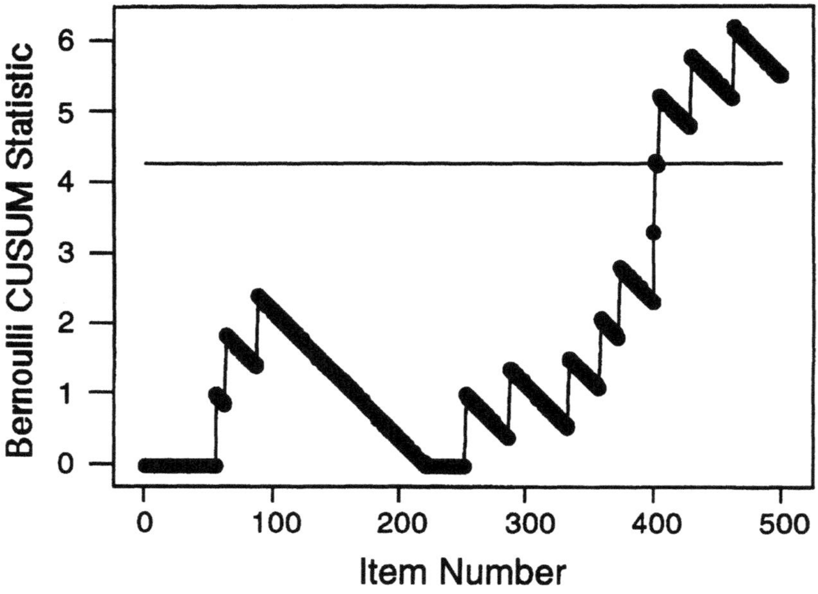

Fig. 4. The Bernoulli CUSUM chart.

As in the cases of the Bernoulli CUSUM and the binomial CUSUM, it will usually be convenient to have $\gamma = 1/m$, where m is a positive integer, so that the SPRT statistic $T_{kj} - \gamma j$ in (12) will take on values which are integer multiples of $1/m$. It will usually be possible to make $\gamma = 1/m$ by a slight adjustment of p_1, as was done before. When $\gamma = 1/m$, m a positive integer, the acceptance limit g in (12) can be chosen to be an integer multiple of $1/m$, and this will insure that the SPRT statistic $T_{kj} - \gamma j$ will exactly hit g when the test accepts H_0. In the development of the

SPRT and the SPRT chart that follows in this paper, it will be assumed that $\gamma = 1/m$ and that g is an integer multiple of γ. If $T_{kj} - \gamma j$ is an integer multiple of γ, then it follows that the rejection limit h can also be taken to be an integer multiple of γ, although $T_{kj} - \gamma j$ may still overshoot h when the test rejects H_0.

11 The Properties of the SPRT Chart

The ANSS, ANOS and ATS of the SPRT chart can be expressed in terms of the *operating characteristic* (OC) function and the *average sample number* (ASN) function or the SPRT. The OC function is the probability of accepting H_0 as a function of p, and the ASN function is the expected sample size as a function of p. We will use notation of the form OC(p) and ANSS(p) to indicate the dependence on p.

Each SPRT either accepts or rejects H_0, and thus the number of SPRTs until a signal has a geometric distribution with parameter $1 - \text{OC}(p)$. Thus, it follows that

$$\text{ANSS}(p) = \frac{1}{1 - \text{OC}(p)}. \tag{13}$$

Using Wald's identity, it follows that

$$\text{ANOS}(p) = \frac{\text{ASN}(p)}{1 - \text{OC}(p)}. \tag{14}$$

When there is a fixed time interval d between samples and the time required to take a sample is negligible, it follows that

$$\text{ATS}(p) = d \cdot \text{ANSS}(p) = \frac{d}{1 - \text{OC}(p)}. \tag{15}$$

Let $\alpha = 1 - \text{OC}(p_0)$ be the probability of a type I error, and let $\beta = \text{OC}(p_1)$ be the probability of a type II error for the test. Then, the ATS when $p = p_0$ is

$$\text{ATS}(p_0) = \frac{d}{\alpha}, \tag{16}$$

and the ATS when $p = p_1$ is

$$\text{ATS}(p_1) = \frac{d}{1 - \beta}. \tag{17}$$

The OC and ASN functions of the SPRT for p can be obtained by modeling the SPRT as a Markov chain (see Reynolds and Stoumbos (1998)). As was discussed previously in the case of the ANOS of the Bernoulli CUSUM chart, it is possible to obtain explicit expressions which do not involve matrices. These expressions, however, are relatively complicated, and thus it would be convenient to have simpler expressions that could be used in practical applications. The CD approximations to the OC and ASN functions will be discussed next, and then the application to the SPRT chart will be discussed.

When the SPRT is used for hypothesis testing, it will usually be desirable to choose the constants g and h such that the test has specified probabilities for type I and type II errors. The CD approximations to the OC and ASN functions use an adjusted value of h, which will be denoted by h^*, in a relatively simple formula. The adjusted value of h^* is

$$h^* = h + \frac{1}{3}(1 - 2p_0),\ 0 < p_0 < 0.5. \tag{18}$$

It is shown in Reynolds and Stoumbos (1998) that, from the CD approximation, the choices for g and h^* to achieve specified α and β are

$$h^* \approx \frac{1}{r_2} \log\left(\frac{1-\beta}{\alpha}\right), \tag{19}$$

and

$$g \approx \frac{1}{r_2} \log\left(\frac{\beta}{1-\alpha}\right) \tag{20}$$

When nominal values are specified for α and β, then h^* and g can be determined by using (19) and (20). Then, the value of h can be obtained from h^* using (18).

The CD approximation to the ASN at p_0 and p_1 can be expressed simply in terms of α and β (see Reynolds and Stoumbos (1998)). For $p = p_0$, this expression is

$$\text{ASN}(p_0) \approx \frac{\alpha \log\left(\frac{1-\beta}{a}\right) + (1-\alpha)\log\left(\frac{\beta}{1-\alpha}\right)}{r_2 p_0 - r_1}, \tag{21}$$

and for $p = p_1$, the expression is

$$\mathrm{ASN}(p_1) \approx \frac{(1-\beta)\log\left(\frac{1-\beta}{\alpha}\right)+\beta\log\left(\frac{\beta}{1-\alpha}\right)}{r_2 p_1 - r_1}. \tag{22}$$

Thus, for given α and β, evaluating the ASN at p_0 and p_1 is relatively easy.

12 A Method For Designing the SPRT Chart

To design the SPRT chart for practical applications, it will be necessary to determine the constants g and h used in each SPRT. In many applications, it will be desirable to specify the in-control average sampling rate and the false alarm rate and design the chart to achieve these specifications. Specifying the sampling interval d and ASN(p_0) will determine the in-control average sampling rate, and specifying ATS(p_0) will determine the false alarm rate. Once these quantities are specified, the design method proceeds as follows. The value of α is determined using (16) and the specified values of d and ATS(p_0). Then, using (21), the value of β can be determined from the specified value of ASN(p_0) and the value of α just determined. Expression (21) cannot be solved explicitly for β in terms of α and ASN(p_0), so the solution for β will have to be determined numerically. Once α and β are determined, (18), (19), and (20) can be used to determine g and h.

13 Comparison of the SPRT Chart with Other Control Charts

The performance of the SPRT chart relative to the p-chart and the binomial CUSUM chart will now be illustrated using an example.

Suppose that the in-control value of p for a process is $p_0 = 0.01$, and the current procedure for monitoring this process is to take samples of $n = 100$ twice per shift (every $d = 4$ hours) and use a p-chart. The p-chart signals if 5 or more defectives are found in a sample. Suppose that items are produced at a rapid rate and an item can be inspected in a relatively short time, so neglecting the inspection time in computations of quantities such as the ATS seems to be reasonable.

The p-chart has an in-control ANSS of 291.35, and this corresponds to an in-control ATS of (4)(291.35) = 1165.4 hours. SSATS values for this p-chart are given in column 2 of Table 2 for various shifts in p (the row corresponding to p = 0.01 has ATS values). This p-chart will not quickly detect small shifts in p. For example, if p increases by 200% from 0.01 to 0.03, the p-chart will require an average of 20.0 hours to detect this increase.

The time required to detect small shifts in p can be reduced by using a binomial CUSUM chart. A binomial CUSUM chart was designed with n = 100 and control limit adjusted to give an in-control ATS as close as possible to the value for the p-chart. As in the example in Section 7, p_1 was specified to be 0.03 and then adjusted slightly to 0.02984, so that $\gamma = 1/55$. Column 3 of Table 2 gives SSATS values for this binomial CUSUM chart. Using this binomial CUSUM chart instead of the p-chart reduces the expected time to detect a shift to 0.03 from 20.0 hours to 12.6 hours.

Table 2. ATS ($p = 0.01$) and SSATS ($p > 0.01$) values for the Shewhart p-chart, binomial CUSUM chart, and SPRT charts

	obs/hr ≈ 25				obs/hr ≈ 12.5	obs/hr ≈ 6.25
	Shewhart p-Chart	Binomial CUSUM Chart	SPRT Chart	SPRT Chart	SPRT Chart	SPRT Chart
p	$d = 4.0$ $n = 100$	$d = 4.0$ $n = 100$	$d = 4.0$ ASN = 99.93	$d = 1.0$ ASN = 23.92	$d = 1.0$ ASN = 12.85	$d = 1.0$ ASN = 6.26
0.01	1165.4	1130.6	1162.8	1194.5	1129.6	1162.6
0.02	76.7	38.1	15.4	15.9	27.1	49.3
0.03	20.0	12.6	4.5	4.4	8.1	16.1
0.04	8.8	7.3	3.0	3.1	4.8	9.6
0.05	5.1	5.1	2.5	2.0	3.5	7.0
0.07	2.8	3.1	2.2	1.3	2.4	4.7
0.10	2.1	2.2	2.0	1.0	1.6	3.2
0.20	2.0	2.0	2.0	0.6	0.9	1.5
0.50	2.0	2.0	2.0	0.5	0.5	0.6
1.00	2.0	2.0	2.0	0.5	0.5	0.5

The time required to detect shifts in p can be reduced even more if sequential inspection is used at each sampling point. The CD approximation design procedure described above was used to design an SPRT chart with $p_1 \approx 0.02984$, in-control ASN approximately 100, and in-control ATS approximately 1165.4 hours (the value for the p-chart). The resulting SPRT chart has an in-control ASN of 99.93 and an in-control ATS of 1162.8 hours. SSATS values for this SPRT chart are given in column 4 of Table 2 for various shifts. This SPRT chart has much better performance than the p-chart or the binomial CUSUM chart. For example, the expected time to detect a shift to 0.03 is only 4.5 hours.

When using a p-chart to monitor a process with p_0 close to zero, it is necessary to use a relatively large value of n to detect shifts (unless the shift is very large). If it is necessary to maintain a reasonable rate of sampling, then this means that the sampling interval d cannot be very small. However, for the SPRT chart, it is not necessary that the in-control ASN be large to have good performance. In the current example, the p-chart is taking samples of $n = 100$ every $d = 4$ hours, and this corresponds to a sampling rate of 25 observations per hour. Another sampling scheme with the same sampling rate per hour would take samples of $n = 25$ every $d = 1$ hours. Using this scheme with the p-chart would increase detection time for all but very large shifts. An SPRT chart was designed using the CD approximation method to have $d = 1.0$, in-control ASN approximately 25, and in-control ATS approximately 1165.4 hours. This will give approximately the same in-control sampling rate and false alarm rate as the other charts. SSATS values for this SPRT chart are given in column 5 of Table 2. Examining these SSATS values shows that reducing the in-control ASN and d has had little effect on the time to detect small and moderate shifts, but has improved performance for large shifts.

The results from Table 2 show that, compared to the p-chart, the SPRT chart can provide tremendous reductions in the expected time required to detect shifts in p. In some applications, it may be desirable to use the increased efficiency of the SPRT chart to reduce sampling costs. This can be accomplished by reducing the in-control ASN of the SPRT chart. For example, suppose that an SPRT chart is designed with $d = 1.0$ and in-control ASN approximately 12.5 (this is half of the sampling rate of the charts in columns 2-5 of Table 2). Using the CD approximation design procedure results in an SPRT chart with an in-control ASN of 12.85 and an in-control ATS of 1129.6 hours. SSATS values for this SPRT chart are given in column 6 of Table 2. Although the SPRT chart in column 6 does not detect shifts as fast as the SPRT chart in column 5, the SPRT chart in column 6 is uniformly faster than the p-chart and the binomial CUSUM chart.

To determine the effect of additional reductions in the average sampling rate of the SPRT chart, an SPRT chart was designed to have one fourth the sampling rate of the charts in Columns 2-5 of Table 2. SSATS values for this SPRT chart are given in column 7 of Table 2. This SPRT chart is better than the p-chart for small and large shifts, but not as good for intermediate shifts.

The conclusion that can be drawn from Table 2 is that the SPRT is much more efficient than the p-chart or the binomial CUSUM chart. This high efficiency can be used either to drastically reduce the time required to detect shifts in p, or to greatly reduce the sampling effort required to provide a given detection ability.

14 Conclusions

The Bernoulli CUSUM chart is a natural CUSUM chart to use in applications in which there is a continuous stream of inspected items. The Bernoulli CUSUM chart will detect shifts in p faster than the binomial CUSUM chart and much faster than the traditional p-chart.

When the inspection process does not run continuously, tremendous reductions in expected detection time can be achieved by using the SPRT chart instead of the p-chart or the binomial CUSUM chart. The SPRT chart allows the sample size at a sampling point to depend on the data at that point, while the p-chart and the binomial CUSUM chart always use a fixed sample size. The SPRT chart achieves the tremendous reductions in detection time by being able to speed up the sampling rate when there is an indication that the value of p may have changed. Another approach to increasing the sampling rate when there is evidence that the value of p may have changed is to use a variable sampling interval and/or variable sample size feature in a control chart. Rendtel (1990) investigated CUSUM charts with variable sampling intervals and variable sample sizes for monitoring p.

The design methods based on the highly accurate CD approximations provide a relatively simple way for practitioners to design the Bernoulli CUSUM chart and the SPRT chart for practical applications. Although the design possibilities for these charts are slightly limited by the discreteness of the distribution of the inspection data, this discreteness is much less of a problem than for the p-chart.

References

1. Bourke, P. D. (1991) "Detecting a Shift in Fraction Nonconforming Using Run-Length Control Charts with 100% Inspection". Journal of Quality Technology **23**, pp. 225-238

2. Gan, F. F. (1993) "An Optimal Design of CUSUM Control Charts for Binomial Counts". Journal of Applied Statistics **20**, pp. 445-460

3. Rendtel, U. (1990) "CUSUM Schemes with Variable Sampling Intervals and Sample Sizes". Statistical Papers **31**, 103-118

4. Reynolds, M. R., Jr. and Stoumbos, Z. G. (1998) "The SPRT Chart for Monitoring a Proportion". IIE Transactions **30**, 545-561

5. Reynolds, M. R., Jr. and Stoumbos, Z. G. (1999a) "A CUSUM Chart for Monitoring a Proportion When Inspecting Continuously". Journal of Quality Technology **31**, 87-108

6. Reynolds, M. R., Jr. and Stoumbos, Z. G. (1999b) "Some Recent Developments in Control Charts for Monitoring a Proportion" in Statistical Process Control for Quality and Productivity Improvement edited by S. H. Park and G. G. Vining. Marcel Dekker, New York, NY. To appear.

7. Reynolds, M. R., Jr. and Stoumbos, Z. G. (1999c) "A General Approach to Modeling CUSUM Charts for a Proportion". Preprint

8. Ryan, T. P. and Schwertman, N. C. (1997) "Optimal Limits for Attributes Control Charts". Journal of Quality Technology **29**, 86-98

9. Siegmund, D. (1979) "Corrected Diffusion Approximations in Certain Random Walk Problems". Advances in Applied Probability **11**, 701-719

10. Siegmund, D. (1985) Sequential Analysis. Springer-Verlag, New York, NY

11. Stoumbos, Z. G. and Reynolds, M. R., Jr. (1996a) "Corrected Diffusion Theory Approximations in Evaluating Properties of SPRT Charts for Monitoring a Process Mean". Nonlinear Analysis **30**, 3987-3996

12. Stoumbos, Z. G. and Reynolds, M. R., Jr. (1996b) "Control Charts Applying a General Sequential Test at Each Sampling Point". Sequential Analysis **15**, 159-183

13. Stoumbos, Z. G. and Reynolds, M. R., Jr. (1997) "Control Charts Applying a Sequential Test at Fixed Sampling Intervals". Journal of Quality Technology **29**, 21-40

14. Woodall, W. H. (1997) "Control Charts Based on Attribute Data: Bibliography and Review". Journal of Quality Technology **29**, 172-183

The Effect of Non-Normality on the Performance of CUSUM Procedures

Thomas P. Ryan[1] and Belinda J. Faddy[2]

[1] Case Western Reserve University, Cleveland, Ohio, USA
[2] University of Newcastle, Newcastle, NSW, Australia

1 Introduction

Cumulative sum (CUSUM) procedures have been used by many practitioners as an alternative to a Shewhart chart. The value of normality-based CUSUM procedures is well-established and CUSUM procedures have been used for decades. In recent years CUSUM procedures have also been proposed for random variables with non-normal distributions (see, e.g., Lucas [16, 17], Wang [32], Gan [8, 9], Radaelli [26], Mathers, Harris, and Lancaster [21], and Edgeman [6]).

The most frequently used CUSUM schemes are normality-based, however, and very little attention has been given to the properties of CUSUM procedures when normality is erroneously assumed. This has probably resulted from the belief that Shewhart charts are robust to non-normality. Messages to this effect can be found in Wheeler and Chambers [33], Schilling and Nelson [29], and many other articles, whereas differing views were given by Moore [23], Ryan [27, 28], Yourstone and Zimmer [35], and Amin, Reynolds, and Bakir [1].

There are some questions that must be addressed, including the following:

(1) Are normality-based CUSUM procedures robust relative to non-normality, and does the robustness (or lack of robustness) differ very much for the various types of CUSUM procedures,

(2) If there is non-robustness, is this a severe problem for the types of distributions that are apt to be encountered most frequently in practice, and

(3) If non-normality is seen to be a problem, how should a practitioner proceed?

We will concentrate primarily on the first question, but will also try to provide some answers to the other two questions.

We consider one-sided and two-sided CUSUM schemes for controlling a process mean, with a two-sided scheme of the form

$$S_{U_i} = max\,[0, (z_i - k) + S_{U_{i-1}}]$$

$$S_{L_i} = min\,[0, (z_i + k) + S_{L_{i-1}}]$$

for the upper and lower cumulative sums, respectively. (We note that some authors have defined S_{L_i} differently, with "*min*" replaced by "*max*" and $(z_i + k)$ replaced by $(-z_i - k)$. The alternative form results in S_{L_i} being non-positive, whereas the form we are using results in it being non-negative. One form is as good as the other.)

The reference value is denoted by k and z is the statistic, possibly standardized, that is being "cusumed". The value of k is generally selected to be one-half of the smallest parameter change, in standard deviation units, that is considered consequential and which the practitioner wishes to detect as quickly as possible.

Unless a fast initial response (FIR) feature is used, the two cumulative sums start from zero and are also reset at zero, such as after the detection and removal of an assignable cause. A search for the latter is initiated whenever either cumulative sum exceeds the reference value, h, with h often set at 5 when z is assumed to be $N(0,1)$. A one-sided CUSUM scheme will have only S_U or S_L.

As stated previously, there are very few results in the literature on the performance of CUSUM schemes designed under the assumption of normality when the actual distribution differs from a normal distribution by varying degrees. Kemp [14] considered the effect of non-normality for a basic CUSUM scheme, and gave expressions that could be used to obtain an upper and lower bound on the in-control ARL. The small number of numerical results cannot be easily interpreted, however, because the data were not assumed to be standardized and the selection of the (k, h) combination was arbitrary. There is also a need to investigate the effect of non-normality for other normality-based CUSUM schemes besides a basic CUSUM scheme, as such schemes are frequently used.

More recently, Hawkins and Olwell [13] discussed the effect of non-normality in their monograph, and gave a few numerical results. They also considered only a basic CUSUM scheme, however, and furthermore considered such a scheme applied only to individual observations.

As in Kemp [14] and Hawkins and Olwell [13], we consider the effect of non-normality only on the in-control average run length (ARL). While it is of interest to also consider the effect of non-normality on ARLs for mean shifts, there is a problem in doing so since a mean shift also implies a variance shift for most non-normal distributions, such as the chi-square distributions that we use in sections 2-4. From a practical standpoint, this problem also exists when normality is assumed since there must be bounds on the possible values of a measurement variable, and a shift in the mean would also certainly imply a change in the variability, even if we started with an approximately normal random variable. Such arguments could be used in contending that normality cannot exist in practice – a position that has been taken by Geary [10], Nester [24], Box and Luceño [2], and many others.

Consequently, we will consider only in-control ARLs. There are two general methods for evaluating ARLs – the integral equation approach and the Markov chain approach. The two methods were compared by Champ and Rigdon [4], who stated "... the Markov chain approach begins by approximating the problem, and then by obtaining the exact solution to the approximate problem, whereas the integral equation approach begins with the exact problem and finds an approximate solution to it". (The Markov chain approach does provide the exact solution for discrete distributions, but such distributions are not used in this paper.) Champ and Rigdon [4] contend that the integral equation approach is preferable. This recommendation stems from their result that the two methods produce the same ARL approximations when the product midpoint rule is used to evaluate the appropriate integral in the integral equation, but there are better methods than the product midpoint rule.

We use the Markov chain approach in this paper for two reasons: (1) we don't need to obtain exact ARLs to evaluate robustness, and (2) difficulties are encountered when the integral equation approach is used with CUSUM schemes other than a basic CUSUM. In particular, Lucas and Crosier [18] in discussing the approximation of FIR CUSUM ARLs noted that the integral equation approach used by previous researchers "does not allow the properties of the FIR feature to be evaluated easily".

The Markov chain approach for evaluating ARLs for a continuous distribution was first studied by Brook and Evans [3], who considered only a one-sided basic CUSUM scheme. The extension to a two-sided basic CUSUM scheme was later given by Woodall [34].

The general idea is to construct a reasonable number of intervals, the union of which will cover the possible values of the appropriate random variable. The "states" for the Markov chain are then represented by one point in each interval, and generally this point is the midpoint of the interval. Clearly the more intervals that are used for a continuous distribution, the better the approximation will be. Some researchers have used trial and error in determining the number of states, looking for the smallest number of states that leads to convergence of the ARLs. Brook and Evans [3] stated that 10 class intervals permit ARLs to be estimated within $\pm$ 1% of the true value, assuming normality. The estimation error for non-normality is apparently unknown. We have used more than 10 intervals, so our estimation error is probably small.

In Section 2 we consider one-sided CUSUM schemes and we examine the robustness of normality-based one-sided schemes when the true distribution is a chi-square or a t-distribution. In Section 3 we consider two-sided schemes and we show the need to look at the ARL properties of one-sided schemes even when a two-sided scheme is to be used. We consider the difficult task of designing robust CUSUM schemes in Section 4.

Minitab macros were written to compute the ARLs in Tables 1-3. Table 4 was constructed using software that can be downloaded from the CUSUM web site (www.stat.umn.edu) that was designed as a companion to Hawkins and Olwell [13].

2 One-Sided Schemes

We consider both one-sided schemes and two-sided schemes. It is of interest to consider the former in addition to the latter for two reasons: (1) the in-control ARLs of two-sided schemes can be somewhat misleading since the departure from the normal-theory ARLs for each of the two one-sided ARLs can be somewhat offsetting (as will be shown in section 3), and (2) there is often a need for a one-sided CUSUM scheme. For instance, Montgomery [22] gave an example in which there is a need to detect an increase in viscosity as quickly as possible, whereas a decrease in viscosity is of no concern.

In addition to a basic CUSUM scheme, we will consider a Shewhart-CUSUM scheme (Lucas, [15]) and a FIR CUSUM (Lucas and Crosier, [18]). The former is useful for detecting a large change, such as a large increase in the mean viscosity, whereas the latter is useful when there is doubt that process repair work that supposedly fixes problems can actually be relied upon. We consider a subgroup size of $n = 5$ and also consider individual observations ($n = 1$).

Since we do not assume division by the appropriate standard deviation, we will use $K = 0.5\sigma_{\overline{x}}$ and $H = 5\sigma_{\overline{x}}$ for an upper scheme and $-5\sigma_{\overline{x}}$ for a lower scheme.

In examining the robustness of each of these three types of CUSUM schemes we consider two types of distributions: chi-square distributions with varying degrees of freedom (*df*) and a *t*-distribution with varying *df*. We will let r denote the degrees of freedom. The chi-square distributions will permit insight into the robustness of CUSUM schemes for skewed distributions, whereas the *t*-distributions will show the effect of a heavy-tailed distribution. It is convenient to consider chi-square distributions because the shape depends only on r. The literature also contains discussions of chi-square distributions as being representative of actual quality control data. For example, Moore [23] points out that data received from W.A. Shewhart resulting from work at Bell Labs had skewness roughly equal to that of a chi-square distribution with $r = 8$.

The general setup when the Markov chain approach is used is well-known and in the interest of conserving space will not be given in detail here (see, e.g., Brook and Evans, [3]). Briefly, transition probabilities $p_{ij} = Pr(E_i \rightarrow E_j)$, which represent the probability of the CUSUM going from state E_i to state E_j, are computed for the various states. For $t + 1$ states, E_0 , E_1, ... E_{t-1} are the *transient states* and E_t, which represents the values of the CUSUM greater than $|H|$, is termed the absorbing state. Once the CUSUM has reached the absorbing state it cannot return to a transient state.

For the lower one-sided scheme, the $t + 1$ states representing the CUSUM values $-H$ to 0 are $E_t = (-\infty, -(t{-}0.5)w)$, ..., $E_1 = [-1.5w, -0.5w)$, $E_0 = [-0.5w, 0]$, with $w = 2H/(2t-1)$ representing the width of each class. The states for an upper one-sided scheme are similarly defined, with the minus signs that define the intervals for the lower one-sided scheme replaced by plus signs.

As stated previously, we will consider a chi-square distribution and a t-distribution. For the former, we will utilize the fact that $n\overline{X}$ is χ^2_{nr} when the individual observations are χ^2_r. This enables us to avoid the problem caused by the fact that the subgroup mean has the distribution of a constant times a chi-square random variable, whereas we need the chi-square distribution itself to partition into intervals. By combining ($k = 0.5$, $h = 5$) under the assumption of a standard normal distribution with the standard deviation of $n\overline{X}$, which is $\sqrt{2nr}$, we obtain $K = 0.5\sqrt{2nr}$ and $H = 5\sqrt{2nr}$.

Brook and Evans [3] approximated the average run length as $\mathrm{ARL} = \mathrm{A} + \mathrm{B}/\mathrm{t} + \mathrm{C}/\mathrm{t}^2$ with A representing the asymptotic (i.e., $t = \infty$) value. Three values of t are chosen, so it is a matter of solving three equations in three unknowns, or linear regression could be used to obtain the solution. (We note at this point that we sometimes obtained a better approximation to a known value (i.e, under normality) for certain CUSUM schemes using a different approximating equation, as have other researchers.) We used $t = 15$, 20, and 25 for the one-sided schemes.

We obtain the **ARL** for each value of t in the usual way as the first element of $\mu = (I - R)^{-1}\mathbf{1}$, with R representing the matrix that results from deleting the last row and last column of the transition probability matrix, and $\mathbf{1}$ represents a column of ones.

2.1 Chi-Square Distribution

Our results for a lower one-sided basic CUSUM scheme (a) and an upper one-sided basic CUSUM scheme (b) for $n = 1$ and $n = 5$ when the actual distribution is a chi-square distribution are given in Table 1.

Table 1. Approximate In-Control ARLs for a One-Sided Basic CUSUM Scheme When the Actual Distribution is Chi-Square

(a) Lower

$n = 1$

r	ARL
4	34112
6	12355
8	7899
10	5766
25	2553
50	1819
100	1465
500	1128
1000	1064
2000	* 1022
10000	970
50000	948
500000	936

$n = 5$

r	ARL
4	2964
6	2298
8	1995
10	1819
25	1391
50	1227
100	1128
500	* 1012
1000	987
10000	948
50000	938
500000	932

(b) Upper

$n = 1$

r	ARL
4	245
6	290
8	323
10	350
25	480
50	559
100	640
500	780
1000	821
2000	** 851
10000	894
50000	914
500000	925

$n = 5$

r	ARL
4	440
6	494
8	531
10	559
25	664
50	729
100	780
500	** 859
1000	879
10000	914
50000	923
500000	927

* denotes ARL that is closest to being 110% of the normal theory value of 930

** denotes ARL closest to being 90% of the normal theory value of 930

We may observe that the extreme positive skewness of a chi-square distribution with small r inflates the in-control ARLs to enormous values, especially when $n = 1$. This can be explained as follows. The smaller the value of r, the closer the mean of the chi-square distribution is to the smallest possible value of the chi-square random variable – a number epsilon greater than zero. Consequently, no value can be very many standard deviations below the mean, unlike the case with normal distributions. Specifically, for a χ_4^2 distribution, no observation can be more than 1.4 standard deviations below the mean. Thus, the CUSUM can be triggered only in a rather narrow range: 0.5–1.4 standard deviations below the mean. Although the probability associated with this interval is greater than for a normal distribution (0.37 versus 0.33) the inability of the random variable to attain values well below the mean makes it harder for the threshold value of the CUSUM to be exceeded.

Of course the ARLs begin to approach the normal theory value as $r \to \infty$, but we see that the convergence is quite slow. We notice some common values in the tables for $n = 1$ and $n = 5$ since the ARLs are determined by the product nr.

If we decide that a departure of 10% from the normal theory value is significant, then clearly r must be quite large in order for the ARL to be within 10%.

We see that the upper one-sided CUSUM is also severely affected by skewness, but not as much as the lower one-sided CUSUM. This is because there is no finite upper bound on possible values of the random variable. Therefore, the difference between these ARLs and the normal theory ARL of 930 is essentially due to the fact that the chi-square distribution has a much heavier tail than a normal distribution when r is small. Specifically, whereas $P(X > \mu + c\sigma) = 0.16$ when $X \sim \chi^2_4$ and 0.15 for a normally distributed random variable when $c = 1$, the corresponding probabilities are 0.05 and 0.02 when $c = 2$. Thus, there is a much greater probability of observing large values that will push the CUSUM statistic close to or above its threshold value for a chi-square random variable with small r than for a normal random variable.

Shewhart-CUSUM schemes have been advocated for detecting small shifts as quickly as possible and detecting large shifts almost as quickly as the corresponding Shewhart chart. Unfortunately, however, since Shewhart schemes are quite sensitive to non-normality, we would expect a Shewhart-CUSUM scheme to be much more adversely affected by non-normality than a basic CUSUM scheme. We can see this by comparing Table 2 with Table 1. (We use 3.5 Shewhart limits, as recommended by Lucas [15].)

Table 2. Approximate In-Control ARLs for a One-Sided Shewhart-CUSUM Scheme When the Actual Distribution is Chi-Square

(a) Lower				(b) Upper			
$n = 1$		$n = 5$		$n = 1$		$n = 5$	
r	ARL	r	ARL	r	ARL	r	ARL
4	34112	4	2964	4	110	4	240
6	12355	6	2298	6	135	6	286
8	7899	8	1995	8	155	8	320
10	5766	10	1818	10	173	10	348
25	2553	25	1367	25	265	25	460
50	1818	50	1176	50	348	50	537
100	1450	100	1052	100	433	100	601
500	1052	500	903	500	601	500	701
1000	970	1000	* 870	1000	652	1000	** 727
4000	* 879	10000	818	4000	** 720	10000	773
10000	847	50000	806	10000	747	50000	785
50000	818	500000	798	50000	773	500000	792

* denotes ARL that is closest to being 110% of the normal theory value of 795.5

** denotes ARL that is closest to being 90% of the normal theory value of 795.5

In comparing Table 2 to Table 1 we see that r for the Shewhart-CUSUM scheme must be twice as large as r for the basic CUSUM scheme in order to be within 10% of the normal theory value. Thus, as expected, the non-normality causes an even greater problem for a Shewhart-CUSUM scheme than for a basic CUSUM scheme.

Notice that this is not true for a lower one-sided scheme with $r = 4$, 6, 8, 10, and 25, however. This is because the Shewhart limit would be set to 0 when $r < 18$ since the computed limit would be negative, and we would thus expect the Shewhart limit to have very little if any effect, in terms of integer-rounded ARLs, when r is slightly greater than 18.

Of course this does not happen with the upper one-sided values since there is no finite upper bound on values of a chi-square random variable. We note that in Table 2(b) that the ARLs are quite small when r is small. Certainly this should be of concern.

A FIR CUSUM scheme is useful when there is doubt that an assignable cause has actually been removed. The CUSUMs are reset at a "headstart value", usually $h/2$, rather than being reset to zero when an assignable cause has been detected and apparently removed. One of the touted features of FIR CUSUM schemes is that the FIR feature does not appreciably reduce the in-control ARL. This is known in the case of normality, but has apparently not been investigated for non-normal distributions. We found this to also be true for a chi-square distribution with the values of r given in Table 1 and Table 2, as the values for r for which the ARLs are within 10% of the normal theory values are the same as for the basic CUSUM scheme. Therefore, we do not show the results for the FIR CUSUM.

2.2 t-distribution

Since a t-distribution is symmetric, the ARLs for an upper one-sided scheme will be the same as the ARLs for a lower one-sided scheme, and the ARLs will be double the ARLs for a two-sided scheme. We consider only $n = 1$. As discussed by Hawkins and Olwell [13], CUSUM schemes for individual observations are the norm and schemes for rational subgroups are the exception, being just the opposite of the case with Shewhart charts. But as discussed by Hawkins and Olwell [13], economies of scale can make it more practical to sample several items at one time than to sample one at a time, so CUSUMs for rational subgroup data are not impractical. Emphasizing $n = 1$ does seem desirable, however.

The ARLs for the basic CUSUM scheme and Shewhart-CUSUM scheme used in the preceding sections are given in Table 3.

Table 3. Approximate In-Control ARLs for a One-Sided Basic CUSUM Scheme and a Shewhart-CUSUM scheme for $n = 1$ when the Actual Distribution is Student's-t

(a) basic CUSUM

r	ARL
3	474
4	503
5	562
6	617
7	657
8	700
9	729
10	752
15	818
20	* 850
25	867
50	900
75	912
100	917
125	919
150	921
175	923
200	924
250	925
300	926
400	927
600	928
800	929
1000	929
2000	930

(b) Shewhart-CUSUM

r	ARL
3	211
4	229
5	263
6	299
7	328
8	362
9	393
10	420
15	519
20	580
25	619
50	705
75	** 736
100	751
125	760
150	766
175	770
200	773
250	778
300	781
400	784
600	788
800	790
1000	791
2000	793

* denotes the ARL that is closest to being 90% of the normal theory value of 930

** denotes the ARL that is closest to being 90% of the normal theory value of 795.5

We see there is not a major problem for the basic CUSUM scheme unless r is small. The case of the Shewhart-CUSUM ARLs is obviously different, however, as the numbers suggest that a heavy-tailed distribution can create serious problems, and even a symmetric distribution with tails somewhat heavier than the normal distribution can cause problems, as we see from Table 3.

As was true for the chi-square distribution, the FIR feature does not create a major problem when it is added to a basic CUSUM scheme when the actual distribution is Student's-t, so the ARLs will not be given here. (See Faddy [7] for tables that are not given here, and for additional discussion.)

3 Two-Sided Schemes

As mentioned previously, Woodall [34] extended the one-sided CUSUM Markov chain approach of Brook and Evans [3] to two-sided CUSUM schemes. In doing so, Woodall pointed out that the number of states required for a two-sided scheme is not the product of the number of states used in each one-sided scheme. This is apparent if we consider that if the upper one-sided CUSUM statistic increases (decreases), the lower one-sided CUSUM statistic must decrease (increase), and similarly for a change in the lower one-sided CUSUM. (Here we assume that $k_1 = k_2 = k$ for the two schemes.) Therefore, if for example, $t_1 = t_2 = t$ and transient states are used in each one-sided CUSUM, the number of states needed for the two-sided scheme will be less than t^2. Specifically, with $k = 0.5$, the number of required states is (from Woodall, [34])

$$2k + (t - 2k)\frac{(t - 2k - 1)}{2} + t = t^2 - \frac{t}{2} + 2$$

which will obviously be less than t^2 when $t > 4$, as in our work. Thus, some saving in computing could be realized using Woodall's result on the number of required states, if the lower and upper one-sided ARLs were not available.

Given the in-control ARL for the lower and upper one-sided schemes, however, we may obtain the in-control ARLs for the two-sided scheme using the relationship

$$\frac{1}{ARL_{two-sided}} = \frac{1}{ARL_{lower}} + \frac{1}{ARL_{upper}} \tag{1}$$

This result, which was given by Van Dobben de Bruyn [31], can also be used for parameter-change ARLs. (We may note that (1) also applies to Shewhart charts. The latter generally have both an upper and a lower control limit, however.)

We can see from (1) that the ARL for a two-sided scheme will be half the ARL for a one-sided scheme whenever $\text{ARL}_{lower} = \text{ARL}_{upper}$, as will be the case for symmetric distributions. Furthermore, if the ARL for each one-sided scheme differs from the normal theory ARL by k%, then the ARL for the two-sided scheme will also be off by k%. So for heavy-tailed or light-tailed symmetric distributions, either one-sided or two-sided ARLs may be used to indicate the degree of robustness of a CUSUM scheme.

This is not the case for asymmetric distributions, however. For such distributions the departure of the one-sided ARLs from the one-sided normal theory ARL can be somewhat offsetting and give a false impression that asymmetry is not a serious problem. For example, assume that the in-control ARLs for the lower and upper one-sided schemes are each off by 10%. These departures must be in opposite directions since a heavy tail on one end of a distribution must cause the tail on the other end to shrink. The heavy tail will increase the ARL, while the lighter tail on the other end will decrease it. For a 10% departure in opposite directions we can easily show that the two-sided ARL will be off by only 1%. And the difference between the two percentages increases as we let the percent by which the one-sided schemes are off increase. For example, if each one-sided scheme is off by 40%, the two-sided scheme will be off by only 16%. And so on.

The chi-square distribution approaches normality (i.e., symmetry) as $r \rightarrow \infty$, so if the actual distribution can be represented by a chi-square distribution with, say, $r = 2000$, a two-sided basic CUSUM scheme will appear to be more robust than it really is.

What happens when the percentages for the upper and lower one-sided schemes differ greatly, as will be the case when r is small? Consider, for example, $r = 8$ and $n = 5$ in Table 2. The ARL for the lower scheme is 115% greater than the normal theory ARL, whereas the ARL for the upper scheme is 42% less than the normal theory ARL. But even though the lower ARL is 1995 and the upper ARL is 531, rather than each being 930, the two-sided ARL is 419, compared to the normal theory value of 465. Thus, the two-sided ARL is off by less than 10%, despite the large percentage differences for the one-sided ARLs.

Why should this be a point of concern when a two-sided scheme is used, as is generally the case? We are also interested in parameter-change ARLs, and if a shift occurs in the direction where the percentage difference is greatest, a CUSUM scheme can perform much differently from its assumed properties under normality.

Thus, it is inadvisable to look at two-sided ARLs in examining the robustness of normality-based CUSUM schemes in the presence of skewed distributions.

4 Designing Robust CUSUM Schemes

Because of the need to look at the two one-sided schemes separately, even when a two-sided scheme is to be used, it is best to use different values of both k and h for the upper and lower schemes. One possibility would be to choose (k_1, h_1) and (k_2, h_2) in such a way as to make the in-control ARLs for the two one-sided schemes each equal to the normal theory in-control ARL for a one-sided scheme (930), while simultaneously having a small mean-shift ARL for the smallest shift that is deemed to be of importance.

The ARL values given in Tables 1-3 suggest the need for designing CUSUM schemes that are robust to non-normality. Hawkins and Olwell [13] imply that one way to combat the nonrobustness problem is to use a larger

h and a smaller k. Obviously a larger value of h will make it harder for extreme values to trigger the CUSUM. Their recommendation results from looking at ARLs for two-sided schemes used in the presence of two symmetric distributions and two asymmetric distributions.

Alternatively, one might proceed by considering the empirical distribution of a large amount of historical data and select a CUSUM scheme for data from certain non-normal distributions, such as the schemes given by Hawkins and Olwell [13]. It seems unlikely, however, that most practitioners would do this, so it seems preferable to design robust CUSUM schemes. (We note that the concept of robust CUSUM schemes is not new, having been proposed by Lucas and Crosier [19], but their robust scheme was designed to be robust to occasional outliers. Similarly, Hawkins [11] used winsorization to robustify CUSUM schemes.)

A price will be paid for using a small value of k if the actual distribution is close to normal, however, as the parameter-change ARLs for shifts that are considered consequential will not be optimal. That is, if we use $k < .50$ we are tuning a CUSUM for detecting less than a one-sigma shift, so the ARL for detecting a one-sigma shift will be larger, for the same in-control ARL, than if the CUSUM had been tuned to detect such a shift.

A potentially more serious problem with this strategy is that unless a very large amount of historical data is available for estimating the parameters, a small k will exacerbate the sampling variability in the CUSUM ARLs that is attributable to parameter estimation. That is, a small k is for detecting a small mean change, but for the purpose of computing CUSUM ARLs small-to-moderate parameter estimation errors essentially become equivalent to a small mean change. Therefore, making k smaller will increase the sensitivity of a CUSUM scheme to estimation errors. Stated differently, estimation errors can become disguised as parameter changes when in-control ARLs are computed, and can also have the effect of adding to or subtracting from an actual parameter change when parameter-change ARLs are computed and there has been a parameter change. Therefore, if a small k is to be used in an attempt to make a CUSUM scheme robust to non-normality, it is imperative that a very large amount of historical data be used for parameter estimation. How large? We will see later than even a calibration sample of approximately 2,000 observations can cause problems. (We should note that this problem is not unique to CUSUM procedures, as Shewhart charts are also affected because of the extreme percentiles that are targeted with 3-sigma limits.)

At the other extreme, what happens when a small amount of historical data is used? Hawkins and Olwell [13] considered a one-sided CUSUM for detecting an increase in a normal mean designed to have an in-control ARL of 1,000. Let m denote the size of the calibration sample (i.e., amount of historical data). A calibration sample size of $m = 50$ was assumed, and the authors showed that the in-control ARL can range from 156 to 11,439 when misestimation of μ by $\overline{X}$ and σ by s is at most $\pm$ one standard error for each estimator. (We should note that this type of analysis differs from the results of published studies of the effects of parameter estimation on Shewhart charts, such as Quesenberry [25], as in those studies there is averaging over the estimation errors, rather than looking at ARLs for specific amounts of misestimation.)

Of course parameter misestimation by more than one standard error will occur quite often, so in-control ARLs when m is not large could be almost literally any number, even if we could specify the distribution properly.

Unquestionably, incorrect ARLs resulting from distribution misspecification can be dwarfed by misrepresentative ARLs resulting from small-to-moderate values of m. This is not to imply that distribution misspecification is not a serious problem, as is shown in Tables 1-3 (which correspond to $m = \infty$).

We will now examine what can happen when m is very large. As an illustration of their software, Hawkins and Olwell [13] use an example of modeling task completion times of workers at a General Motors plant that was given by Desmond and Chapman [5]. Here m=1,995. Hawkins and Olwell [13] state "the inverse Gaussian distribution is highly flexible for modeling positive skew processes". Other general reasons for considering the inverse Gaussian (IG) distribution as a modeling distribution are given by Hawkins and Olwell [12, 13].

The shape of the distribution depends on the mean and scale parameters, however, so distribution fitting from the family of inverse Gaussian distributions would be required. It seems unlikely that practitioners will engage in distribution fitting from this family unless appropriate software is available. Furthermore, distribution fitting is problematic unless there is a very large amount of historical data.

Nevertheless, Desmond and Chapman [5] showed that their data are well-fit by an inverse Gaussian distribution with $\hat{\mu} = 42.6257$ and $\hat{\lambda} = 66.282$. The distribution is given by

$$f(x) = \left(\frac{\lambda}{2\pi x^3}\right) exp\left(\frac{-\lambda(x-\mu)^2}{2\mu^2 t}\right) \qquad (x > 0; \mu > 0; \lambda > 0)$$

The maximum likelihood estimators are (Tweedie [30])

$$\hat{\mu} = \overline{x}$$

and

$$\hat{\lambda} = \frac{n}{\sum_{i=1}^{n} \left(\frac{1}{X_i} - \frac{1}{\overline{X}}\right)}$$

The distribution of $\overline{X}$ is known to be $IG(\mu, n\lambda)$ and $(\hat{\lambda})^{-1} \sim (\lambda n)^{-1} \chi^2_{n-1}$. Since $Var(X) = \mu^3/\lambda$, $Var(\overline{X}) = \mu^3/n\lambda$, so the standard error of $\overline{X}$ is $\sqrt{\mu^3/n\lambda}$ and the standard error of $(\hat{\lambda})^{-1}$ is $\sqrt{2(n-1)/\lambda^2 n^2}$.

These results will allow us to later perform the same type of sensitivity analysis as was performed by Hawkins and Olwell [13] for a normal distribution. That is, we will consider estimation errors of at most ± one standard error for each parameter, and see how these estimation errors affect the in-control ARL.

Using the notation of Hawkins and Olwell, the reference values for an upward (+) and a downward (−) mean shift for the inverse Gaussian distribution are given, respectively, as

$$k^{+} = \frac{2\mu\mu^{+}}{\mu + \mu^{+}}$$

and

$$k^{-} = \frac{2\mu\mu^{-}}{\mu + \mu^{-}}$$

(The inverse Gaussian distribution is a member of the exponential family, and these expressions are obtained using the general forms for members of the exponential family originally given by Van Dobben De Bruyn [31].)

Increases and decreases in both parameters are of interest in this application, but we will consider only an increase in the mean. Hawkins and Olwell [13] use 75 as the out-of-control mean that the CUSUM (of individual observations) is to be tuned to detect. Since 70 is slightly less than one estimated standard deviation (using $\hat{\lambda}$ in place of λ) above $\overline{x}$; we will instead use 75 so as to be closer to one standard deviatio above $\overline{x}$. This forces k^{+} to be 54.358 and if we specify an in-control ARL of 930, we have $h = 325.1459$.

The estimated standard errors for $\hat{\mu}$ and $(\hat{\lambda})^{-1}$ are 0.5857 and 0.000478. Therefore, we will use 42.6257 ± 0.5857 and $(66.282)^{-1}$ ± 0.000478. The results are given in Table 4.

Table 4. Approximate In-Control ARLs for a basic CUSUM Designed to Detect a One-Sigma Increase in the Mean of an Inverse Gaussian Distribution, with Nine Combinations of Parameter Values

λ/μ	42.04	42.6257	43.2114
64.2483	1034	837	684
66.282	1159	**930**	755
68.4487	1308	1041	839

The numbers in Table 4 are discouraging. Although there is not the enormous variation in the in-control ARLs that Hawkins and Olwell [13] found using $m = 50$ for a normality-based scheme, the range of ARLs is about half an order of magnitude, despite the fact that almost 2,000 observations were used to estimate the parameters! Thus, even if we knew that the true distribution were inverse Gaussian, we still have a major problem.

5 Conclusions

We have seen that a skewed distribution and a heavy-tailed distribution can have a serious effect on in-control ARLs, as CUSUM procedures are clearly *not* robust to non-normality. The ARLs given in Tables 1-3 are based on the assumption that the parameters are known without error. The nonrobustness problem shown by these tables could be either exacerbated or lessened by parameter estimation errors, depending on the direction and magnitude of the errors of estimation. Users of CUSUM procedures based on normality should be aware of these problems. CUSUM procedures are inarguably superior to Shewhart charts, however, and this same combination of problems also exists for Shewhart charts.

CUSUM procedures are also less sensitive to non-normality than Shewhart charts. Hawkins and Olwell [13] conjecture that CUSUM charts are less affected by non-normality than Shewhart charts. We could show this for the distributions that we are considering in this paper. We would logically expect a Shewhart-CUSUM procedure to fall between a Shewhart chart and a CUSUM procedure in terms of robustness, and we could also show this to be true. For example, assuming a χ^2_8 distribution with $n = 5$, Ryan [27] showed that the upper tail area for an $\overline{X}$-chart constructed under the assumption of normality is .0049. Using $1/.0049 = 204$ as the in-control ARL for an "upper one-sided $\overline{X}$-chart", we see that this is smaller than the corresponding ARL (320) from Table 2, which in turn is less than 531 from Table 1. So the ordering of ARLs is as we would expect.

In a given application with unknown parameter values and an unknown distribution, however, the ARL properties of any of the CUSUM procedures may be much different from the ARL properties that might be anticipated.

Research is needed to determine desirable combinations of (k, h) values in the presence of non-normality. Ideally, such combinations should produce relatively stable ARL values for the calibration sample sizes that are apt to be used, and to also produce relatively stable ARL values over a range of possible non-normal distributions for a given application. This is clearly a daunting task, however, as we have seen that decreasing k can cause problems. So it appears as though we are faced with choosing between the consequences of distribution non-robustness that occurs if h is not increased and the estimation non-robustness that occurs when k is increased. Of course we could achieve robustness to both problems by making both k and h large, but then the parameter-change ARLs would almost certainly be unacceptably large.

References

1. Amin, R.W., M. R. Reynolds, Jr., and S. Bakir (1995) Nonparametric control charts based upon the sign statistic. Communications in Statistics - Theory and Methods **24**, 1597-1623

2. Box, G. E. P. and A. Luceño (1997) Statistical Control by Monitoring and Feedback Adjustment. New York: Wiley
3. Brook, D. and D. A. Evans (1972) An approach to the probability distribution of CUSUM run length. Biometrika **59**(3), 539-549
4. Champ, C. W. and S. E. Rigdon (1991) A comparison of the Markov chain and the integral equation approaches for evaluating the run length distribution of quality control charts. Communications in Statistics – Simulation and Computation **20**(1), 191-204
5. Desmond, A. F. and G. R. Chapman (1993) Modeling task completion data with inverse Gaussian mixtures.Applied Statistics **42**(4), 603-613
6. Edgeman, R. L. (1996) SPRT and CUSUM results for inverse Gaussian processes. Communications in Statistics – Theory and Methods **25**(11), 2797-2806
7. Faddy, B. J. (1996) The Effect of Non-Normality on the Performance of Cumulative Sum Quality Control Schemes. Honours Thesis, Department of Statistics, University of Newcastle, Australia
8. Gan, F. F. (1993) An optimal design of CUSUM control charts for binomial counts. Journal of Applied Statistics **20**, 445-460
9. Gan, F. F. (1994) Design of optimal exponential CUSUM control charts. Journal of Quality Technology **26**, 109-124
10. Geary, R.C. (1942) Testing for normality. Biometrika **34**, 209-242
11. Hawkins, D. M. (1993) Robustification of cumulative sum charts by Winsorization. Journal of Quality Technology **25**, 248-261
12. Hawkins, D. M. and D. H. Olwell (1997) Inverse Gaussian cumulative sum control charts for location and shape. The Statistician **46**(2), 323-335
13. Hawkins, D. M. and D. H. Olwell (1998) Cumulative Sum Charts and Charting for Quality Improvement. New York: Springer-Verlag
14. Kemp, K. W. (1967) An example of errors incurred by erroneously assuming normality for Cusum schemes. Technometrics **9**, 457-464
15. Lucas, J. M. (1982) Combined Shewhart-CUSUM quality control schemes. Journal of Quality Technology **8**(1), 1-12
16. Lucas, J. M. (1985) Counted data CUSUMs. Technometrics **27**, 129-144
17. Lucas, J. M. (1989) Control schemes for low count levels. Journal of Quality Technology **21**, 199-201
18. Lucas, J. M. and R. B. Crosier (1982a) Fast Initial Response for CUSUM quality control schemes: Give your CUSUM a head start. Technometrics **24**(3), 199-205
19. Lucas, J. M. and R. B. Crosier (1982b) Robust CUSUM. Communications in Statistics – Theory and Methods **11**, 2669-2687
20. Lucas, J. M. and M. S. Saccucci (1990) Exponentially weighted moving average control schemes: properties and enhancements (with discussion). Technometrics **32**, 1-29
21. Mathers, C. D., R. S. Harris, and P. A. L. Lancaster (1994) A cusum scheme based on the exponential distribution for surveillance of rare congenital malformations. Australian Journal of Statistics **36**, 21-30
22. Montgomery, D. C. (1996) Introduction to Statistical Quality Control, 3rd edition. New York: Wiley
23. Moore, P. G. (1957) Normality in quality control charts. Applied Statistics **6**(3), 171-179
24. Nester, M. (1996) An applied statistician's creed. Applied Statistics **45**(4), 401-410

25. Quesenberry, C. P. (1993) The effect of sample size on estimated limits for $\overline{X}$ and X control charts. Journal of Quality Technology **25**(4), 237-247
26. Radaelli, G. (1994) Poisson and negative binomial dynamics for counted data under CUSUM-type charts. Journal of Applied Statistics **21**, 347-356
27. Ryan, T. P. (1989)Statistical Methods for Quality Improvement. New York: Wiley
28. Ryan, T. P. (1997) "Individual contribution" to: A discussion of statistically-based process monitoring and control. Journal of Quality Technology **29**, 150-151
29. Schilling, E. G. and P. R. Nelson (1976) The effect of non-normality on the control limits of $\overline{X}$-charts. Journal of Quality Technology **8**(4), 183-188
30. Tweedie, M. C. K. (1957) Statistical properties of inverse Gaussian distributions, I. Annals of Mathematical Statistics [**28**, 362-377
31. Van Dobben de Bruyn, C.S. (1968) Cumulative Sum Tests: Theory and Practice. Griffin's Statistical Monographs and Courses No. 24. New York: Hafner
32. Wang, A.-L. (1990) Exponential cusum scheme for detecting a decrease in mean. Sankhya, Series B **52**, 105-114
33. Wheeler, D. J. and D. S. Chambers (1986)Understanding Statistical Process Control. Knoxville, TN: SPC Press
34. Woodall, W. H. (1984) On the Markov chain approach to the two-sided CUSUM procedure. Technometrics **26**(1), 41-46
35. Yourstone, S. A. and W. J. Zimmer (1992) Non-normality and the design of control charts for averages. Decision Sciences **23**, 1099-1113

Process Control for Non-Normal Populations Based on an Inverse Normalizing Transformation

Haim Shore

Department of Industrial Engineering, Faculty of Engineering Science, Ben-Gurion University of the Negev, P.O.B. 653, Beer-Sheva 84105, Israel

Abstract. When the process distribution is non-normal, traditional Shewhart charts may not be applicable. A common practice in such cases is to normalize the process data by applying the Box-Cox power transformation. The general effectiveness of this transformation implies that an inverse normalizing transformation (INT) may also be effective, namely: a power transformation of the standard normal quantile may deliver good representation for the original process data without the need to transform them. Using the Box-Cox transformation as a departure point, three INTs have recently been developed and their effectiveness demonstrated. In this paper we adapt one of these transformations to develop new schemes for process control when the underlying process distribution is non-normal. The adopted transformation has three parameters, and these are identified either by matching of the median, the mean and the standard deviation (Procedure I), or by matching of the median, the mean and the mean of the log (of the original monitoring statistic; Procedure II). Given the low mean-squared-errors (MSEs) associated with sample estimates of these parameters, the new transformation may be used to derive control limits for non-normal populations without resorting to estimates of third and fourth moments, known for their notoriously high MSEs. Implementation of the new approach to monitor processes with non-normal distributions is demonstrated.

1 INTRODUCTION

Statistical quality control methods often make two commonly accepted assumptions: That the quality data on hand contain statistically independent (non-autocorrelated) observations, and that available data have been derived from normal populations. While these assumptions are often valid (for instance, when data are collected from periodic samples taken far apart in time), they become questionable in certain situations. Typical examples for the latter are real-time process control, which often deals with autocorrelated observations, or the use of control charts for individual measurements when the monitored process has a skewed distribution. If the independence and the normality assumptions are invalidated, alternative methodologies need to be pursued that take account of the possible existence of autocorrelations in the data and of possible departure from normality in the underlying process distribution.

The first problem, that of autocorrelated processes, has been addressed extensively in the literature both with regard to statistical process control (SPC) and with regard to process capability analysis (PCA). Examples for the former are Hunter (1995) and Runger and Willemain (1995). Examples for the latter are Kotz and Johnson (1993), Pignatiello and Ramberg (1993) and more recently Shore (1997). The latter reference, in particular, provides some remedies for the existence of autocorrelations when capability indices need to be calculated.

The second problem, that of the analysis of non-normal quality data, in general, and statistical process control for non-normal populations, in particular, have also attached intensive research effort. Examples are Ferrel (1958), Burr (1967), Schilling and Nelson (1976), Nelson (1979), Balakrishnan and Kocherlakota (1986), Choobineh and Ballard (1987), Chan, Hapuarachchi and Macpherson (1988), Chan, Cheng and Spiring (1988), Clements (1989), Shore (1991), Wheeler (1991), Kocherlakota et al (1992), Kotz and Johnson (1993, and references therein), Pearn and Kotz (1994), Bai and Choi (1995), Grimshaw and Alt (1997) and Shore (1998b). The latter reference surveys the various solutions suggested in these publications, and offers a general classification scheme for these efforts.

Perhaps the most common approach to deal with non-normal data is the use of the Box-Cox transformation. This transformation is widely applied to achieve a normalizing effect for non-normal data, and it may be found as standard procedure in most statistical packages available on the market. The basis for this transformation, as articulated by Box and Cox (1964) and also from a personal message (Box, 1993), is the empirical observation that a power transformation is equivalent to finding the "right" scale for given data. Furthermore, expressing data by their "right" scale can simultaneously achieve three objectives: Simplicity of structure of descriptive models, homogeneity of the error-variance and normalization (Box and Cox, 1964, pp. 211, 213). In the years that have passed since the introduction of the Box-Cox transformation it has proved to be effective in achieving normality for a variety of published cases involving widely differing source distributions. Yet, the obvious implication of this general effectiveness, namely, the potential of the Box-Cox transformation to serve as a starting point for developing general representations for statistical distributions, has never been seriously observed.

Recently (Shore, 1998c), we have used the Box-Cox transformation as a departure point to develop inverse normalizing transformations. These transformations express quantiles of a non-normal r.v. as power transformations of corresponding standard normal quantiles. Thus, they may provide standard form for approximate general representations of non-normal distributions. Although such an endeavor has been attempted in the past (for example, Cornish-Fisher, 1937), the new transformations have several unique properties:

A. They are general in the sense that the new transformations are not restricted by distributional assumptions (like the requirement of the Cornish-Fisher expansions for asymptotic normality).

B. The new transformations use power transformations of the standard normal quantile (with non-integer exponent) to express in a simple form the corresponding quantile of a non-normal (presumably unknown) distribution. Thus, the transformations are associated with a standard (distribution-free) procedure that is simple to apply.

C. The scale parameter in all the new transformations is equal to the median. Given the low sampling uncertainty associated with estimators of the median, this is of paramount importance for SPC implementation.

D. When the new transformations are fitted to given sets of quantile values (using non-linear least-squares), extensive numerical comparisons have shown (Shore, 1998c) that the basic distributional properties of the underlying distribution (namely, its first four moments) are preserved. This is valid for distributions with skewness that ranges from zero (the normal) to over eleven (Weibull). Thus, the new transformations are flexible enough to represent distributions with widely different shapes.

E. When fitted to known statistical distributions by moment matching, the new transformations require specification of moments of second-degree at most. Thus, the large MSEs associated with sample estimates of higher moments (third- and fourth-degree moments) are avoided. In fact, one of the fitting procedures adopted here requires sample estimates of first-degree moments only. This implies that sample estimates needed in constructing the control scheme will display low sampling uncertainty.

The rest of the paper is structured as follows: In the second section we introduce the three new transformations developed recently (Shore, 1998c). For the three-parameter transformation, adapted here for the purpose of process control, we detail three fitting procedures, and numerically demonstrate the effectiveness of the transformation in depicting variously-shaped distributions. The third section presents process control schemes for non-normal distributions, based on the adopted transformation, and numerically demonstrates properties of one scheme. The last section 4 discusses the results.

2 PRESENTING THE TRANSFORMATIONS

The Box-Cox single-parameter normalizing transformation of a non-negative r.v., X, is defined by (Box and Cox, 1964):

$$X^{(\lambda)} = \begin{cases} (X^{\lambda} - 1) / \lambda, \ \lambda \neq 0 \\ \mathrm{Log}(X), \ \lambda = 0. \end{cases} \qquad (1)$$

Using this transformation as a departure point, and denoting the standard normal variate by Z, we may express the quantile relationship between Z and X by the following inverse normalizing transformation (Shore, 1998c):

$$x = H(z;\alpha) = (M)(1+az)^{b}\mathrm{Exp}[cz] =$$

$$= (M)\mathrm{Exp}[b\mathrm{Log}(1+az)+cz], \ z > -1/a \qquad (2)$$

where $H(z;\alpha)$ denotes the quantile relationship between Z and X, $\alpha = \{a, b, c\}$ is a three-parameter vector, to be determined, M is the median (z=0), and {x,z} are corresponding quantiles of X and Z, namely: $F(x) = \Phi(z) = P$, where F(.) and Φ(.) are the distribution functions of X and Z, respectively. Note, that for: b=1; c=0, X is normally distributed, while for b=0, X follows the log-normal distribution.

When transformation (2) is fitted to differently-shaped distributions, we find that the parameters "a" and "c" are very close in their absolute values, however they have opposite signs. This gives rise to the second inverse normalizing transformation (Shore, 1998c):

$$x = H(z;\alpha) = (M)(1+az)^{b}\ \mathrm{Exp}[-az] =$$

$$= (M)\mathrm{Exp}[b\mathrm{Log}(1+az) - az], \ z > -1/a \qquad (3)$$

A third inverse normalizing transformation is (Shore, 1998c):

$$x = H(z;\alpha) = (M)\mathrm{Exp}[bz^{(\lambda)}] \qquad (4)$$

where $z^{(\lambda)}$ is a Box-Cox transformation of (1+az), namely:

$$z^{(\lambda)} = [(1+az)^{\lambda} - 1] / \lambda, \ z > -1/a \qquad (5)$$

The last two expressions (eqs. 4 and 5) may be used to derive a normalizing transformation that turns out to be appreciably more effective than (1).

Extensive numerical comparisons of the above transformations are given in Shore (1998c). As earlier noted, all of these transformations deliver good fit to distributions with widely different shapes, and all first four moments of the source distributions are well preserved by the fitted transformation. We will demonstrate only the accuracy of the second transformation (eq. 3), which will subsequently be used to construct process control schemes for non-normal populations. Two fitting procedures will be demonstrated: Fitting via a set of quantile values, using a non-linear least-squares procedure, and fitting via moment-matching.

To fit (3) via quantile matching, we have selected a representative set of commonly encountered distributions (gamma, Weibull, Rayleigh and Extreme-value), and for each distribution in the testing set seventy-one quantile values were identified, corresponding to z: -4.0(0.1)3.0. These were then incorporated in a non-linear least-squares procedure to determine the transformation's parameter values. The procedure used is incorporated in the package "Statistics'NonlinearFit'", a standard package distributed with Mathematica®. Once the parameters are determined, we have examined the suitability of the fitted transformation by calculating numerically its mean, variance and the skewness (Sk) and kurtosis (Ku) measures (for the normal, Sk=Ku=0). Numerical integration, needed to calculate the moments, extends from $P=2.866(10)^{(-7)}$ (z= -5) to $P=1-2.866(10)^{(-7)}$ (z= 5). To enable calculation of these moments, irrespective of the value of the parameter "a", the following transformation was used:

$$x= H(z;\alpha)= (M)\mathrm{Exp}\{(1/2)b\mathrm{Log}[(1+az)^2] - az\} \quad (3a)$$

Table I provides the identified parameters' values and the values of the numerically computed moments. Also given are the 0.135 and 99.865 percentile values. Note, that the constraint imposed on transformations (3) and (5), namely: z>-1/a, does not constitute a real problem in application since for distributions that are not extremely skewed the parameter "a" is very small, while for extremely skewed distributions quantile values that correspond to z<-1/a are extremely small. For example, for gamma with parameters: α= 1/4, β=2, we have obtained a= 0.4861, so that we need: z>-1/a= -2.057. The corresponding limiting quantile value of gamma is $2(10)^{(-7)}$. Similarly, for Weibull with parameters: α=0.4, β=5, we have obtained a= 0.3496, so that we need: z>-1/a= -2.86. The corresponding limiting quantile value of Weibull is $1(10)^{(-6)}$. Thus, values of z that violate the lower limit posed by the constraint would rarely be needed in practice.

To fit (3) via moment matching, and also prepare the ground for the control scheme developed in the next section, note first that apart from the median eq. 3 contains two parameters: "a" and "b". These parameters may be determined by matching of the mean and the standard deviation, and indeed good accuracy is achieved when this fitting procedure is pursued (skewness and kurtosis are well preserved). However, since the MSEs associated with sample estimates of the variance are higher than those associated with sample estimates of the mean, a fitting procedure based on matching of only first-degree moments is suggested. Define: LT= Log[X] and let $\mu_i(LT)$ denote the ith non-central moment of LT (moment about zero). Then from (3) we obtain for the first moment of LT:

$$\mu_1[LT]= \mathrm{Log}[M] + b \int [\mathrm{Log}(1+az)]\phi(z)dz \quad (6)$$

where $\phi(z)$ is the standard normal density function, integration extends from

Max[-5, -1/a] to 5, and we have set: $E(Z)=0$. Also for the mean of X we have:

$$\mu= M \int \mathrm{Exp}[b\mathrm{Log}(1+az) - az]\, \phi(z)dz \quad (7)$$

with the same integration limits. From (6) and (7), the parameters "a" and "b" may easily be identified. If only sample data are available, introducing sample estimates for M, $\mu_1[LT]$ and μ will provide the required values.

Table I displays the parameters' values obtained from this fitting procedure, together with the numerically calculated moments (with integration limits as defined above) of the resulting transformation. Also given are the 0.135 and 99.865 percentile values. We realize that except for distributions with extreme skewness (gamma with skewness of 4 and Weibull with skewness of 11.35) for all other distributions all first four moments are well preserved, notwithstanding that only first-degree moments have determined the parameters' values. This fact is very desirable since it implies that the control charts, based on transformation (3) and the allied first-degree moment-matching procedure, will be less susceptible to sampling inaccuracies relative to current approaches (we will return to this point in the discussion of the concluding section).

3 CONTROL SCHEMES FOR NON-NORMAL POPULATIONS

The inverse normalizing transformation (3) with its allied fitting procedures imply that in order to detect changes in a non-normal process distribution three process-parameters need to be monitored:

Table 1. Parameters' values for eq. 3 and the resulting moments (Var., Sk and Ku are the variance, the skewness and the kurtosis measures). For the parameters ("a" and "b"), upper entries refer to quantile matching (using NL-LS), and lower entries refer to matching of the means and of the means of the log-transformed variable (median assumed known). Upper entries for moments and quantiles are the exact values.

Dis.	a	b	M (ex.)	Mean	Var.	Sk	Ku	P= .00135	P= .99865
Gamma			13.34	14.00	28.00	.7559	.8571	3.206	35.25
α= 7,	.0996	4.859		14.00	27.93	.7578	.8609	3.201	35.26
β=2	.1005	4.833		--	28.00	.7556	.8532	3.182	35.27
Gamma			5.348	6.000	12.00	1.155	2.000	.423	21.74
α= 3,	.1541	4.909		5.999	11.96	1.157	2.005	.404	21.75
β=2	.1548	4.898		--	12.00	1.159	2.019	.400	21.77
Gamma			.08735	.5000	1.000	4.000	24.00	0	8.231
α= 1/4,	.4861	6.742		.4379	.8971	5.154	42.87	.0019	8.738
β=2	.4873	6.808		--	1.009	5.340	47.14	0	9.339
Weibull			3.610	4.790	18.19	1.676	4.040	.0141	26.79
α=1.12	.2733	4.714		4.784	18.16	1.675	3.982	.0025	26.75
5, β=5	.2658	4.792		--	18.20	1.707	4.213	.0038	27.01
Weibull			3.466	5.000	25.00	2.000	6.000	.0067	33.04
α=1,	.2823	5.054		4.992	24.96	1.994	5.877	.0006	33.00
β=5	.2642	5.258		--	25.01	2.047	6.333	.0019	33.79
Weibull			3.162	5.665	50.99	2.815	12.74	.0013	52.97
α=0.8,	.2992	5.799		5.655	50.89	2.794	12.29	0	52.92
β=5	.2795	6.035		--	51.04	2.915	13.82	.0001	53.94
Weibull			2.000	16.62	2723	11.35	287.6	0	561.9
α=0.4,	.3496	9.322		16.62	2678	10.43	204.3	0	561.1
β=5	.3018	10.17		--	2768	12.70	347.5	0	570.8

Table I (Cont'd)

Dis.	a	b	M (ex.)	Mean	Var.	Sk	Ku	P= .00135	P= .99865
Rayleigh			2.355	2.507	1.717	.6311	.2451	.1039	7.271
σ=2	.2323	3.446		2.511	1.703	.6405	.2649	.0773	7.255
	.2389	3.398		--	1.717	.6223	.2127	.0662	7.217
Extreme Value			3.366	3.577	1.645	1.139	2.400	1.112	9.607
α=3,	.0149	24.94		3.576	1.701	1.080	2.086	1.127	9.570
β=1	-.346	-.003		--	1.647	1.133	2.361	1.189	9.616

- The median, the mean and the standard deviation (correspondingly with the first fitting procedure, henceforth denoted procedure I)
- The median, the mean and the mean of the log of the monitoring statistic (correspondingly with the second fitting procedure, henceforth denoted procedure II)

Accordingly, the following control schemes are suggested:

Control scheme A (fitting procedure I):

In the framework of traditional Shewhart control charts, construct and maintain control charts for the median, the average and the standard deviation (or range). The robustness to deviation from normality of these control charts has been extensively documented in the literature (refer to the aforecited references). For extremely skewed process distributions, one may use the X and R charts for skewed populations, suggested by Bai and Choi (1995). Guidelines for the median control chart may be found in textbooks for statistical quality control (for example, Montgomery, 1996).

Control scheme B (fitting procedure I):

Based on sample estimates of the median, the mean and the standard deviation, fit transformation (3) (identify the parameters), and calculate the end-points of the process distribution by introducing into the fitted transformation: $z = \pm 3$. These end-points will comprise the control limits for a control chart for individual observations. An outlying observation will indicate an out-of-control situation.

Control scheme C (fitting procedure II):

In the framework of traditional Shewhart control charts, construct and maintain control charts for the median, for the average of *X* (the process variable), for the average of *LT* (the log of the process variable), and for the standard deviation (or range). Again, for extremely skewed process distributions, the X and R charts for skewed populations, suggested by Bai and Choi (1995), may be employed.

Control scheme D (fitting procedure II):

Based on sample estimates of the median, the mean of *X* and the mean of *LT*, fit transformation (3) (identify the parameters), and calculate the end-points of the process distribution by introducing into the fitted transformation: $z= \pm 3$. These end-points will serve as control limits for a control chart for individual observations. An outlying observation will indicate an out-of-control state.

We will now demonstrate the effectiveness of the latter control scheme (D). It is assumed that the underlying process distributions are those given in Table I, however we limit ourselves only to the gamma and the Weibull distributions for which changes in the mean can be easily effected by respective changes in one of the distribution's parameters. Thus, for the gamma distribution to change the mean from μ to $(1+\delta)\mu$, the parameter α is changed to $(1+\delta)\alpha$. Similarly, for the Weibull distribution the parameter β is changed to $(1+\delta)\beta$.

To examine the effectiveness of control scheme D, ARLs associated with deviations in the mean are calculated for the selected distributions. The ARLs are calculated by re-fitting eq.3 (using the changed process-parameter), and thence identifying values of z corresponding to current (pre-change) control limits. From the z values the respective ARLs may be computed. Table II gives ARLs associated with –50%, -20%, -10%, 0%, 10%, 20% and 50% deviations. Only for the first two gamma distributions were the lower control limits large enough to allow reliable calculation of the ARLs associated with downward shifts in the mean. When the LCL is near zero there is no meaning to calculating ARLs realting to the LCL. Therefore, for the rest of the distributions no ARLs are given for negative δ.

Examining Table II we realize that the suggested scheme D is effective in detecting large shifts, however it is relatively insensitive to small to moderate shifts. Such a phenomenon is expected with respect to control charts for individual observations. Also note the accuracy of the ARLs calculated from the re-fitted transformation ("approximate values"), relative to the exact values obtained by using exact expressions of the respective distribution ("exact values"). The proximity of the approximate and exact values further corroborates the assertion, made earlier, that the new approach may be applied to non-normal process distributions with widely differing shapes.

Table 2. Exact (upper entries) and approximate (lower entries) control limits, and the ARLs associated with a deviation of δ in the value of the mean. Upper entries are the exact ARLs associated with the approximate control limits, and lower entries are the approximate ARLs (derived from the fitted transformation). Under δL=0 are ARLs relating to the LCL, and under δU=0 are ARLs relating to the UCL

						ARL				
Dis.	**LCL**	**UCL**	**δ= -50%**	**δ= -20%**	**δ= -10%**	**δL=0%**	**δU=0%**	**δ=10%**	**δ=20%**	**δ=50%**
Gamma α= 7, β=2	3.206 3.182	35.25 35.27	7.557 7.587	96.43 95.31	263.9 257.6	772.8 741.6	745.9 741.4	364.8 363.0	190.5 189.8	37.93 37.83
Gamma α= 3, β=2	.4230 .4012	21.74 21.77	16.67 16.96	162.1 151.7	369.1 331.6	863.3 740.7	750.3 740.7	477.4 472.6	313.3 310.8	103.8 103.4
Weibull α=1.125, β=5	.0141 .0038	26.79 27.01	n.a.	n.a.	n.a.	n.a.	788.4 708.1	417.8 382.9	237.8 222.3	70.6 68.6
Weibull α=1, β=5	.0067 .0019	33.04 33.79	n.a.	n.a.	n.a.	n.a	802.7 692.7	465.9 412.9	279.2 253.6	90.53 86.32
Weibull α=0.8, β=5	.0013 1E-4	52.97 53.94	n.a.	n.a.	n.a.	n.a	815.9 664.2	556.4 465.9	363.7 313.9	138.7 127.3
Weibull α=0.4, β=5	0 0	561.9 570.8	n.a.	n.a.	n.a.	n.a	774.9 619.4	733.1 589.4	584.9 481.5	339.5 294.6

4. CONCLUDING REMARKS

Current tendency to practice process control in real time has exacerbated the problem of monitoring processes with non-normal distributions. This has underscored the need to develop universally applicable approaches that circumvent the need to identify exactly the underlying process distribution, and at the same time avoid the large MSEs associated with currently available methods of distribution fitting. Thus, two desirable and seemingly conflicting requirements are requested: that only low degree moments participate in the fitting procedure, and that the basic properties of the underlying distribution (as exhibited by the first four moments) will be preserved in the fitted expression.

The new approach seems to fulfill both requirements. As we have repeatedly shown (Shore, 1995a, 1995b, 1996, 1998a, 1998b), fitting with low degree sample moments may provide better representation of the underlying distribution relative to three- or four-moment fitting procedures. In this paper we have pursued this assertion, and adopted, for the purpose of process control for non-normal populations, new inverse normalizing transformations that require low degree moments (first or second degree, at most), and yet preserve well all first four moments of the underlying (presumably unknown) process distribution. As a result, far-tail quantiles, needed to determine control limits, may be approximated by the fitted transformation with error-deviations that are generally acceptable. This property of the new transformations has also been demonstrated by an approximate calculation of the ARLs, associated with downward- and upward-shifts of the mean.

The new transformations with their allied fitting procedures may prove useful also in other applied areas of quality and reliability, for example: In calculating process capability indices for non-normal populations (refer also to an alternative approach introduced in Shore, 1998b).

Acknowledgment

Mathematica is a registered trademark of Wolfram Research.

REFERENCES

1. Bai, D. S., and Choi, I. S. (1995) "X and R control charts for skewed populations". Journal of Quality Technology **27(2)**, 121-131
2. Balakrishnan, N., and Kocherlakota, S. (1986) "Effects of non-normality on X charts: Single assignable cause model". Sankya. **B 48**, 439-444
3. Box, G. E. P., and Cox, D. R. (1964) "An analysis of transformations". Journal of the Royal Statistical Society B **26**, 211-243
4. Burr, I. W. (1967) "The effect of non-normality on constants for X and R charts". Industrial Quality Control. **34**, 563-569

5. Chan, L. K., Cheng, S. W., and Spiring, F. A. (1988) "The robustness of process capability index C_p to departure from normality". In Statistical Theory and Data Analysis, II (K. Matusita, ed.), North-Holland, Amsterdam, 223-229

6. Chan, L. K., Hapuarachchi, K. P., and Macpherson, B. D. (1988) "Robustness of X and R charts". IEEE Transactions on Reliability. **37**, 117-123

7. Choobineh, F., and Ballard, J. L. (1987) "Control limits of QC charts for skewed distributions using weighted variance". IEEE Transactions on Reliability. **36**, 473-477

8. Clements, J. A. (1989) "Process capability calculations for non-normal distributions". Quality Progress. **22 (2)**, 49-55

9. Cornish, E. A., and Fisher, R. A. (1937:. "Moments and cumulants in the specification of distributions". Review of the International Statistics Institute. **5**, 307

10. Ferrell, E. B. (1958) "Control charts for lognormal universe". Industrial Quality Control **15**, 4-6

11. Grimshaw, S. D., and Alt, F. B. (1997) "Control charts for quantile function values". Journal of Quality Technology **29 (1)**, 1-7

12. Hunter, S. (1995) "Just what does an EWMA do?". ASQC Statistics Division Newsletter, **16 (1)**, 4-12

13. Kocherlakota, S., Kocherlakota, K., and Kirmani, S. N. U. A. (1992) "Process capability indices under non-normality". International Journal of Mathematical Statistics **1**

14. Kotz, S. and Johnson, N. L. (1993) Process Capability Indices. Book. Chapman and Hall, London

15. Montgomery, D. C. (1996) Introduction to statistical quality control. 3rd Ed., John Wiley & Sons

16. Nelson, R. P. (1979) "Control charts for Weibull processes with standards given". IEEE Transactions on Reliability **28**, 283-287

17. Pearn, W. L., and Kotz, S. (1994) "Application of Clements' method for calculating second- and third-generation process capability indices for non-normal Pearsonian populations". Quality Engineering **7 (1)**, 139-145

18. Pignatiello, J. J., and Ramberg, J. S. (1993) "Process capability indices: Just say "no". Transactions of ASQC 47th Annual Quality Congress, 92-104

19. Runger, G. C., and Willemain, T. R. (1995) "Model-based and model-free control of autocorrelated processes". Journal of Quality Technology **27**, 283-292

20. Schilling, E. G., and Nelson, P. R. (1976) "The effect of non-normality on the control limits of X charts". Journal of Quality Technology **8**, 183-188

21. Shore, H. (1991): "New control limits that preserve the skewness of the monitoring statistic". In The ASQC 45th Annual Quality Congress Transactions, 77-83

22. Shore, H. (1995a): "Fitting a distribution by the first two sample moments (partial and complete)". Computational Statistics and Data Analysis **19**, 563-577

23. Shore, H. (1995b) "Some applications of a new two-moment distributional fitting procedure". In the proceedings of The 13th International Conference on Production Research (August, 1995, Jerusalem). 704-706

24. Shore, H. (1996) "A new estimate of skewness with MSE smaller than that of the sample skewness". Communications in Statistics (Simulation and Computation). **25 (2)**, 403-414

25. Shore, H. (1997) "Process capability analysis when data are autocorrelated". Quality Engineering **9 (4)**, 615-626

26. Shore, H. (1998a) "Approximating an unknown distribution when distribution information is extremely limited". Communications in Statistics (Simulation and Computation) **27 (2)**, 501-523

27. Shore, H. (1998b) "A new approach to analysing non-normal quality data with application to process capability analysis". International Journal of Production Research **36 (7)**, 1917-1933

28. Shore, H. (1998c) "Inverse normalizing transformations and an extended normalizing transformation". Submitted

29. Wheeler, D. J. (1991) "Shewhart's charts: Myths, facts and competitors". In The ASQC 45th Annual Quality Congress Transactions, 533-538

On Nonparametric Multivariate Control Charts Based on Data Depth

Zachary G. Stoumbos[1], L. Allison Jones[2], William H. Woodall[3] and Marion R. Reynolds, Jr.[4]

[1] Rutgers, The State University of New Jersey, Newark, NJ 07102-1895
[2] The University of Miami, Coral Gables, FL 33124-8237
[3] The University of Alabama, Tuscaloosa, AL 35487-0226
[4] Virginia Polytechnic Institute and State University, Blacksburg, VA 24061-0439

Abstract. Most traditional multivariate control charts for monitoring multiple process quality variables are designed and evaluated under the assumption that the process data follow a multivariate normal distribution with known covariance matrix. However, in many practical situations, this assumption of multivariate normality may not hold or may be difficult to verify. Recently, nonparametric Shewhart-type control charts that are based on the notion of simplicial data depth have been proposed for monitoring the locations and scales of multiple process variables. Although these Shewhart charts have potential for addressing the hard problem of multivariate nonparametric quality control, there exist some important unresolved issues with these charts. This paper investigates the effect of the size of the reference sample and the size of the subgroups on the statistical properties of Shewhart-type control charts that are based on simplicial data depth. The sizes of the reference sample and subgroups are the primary factors which determine whether these charts can be designed to achieve sufficiently large in-control average run lengths for practical applications.

1 Introduction

Control charts are graphical devices extensively employed to monitor industrial processes, with the objective of detecting any change in a process that might result in a deterioration in product quality. The traditional approaches to control charts have been parametric and univariate, with the most frequently investigated charts being those designed to monitor the mean of a single normally distributed process characteristic. However, with modern data acquisition technologies and on-line computers, it is now common practice to simultaneously monitor several, usually correlated, quality variables. Using separate univariate charts for each variable does not account for correlation between variables, so a more suitable approach is to use a multivariate control chart procedure.

Recently, a number of multivariate extensions to univariate Shewhart, cumulative sum (CUSUM), and exponentially weighted moving average (EWMA) charts have been investigated. The survey papers of Alt and Smith (1988), Wierda (1994), Lowry and Montgomery (1995), and Mason et al. (1997) may be consulted for good discussions on multivariate control charts and extensive lists of references. A potential drawback of most multivariate procedures developed to this date is that they are based on the assumption of multivariate normal process data. However, measurements may not follow this distribution, and in the case of small samples—quite common in statistical process control (SPC) practices—taken from a non-normal population, a reasonable approximation of the sampling distribution of a parametric control statistic may be difficult to obtain. Moreover, if the production run is short, there may be insufficient data to obtain good estimates of all parameters involved in designing a parametric multivariate control chart. Coleman (1997) stressed that he "would never believe the multivariate normal assumption for industrial data", and that even if he wanted to, he "cannot believe that there are tests for multivariate normality with sufficient power for practical sample sizes" that he "would even bother to use them". Coleman (1997) concluded that, "distribution-free multivariate SPC is what we need".

Nonparametric control charts require few restrictive assumptions on the process data and are generally insensitive to potential outliers. The literature on nonparametric control charts is much less extensive than that on parametric charts, and it has been limited almost exclusively to the univariate problem (see, e.g., Bakir and Reynolds (1979), Bhattacharya and Frierson (1981), Park and Reynolds (1987), King and Longnecker (1990), Amin and Searcy (1991), Hackl and Ledolter (1991, 1992), Amin et al. (1995), Pappanastos and Adams (1996), Willemain and Runger (1996), Janacek and Meikle (1997), and Bakir (1999)). Very little work has been conducted on developing multivariate nonparametric control charts. Hayter and Tsui (1994) proposed a multivariate nonparametric control scheme, the M procedure, for monitoring the process location-parameter vector. The M procedure, however, ignores any correlation structure among the multivariate components. Kapatou (1996) used EWMAs of the classical sign and signed-rank statistics to develop two multivariate control charts for monitoring the process location vector. While Kapatou's charts are robust compared to purely parametric counterparts, they are not nonparametric in the usual sense because they require the estimation of nuisance parameters related to the process covariance structure.

Liu (1995) used the concept of data depth (Tukey (1975) and Liu (1990)) to construct a Shewhart-type multivariate nonparametric control chart, for the simultaneous detection of changes in the process location and/or scale. This chart was designed as a simplicial-depth-based multivariate generalization of the univariate $\overline{X}$ chart. The main idea behind Liu's (1995) control chart is to "reduce each multivariate measurement to a univariate index—namely, its relative center-outward ranking induced by a data depth" measure.

Liu (1995) presented an innovative application of the idea of simplicial data depth to the long-standing problem of distribution-free multivariate SPC. This topic is of great interest to quality control researchers and practitioners alike. In fact, Cheng et al. (1997) decsribed use of Liu's (1995) chart to develop a unified threshold system of airline safety measures for the Federal Aviation Administration (FAA), and then to implement and monitor these measures for the FAA. Although Liu's (1995) chart has potential for addressing the hard problem of multivariate nonparametric SPC, there exist some important unresolved issues with this chart that must be addressed. The purpose of this paper is to investigate certain statistical properties of Liu's (1995) chart, in order to address some of the unresolved issues that are of central importance to its design and practical implementation.

The remaining part of this paper is structured as follows. Some useful geometrical concepts are defined in Section 2, and the notion of simplicial depth is discussed in Section 3. In Section 4, the Shewhart control chart based on data depth is presented. Some statistical properties of this chart are investigated in Section 5, in order to address some issues that are critical to its design and practical usage. Finally, Section 7 contains some concluding remarks. Several technical results on stochastic geometry have been deferred to the Appendix.

2 Some Geometrical Concepts

Liu (1990) proposed the nonparametric concept of simplicial data depth as a multivariate generalization of the univariate idea of ranking data according to a sign statistic. For a given point y in the p-dimensional Euclidean space $\mathbf{R}^p$, simplicial depth is a measure of how central y is within a random sample $\{Y_1,...,Y_m\} \subset \mathbf{R}^p$. In particular, for $p = 1$, the simplicial depth of y is the proportion of intervals, among all intervals formed by pairs of points from $\{Y_1,...,Y_m\} \subset \mathbf{R}$, that contain y. Analogously, for $p = 2$, the simplicial depth of y is the proportion of triangles, among all triangles formed by triplets of points from $\{Y_1,...,Y_m\} \subset \mathbf{R}^2$, that include y. In the context of SPC, simplicial depth can be used to determine whether the process data are consistent with an in-control reference sample. To precisely define simplicial depth in higher dimensions, some geometrical concepts are needed. These concepts are presented in this section, and a detailed definition of simplicial depth is provided in the following section.

A set $C \subset \mathbf{R}^p$ is said to be *convex* if the line segment joining any two points of C also belongs to C. In other words, if y_1 and $y_2 \in C$, then $ay_1 + (1-a)y_2$ must also belong to C, for each $a \in [0,1]$. A point y in a convex set $C \subset \mathbf{R}^p$ is called an *extreme point* of C if $y = ay_1 + (1-a)y_2$, with y_1, $y_2 \in C$ and $a \in (0,1)$, implies that $y = y_1 = y_2$.

The *convex hull* of a set $S \subset \mathbf{R}^p$, denoted $h(S)$, is the smallest convex set in $\mathbf{R}^p$ that contains S. The convex hull of a finite set of m points $\{y_1,..., y_m\}$ in $\mathbf{R}^p$ is a closed and bounded polyhedron called a *polytope*. The extreme points of $h(y_1,..., y_m)$ are the elements of $\{y_1,..., y_m\}$ that *minimally* define the *polytope*.

A *simplex* in p dimensions is the convex hull of a set of $(p+1)$ noncoplanar points in $\mathbf{R}^p$, i.e., points not all lying on the same hyperplane. For $p = 1$, this is a closed line segment, for $p = 2$, it is a closed triangle, and for $p = 3$, it is a closed tetrahedron. Further, a random set of $(p+1)$ points from a p-variate continuous distribution are noncoplanar with probability 1. Thus, the convex hull of a random set of $(p+1)$ such points is a simplex with probability 1. Clearly, the $(p+1)$ vertices of a p-dimensional simplex are its extreme points. The extreme points of a random sample play a critical role in investigating the properties of simplicial depth (Stoumbos and Jones (2000)). Some useful results on the extreme points of a random sample are provided in the Appendix. More information on the concepts defined in this section can be obtained in Bazaraa et al. (1993).

3 Simplicial Depth

The *simplicial depth* of a given point $y \in \mathbf{R}^p$, with respect to a continuous distribution F, is defined to be

$$D_F(y) = P_F\left(y \in s[Y_1,...,Y_{p+1}]\right),$$

where $s[Y_1,...,Y_{p+1}]$ is the p-dimensional simplex whose vertices are the random observations $Y_1,...,Y_{p+1}$ from F. $D_F(y)$ measures how central the point y is within the distribution F. When F is unknown, a reference sample $\{Y_1,...,Y_m\}$ from F can be used to compute the *sample* simplicial depth, which is defined to be

$$D_{F_m}(y) = \binom{m}{p+1}^{-1} \sum_{1 \le i_1 < ... < i_{p+1} \le m} I\left(y \in s[Y_{i_1},...,Y_{i_{p+1}}]\right), \tag{1}$$

where F_m denotes the empirical distribution function of $\{Y_1,...,Y_m\}$, $m \ge (p+1)$, and $I(\cdot)$ is the indicator function

$$I\left(y \in s[Y_{i_1},\ldots,Y_{i_{p+1}}]\right)=\begin{cases}1, & \text{if } y \in s[Y_{i_1},\ldots,Y_{i_{p+1}}]\\ 0, & \text{otherwise.}\end{cases}$$

The summation in (1) runs over all subsets of $(p+1)$ points from $\{Y_1,\ldots,Y_m\}$. For example, for $y \in \mathbf{R}^2$, $D_{F_m}(y)$ is the proportion of closed triangles, among all closed triangles formed by triplets of points from $\{Y_1,\ldots,Y_m\} \subset \mathbf{R}^2$, that include y. The value $D_{F_m}(y)$ is a multivariate measure of how central (or outward) the point y is within the data cloud of $\{Y_1,\ldots,Y_m\}$. Since the emphasis here is nonparametric, the developments presented in the remainder of this paper will stem from the sample simplicial depth $D_{F_m}(y)$ in (1). Moreover, in the balance of this paper, for brevity, the modifier "sample" will be suppressed, and $D_{F_m}(y)$ will be referred to merely as simplicial depth.

An investigation of the basic properties of simplicial depth was conducted by Liu (1990), who showed that if F is absolutely continuous, then as $m \to \infty$, $D_{F_m}(y)$ converges uniformly and strongly to $D_F(y)$. Liu (1990) also showed that $D_F(y)$ is *affine invariant*; i.e., if y is transformed to $Ay+b$, where A is a $p \times p$ non-singular matrix and $b \in \mathbf{R}^p$, then

$$P_F\left(Ay+b \in s[AY_1+b,\ldots,AY_{p+1}+b]\right)=D_F(y).$$

The affine invariance will render the simplicial-depth-based control chart, which is described in the next section, coordinate-free.

4 The Shewhart Chart Based on Simplicial Depth

Liu (1995) gave the name "*Q* chart" to the Shewhart-type control chart that she proposed. This chart should not be confused with Quesenberry's homonymous family of short-production-run control charts (see, e.g., Quesenberry (1991, 1995) and references therein). In fact, to avoid any confusion, for the remainder of this paper, the simplicial-depth-based Shewhart chart will be descriptively referred to as the *simplicial depth* (SD) chart. We now proceed to describe the SD chart.

Consider a continuous p-variate quality vector whose distribution is F when the process is in control. Assume that if an assignable cause occurs, then any resulting change in the process will be reflected by a location change and/or a scale increase and characterized as a departure from F to a continuous out-of-control distribution G. Let $\{Y_1,\ldots,Y_m\}$ denote a reference sample of m independent p-variate data points, with $m \geq (p+1)$, initially taken from F in order to design an SD chart. Suppose that after the reference sample is obtained, independent random samples $\boldsymbol{X}_1 = (\boldsymbol{X}_{11}, ..., \boldsymbol{X}_{1n})$, $\boldsymbol{X}_2 = (\boldsymbol{X}_{21}, ..., \boldsymbol{X}_{2n})$, ... of $n \geq 1$ p-variate observations each are taken from the process at specified sampling points $i = 1, 2, ...$, respectively, in order to determine if the process is in control.

Now, for a given p-variate observation $\boldsymbol{X}_{ij}$, define the statistic,

$$r_m(\boldsymbol{X}_{ij}) = \frac{1}{m+1}\left[1 + \sum_{k=1}^{m} I\left(D_{F_{m+1}}(\boldsymbol{Y}_k) \leq D_{F_{m+1}}(\boldsymbol{X}_{ij})\right)\right], \tag{2}$$

where $I(\cdot)$ is the indicator function

$$I\left(D_{F_{m+1}}(\boldsymbol{Y}_k) \leq D_{F_{m+1}}(\boldsymbol{X}_{ij})\right) = \begin{cases} 1, & \text{if } D_{F_{m+1}}(\boldsymbol{Y}_k) \leq D_{F_{m+1}}(\boldsymbol{X}_{ij}) \\ 0, & \text{otherwise,} \end{cases}$$

and where $D_{F_{m+1}}(\boldsymbol{Y}_k)$ and $D_{F_{m+1}}(\boldsymbol{X}_{ij})$ in (2) are calculated using (1) with respect to the *expanded reference sample* (ERS) $\{\boldsymbol{Y}_1, ..., \boldsymbol{Y}_m, \boldsymbol{X}_{ij}\}$. The statistic $r_m(\boldsymbol{X}_{ij})$ denotes the fraction of observations in the ERS that have simplicial depths not exceeding $D_{F_{m+1}}(\boldsymbol{X}_{ij})$.

The *control statistic* of the SD chart is a function of the simplicial depth from (1) through the statistic $r_m(\boldsymbol{X}_{ij})$ from (2), and is defined to be the average

$$\bar{r}_m(\boldsymbol{X}_i) = \frac{1}{n}\sum_{j=1}^{n} r_m(\boldsymbol{X}_{ij}), \text{ for } i = 1, 2, \ldots . \tag{3}$$

A small value of $\bar{r}_m(\boldsymbol{X}_i)$ indicates that only a small proportion of the points in $\{\boldsymbol{Y}_1, ..., \boldsymbol{Y}_m\}$ are more outlying than most observations in sample $\boldsymbol{X}_i$, thus placing these observations at the outskirts of $\{\boldsymbol{Y}_1, ..., \boldsymbol{Y}_m\}$. That is, a small value of $\bar{r}_m(\boldsymbol{X}_i)$ suggests a possible shift in location and/or an increase in scale. Specifically, at the ith sampling point, the SD chart uses sample $\boldsymbol{X}_i$ in the following decision rule:

1. if $\bar{r}_m(\boldsymbol{X}_i) \leq g$, then stop sampling and declare the process out-of-control;

2. if $g < \bar{r}_m(\boldsymbol{X}_i)$, then declare the process in-control and wait until the $(i+1)$st sampling point to observe sample $\boldsymbol{X}_{i+1}$ and compute $\bar{r}_m(\boldsymbol{X}_{i+1})$.

A graphical illustration of an SD chart is given in Figure 1.

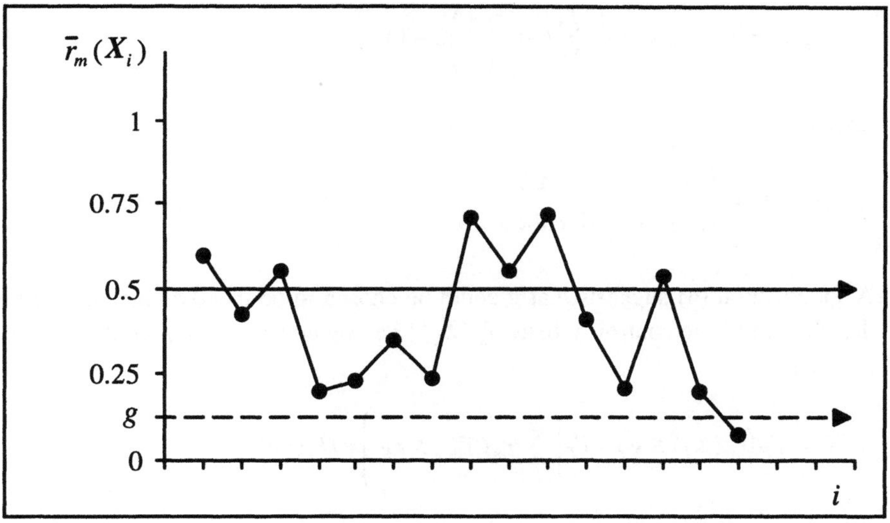

Fig. 1. A graphical illustration of an SD chart.

The control limit g is selected so that the SD chart will achieve a specified in-control *average run length* (ARL). The ARL is the expected number of samples required for the SD chart to produce a signal. The ARL computed under the out-of-control distribution G is a measure of the chart's ability to detect changes from the in-control distribution F to G. When no shift has occurred at the process onset, the ARL computed under F is a measure of *false alarm rate* (FAR). Specifically, the FAR is the long-run average number of false signals (false alarms) generated by the SD chart per sample. The FAR can be expressed as the reciprocal of the in-control ARL. For example, if the in-control ARL is 250, then this corresponds to an average of one false alarm every 250 samples, or 1/250 = 0.004 false alarms per sample. Conditioning on the reference sample $\{Y_1,...,Y_m\}$, the in-control ARL of the SD chart is the mean of a geometric distribution, which is

$$\text{ARL} = \frac{1}{\alpha}, \tag{4}$$

where α is the probability of a false alarm. That is, since the SD chart is a Shewhart control chart, FAR = α.

Liu and Singh (1993) showed that when F is absolutely continuous, as $m \to \infty$, the statistic $r_m(X_{ij}) \to U[0,1]$ in law along almost all samples $\{Y_1,...,Y_m\}$, where $U[0,1]$ denotes the uniform distribution in $[0,1]$. Furthermore, a result in Feller (1971) states that the cumulative distribution function (cdf) of the sum S_n of n independent $U[0,1]$ random variables is

$$U_n(s) = P(S_n \le s) = \frac{1}{n!}\sum_{k=0}^{n}(-1)^k \binom{n}{k}(s-k)_+^n, \tag{5}$$

where

$$(s-k)_+^n = \begin{cases} 0, & \text{if } s \le k \\ (s-k)^n, & \text{if } s > k. \end{cases}$$

Also, equation (4) suggests that g could be chosen to be the αth quantile of the distribution of the control statistic $\bar{r}_m(\boldsymbol{X}_i)$; i.e., by equation (5), g could be taken to satisfy,

$$\alpha = P_F\left(\bar{r}_m(\boldsymbol{X}_i) \le g\right) = P_F\left(\sum_{j=1}^{n} r_m(\boldsymbol{X}_{ij}) \le ng\right) \approx U_n(ng),$$

where the last relation in this expression is an approximate equality because it is based on the asymptotic distribution of the control statistic. Thus, the quantile $s_{n,\alpha}$, such that $U_n(s_{n,a}) = \alpha$, needs to be determined from (5), in order to compute $g = s_{n,\alpha}/n$. While the cdf in equation (5) is a seemingly complicated piecewise polynomial, as long as $\alpha \le 1/n!$, $s_{n,\alpha} = (n!\alpha)^{1/n}$. Hence, if $\alpha \le 1/n!$, the control limit would be taken to be

$$g = \frac{(n!\alpha)^{1/n}}{n}. \tag{6}$$

For typical SPC applications, were n is usually chosen to be 3-5 observations and the in-control ARL exceeds 5! = 120, the restriction $\alpha \le 1/n!$ imposed on (6) is irrelevant for all practical purposes. For example, if the sample size n = 3 is to be used in an SD chart, and an in-control ARL of 200 is desired for this chart, then from (4), α = 1/200 = 0.005 < 1/3!, and the control limit would be chosen to be $g = ((3!)(0.005))^{1/3}/3 = 0.1036$. However, if for a certain practical situation, $\alpha > 1/n!$, then $s_{n,\alpha}$ would have to be determined using simple trial-and-error, or numerically, using a simple application of a Newton-Raphson procedure.

A point worth mentioning is that, since the control statistic $\bar{r}_m(\boldsymbol{X}_i)$ in (3) is discrete for $m < \infty$, a particular FAR may be unattainable. While this is a common complication with control charts based on any discrete distribution, in the case of the SD chart, it is a minor problem because m will usually be large enough so that an FAR very near the desired FAR will be attainable, especially for n > 1. The determination of the size of the reference sample in the design of SD charts is the focus of the next section.

Before concluding the description of the SD chart, it should be noted that, from the $U[0,1]$ distribution, as $m \to \infty$, the asymptotic in-control median of the control statistic $\bar{r}_m(X_i)$ is 0.5. This suggests that a reference line be drawn at the value 0.5, which can help the user identify any trends or patterns that could potentially develop on an SD chart while it is being implemented (see Figure 1).

5 The Effect of the Reference Sample

The geometrical nature of simplicial depth renders the SD chart easy to visualize and interpret. The SD chart is nonparametric, and thus it can be implemented without any restrictive parametric assumptions on the process model. Moreover, the SD chart simultaneously detects changes in the process location and scale. The practical implementation, however, of the SD chart requires obtaining a reference sample whose size requirement has been an unresolved issue. Liu (1995) recommended employing a reference sample of $m = 500$ observations (in the bivariate case) or more (in higher-dimensional cases) in designing an SD chart, but these are heuristic recommendations that were neither scientifically founded nor accounted for the effect of using samples of various sizes $n \geq 1$.

In this section, the effect of the reference sample on some statistical properties of the SD chart is investigated, and useful information is provided regarding the reference-sample size that should be employed in designing an SD chart with a desirable FAR. This is the main contribution of the present paper.

The extreme points of a random sample play a central role in investigating the properties of simplicial depth. In particular, Stoumbos and Jones (2000) showed that for a reference sample of $m \geq (p+1)$ points from a p-variate continuous distribution, the minimum simplicial depth that any point can have is $(p+1)/m$ (with probability (w.p.) 1). They also showed that a point assumes this minimum depth (w.p. 1) if and only if it is an extreme point. Finally, Stoumbos and Jones (2000) established that there exist at least $(p+1)$ extreme points (w.p. 1) that share the minimum simplicial depth of $(p+1)/m$. Hence, from (2), for a reference sample of m observations $\{Y_1,...,Y_m\}$, lower bounds to the statistic $r_m(X_{ij})$ from (2) are given (w.p. 1) by

$$r_m(X_{ij}) \geq \frac{E_{i,j}^{(m+1)}}{m+1} \geq \frac{p+1}{m+1}, \text{ for } i = 1, 2, ..., \text{ and } j = 1, ..., n, \tag{7}$$

where $E_{i,j}^{(m+1)}$ denotes the number of extreme points in the ERS $\{\boldsymbol{Y}_1,...,\boldsymbol{Y}_m,\boldsymbol{X}_{ij}\}$. The $E_{i,j}^{(m+1)}$'s are identically distributed and (conditionally) independent random variables whose distribution depends on *m*, *p*, and the particular underlying process distribution. Then, from (3) and (7), lower bounds to the control statistic $\bar{r}_m(\boldsymbol{X}_i)$ are given (w.p. 1) by

$$\bar{r}_m(\boldsymbol{X}_i) \geq \frac{1}{n(m+1)}\sum_{j=1}^{n} E_{i,j}^{(m+1)} \geq \frac{p+1}{m+1}, \text{ for } i = 1, 2, \ldots. \tag{8}$$

The lower bounds in (8) provide useful information regarding the choices of reference-sample size *m* and control limit *g* in the design of SD charts. In particular, for $\alpha \leq 1/n!$, using (6) and (8), it can be shown that the absolute smallest control limit that gives a positive FAR and the absolute smallest positive FAR that can be used in the design of an SD chart are (w.p. 1)

$$g = \frac{p+1}{m+1} \quad \text{and} \quad \alpha \approx \frac{1}{n!}\left(\frac{n(p+1)}{m+1}\right)^n,$$

respectively, where the right expression is an approximate equality because it is based on the asymptotic distribution of the control statistic. For example, for p = 2, n = 3, and m = 100, the smallest control limit that could possibly generate a signal is $g = (p+1)/(m+1) = 3/101 = 0.03$, and the largest achievable in-control ARL is $1/\alpha = 1/0.00012 = 8{,}479.84$. This ARL can be extremely misleading, because the left inequality in (8) places much stricter limitations on the choices of *m* and *g*, and hence on the FAR of the SD chart. Actually, the smallest control limit that can be used for the SD chart to be able to produce a signal at the *i*th sampling point is

$$g = \frac{1}{n(m+1)}\sum_{j=1}^{n} E_{i,j}^{(m+1)} \tag{9}$$

(w.p. 1). This control limit can be much greater than $(p+1)/(m+1)$, depending on the process data. Since the control limit in (9) is a random variable, the distribution of $E_{i,j}^{(m+1)}$ must be considered in this investigation.

When the data follow a multivariate uniform or normal distribution, as $m \to \infty$, $E_{i,j}^{(m+1)} \to N\left(E(E^{(m+1)}), Var(E^{(m+1)})\right)$, where $N(\mu,\sigma^2)$ is the standard notation for the normal distribution with mean μ and variance σ^2 (Groeneboom (1988) and Hueter (1994, 1996)). This fact was empirically verified by Stoumbos and Jones (2000), who also found, empirically, the convergence to normality to hold for other distributions, such as the exponential distribution (for excellent discussions on the limiting distribution of the number of extreme points, see Aldous et al. (1991) and Hueter (1996, 1999)).

A common approach to characterizing a distribution is by its mean. Efron (1965) derived exact integral expressions for the expected number of extreme points in random samples from bivariate and trivariate normal distributions and bivariate and trivariate uniform distributions over the unit circle and sphere. These expressions can be evaluated numerically, and for the convenience of the reader, are supplied in the Appendix.

Clearly, the FAR of the SD chart is a random variable, because it depends on the data observed in the reference sample. Table 1 contains values for the expected number of extreme points, $E(E^{(m+1)})$, and the *ARL at the expected minimum positive FAR*, computed using the expression,

$$\text{ARL at the expected minimum positive FAR} \approx n!\left(\frac{m+1}{nE(E^{(m+1)})}\right)^{n}, \qquad (10)$$

(obtained for $\alpha \leq 1/n!$, using (6) and (8)), for various sample sizes and sizes of ERSs from the bivariate normal distribution. Note that the expression in (10) is an approximate equality because it is based on the asymptotic distribution of the control statistic. Tables 2-4 have been constructed in the same manner as Table 1, except that the values provided in Table 2 pertain to the trivariate normal distribution, and the values provided in Tables 3 and 4 respectively pertain to the bivariate uniform distribution over the unit circle and the trivariate uniform distribution over the unit sphere.

Throughout Tables 1-4, the results were computed using Efron's (1965) exact integral expressions, which are given in equations (A1)-(A4) from the Appendix. The accuracy of equation (10) was assessed using simulation studies for the bivariate normal distribution and for ERSs of $(m+1)=250$ observations and sample sizes of n = 1, 3, and 5 observations. In all of these simulations studies, equation (10) was found to be within 3% of the simulation results, and thus, should be regarded as highly accurate. Additional bivariate simulation studies were not conducted, because the computation of simplicial depths is extremely computer intensive and time consuming, requiring many days of simulations. The simulation studies were restricted to the bivariate case because efficient and reliable computer programs that compute simplicial depths in three (or more) dimensions have yet to be developed.

Table 1. Values for the expected number of extreme points $E(E^{(m+1)})$ and ARL at the expected minimum positive FAR for ERSs of $(m+1)$ bivariate normal observations

		ARL at Expected Minimum Positive FAR Computed Using Equation (10)				
$(m+1)$	$E(E^{(m+1)})$	$n=1$	$n=2$	$n=3$	$n=4$	$n=5$
50	7.9728	6.27	19.66	54.81	145.02	372.51
100	8.8892	11.25	63.28	316.37	1,501.49	6,918.64
150	9.3913	15.97	127.56	905.48	6,101.36	39,916.35
250	9.9937	25.02	312.89	3,478.74	36,712.82	376,174.50
500	10.7648	46.45	1,078.69	22,267.80	436,339.35	8,301,328.62
750	11.1944	67.00	2,244.36	66,829.86	1,888,927.90	51,836,581.74
1,000	11.4906	87.03	3,786.91	146,473.96	5,377,765.58	191,698,830.35
1,250	11.7158	106.69	5,691.77	269,900.64	12,148,607.72	530,915,329.09
1,500	11.8970	126.08	7,948.41	445,402.96	23,691,474.69	1,223,509,221.73
1,750	12.0483	145.25	10,548.67	680,972.02	41,727,943.66	2,482,566,313.80
2,000	12.1780	164.23	13,485.95	984,361.40	68,201,521.13	4,587,855,229.11
2,500	12.3920	201.74	20,350.20	1,824,674.90	155,298,988.02	12,832,991,630.63

Tables 1-4 demonstrate that the reference-sample size m required to design an SD chart with a reasonable in-control ARL is highly dependent on the sample size n. In particular, if individual observations are to be sampled (i.e., $n = 1$), m must be much larger than 500, which Liu (1995) recommended for the bivariate case. For example, for $n = 1$ and $m = 500$ observations from a bivariate normal distribution, the ARL at the expected minimum positive FAR is 46.53. For $n = 1$ and $m = 500$ observations from the bivariate uniform distribution on the circle, the ARL at the expected minimum positive FAR is merely 18.86. As a point of reference, typical SPC practices use in-control ARLs of at least 370.4, which is the in-control ARL of the two-sided Shewhart $\overline{X}$ chart with the "standard" three-sigma control limits. More importantly, however, Tables 1-4 reveal that the demands on the reference-sample size rapidly decrease as the sample size increases, and Liu's (1995) recommendation of $m = 500$ data points becomes reasonable for the bivariate case, when subgrouping is feasible. In fact, for $n = 3$ and $m = 500$ observations from a bivariate normal distribution, the ARL at the expected minimum positive FAR is 22,388.12, while for $n = 5$ and $m = 500$ observations from this distribution, it is 8,376,220.54. Also, for $n = 3$ and $m = 500$ data points from the bivariate uniform distribution on the circle, the ARL at the expected minimum positive FAR is 1,491.64, while for $n = 5$ and $m = 500$ data points from the same distribution, it is 91,719.20. Note that the values discussed in this paragraph, which have all been computed for $m = 500$ observations, are not included in Tables 1-4, as these tables pertain to ERSs of $(m+1)$ observations.

Table 2. Values for the expected number of extreme points $E(E^{(m+1)})$ and ARL at the expected minimum positive FAR for ERSs of $(m+1)$ trivariate normal observations

		ARL at Expected Minimum Positive FAR Computed Using Equation (10)				
$(m+1)$	$E(E^{(m+1)})$	$n=1$	$n=2$	$n=3$	$n=4$	$n=5$
100	13.2002	7.58	28.70	96.62	308.78	958.15
250	16.5355	15.12	114.29	767.98	4,898.44	30,334.65
500	19.1075	26.17	342.38	3,981.89	43,958.22	471,158.73
750	20.6266	36.36	661.05	10,682.86	163,872.02	2,440,608.53
1,000	21.7100	46.06	1,060.84	21,717.37	422,018.06	7,962,160.53
1,500	23.2439	64.53	2,082.26	59,722.09	1,625,928.00	42,977,759.06
2,000	24.3366	82.18	3,376.83	123,338.08	4,276,129.60	143,939,736.34
2,500	25.1865	99.26	4,926.23	217,322.15	9,100,389.09	369,991,783.77
3,000	25.8823	115.91	6,717.48	346,052.52	16,921,699.73	803,382,022.82
4,000	26.9826	148.24	10,988.07	723,960.30	45,276,651.02	2,749,223,191.18
5,000	27.8380	179.61	16,129.98	1,287,606.41	97,566,067.23	7,177,788,019.05
6,000	28.5381	210.25	22,101.58	2,065,225.87	183,179,992.84	15,774,829,574.98

Tables 1-4 also demonstrate that, while the reference-sample size requirements increase rapidly with the number of variables, analogous observations to the ones made in the previous paragraph for bivariate measurements can also be made about trivariate measurements. In particular, from Tables 2 and 4, it is evident that, when subgrouping is appropriate, depending on the sample size n being used, reference samples of $600 \leq m \leq 1{,}000$ trivariate data points should be sufficiently large for most practical situations, and especially those with $n \geq 3$.

While the SD chart appears to have limited potential for practical applications that require individual observations because of the large reference-sample size demands established in this paper, these demands significantly decrease when subgrouping is feasible. Nevertheless, some practitioners may also be sceptical about the SD chart's usefulness even when subgrouping is feasible, since they may regard reference samples in excess of 500 data points problematic for practical use, because typical SPC practices traditionally use reference samples of 100-200 observations (see e.g., Ryan (1989) and Montgomery (1996)). However, it is known (see Ghosh et al. (1981) and Quesenberry (1993)) that even in the simple case of the $\overline{X}$ chart with independent data and unknown variance σ^2, a surprisingly large reference-sample size of more than 500 observations is required before the FAR for the case of the estimated σ^2 is similar to the FAR for the case of known σ^2. Recently, Lu and Reynolds (1999) showed that reference samples of more than 1,000 data points are needed in designing univariate EWMA charts for serially correlated data, so that these charts will attain an in-control ARL of approximately 370.4, when several parameters must be estimated and the process distribution is normal. Clearly, the demands on the reference-sample size for a multivariate chart, such as Hotelling's (1947) T^2 chart, which can require the estimation of multiple parameters related to the location and covariance structures, will be much greater (see Rigdon (1995)). When considering these facts, the requirements on the reference-sample size of the SD chart should seem quite reasonable, especially for $n \geq 3$.

Table 3. Values for the expected number of extreme points $E(E^{(m+1)})$ and ARL at the expected minimum positive FAR for ERSs of $(m+1)$ bivariate uniform observations over the unit circle

		ARL at Expected Minimum Positive FAR Computed Using Equation (10)				
$(m+1)$	$E(E^{(m+1)})$	$n=1$	$n=2$	$n=3$	$n=4$	$n=5$
100	15.1713	6.59	21.72	63.64	176.96	477.78
250	20.9205	11.95	71.40	379.22	1,911.80	9,357.75
500	26.5407	18.84	177.45	1,485.80	11,808.67	91,120.98
750	30.4659	24.62	303.02	3,315.36	34,432.05	347,192.87
1,000	33.5842	29.78	443.30	5,866.53	73,693.53	898,780.89
2,000	42.4288	47.14	1,110.99	23,275.30	462,858.58	8,936,707.15
3,000	48.6222	61.70	1,903.46	52,197.36	1,358,686.64	34,337,309.36
4,000	53.5486	74.70	2,789.93	92,623.72	2,918,889.02	89,307,741.54
5,000	57.7068	86.64	3,753.67	144,549.51	5,283,764.59	187,519,507.68
7,500	66.0972	113.47	6,437.64	324,655.47	15,541,213.52	722,309,435.17
10,000	72.7737	137.41	9,441.06	576,585.14	33,425,087.36	1,881,298,969.56
15,000	83.3364	179.99	16,198.79	1,295,855.03	98,400,322.13	7,254,588,432.68

Finally, Tables 1-4 demonstrate that the reference-sample size m required to design an SD chart with a specified in-control ARL of $1/\alpha$ depends on the particular type of underlying process distribution. Although equations (A1)-(A4) were derived for specific normal and uniform distributions, in practical situations where the SD chart would be used, the process distribution would be unknown. Thus, while the expectations in (A1)-(A4) can be used in (10) to form the useful guideline,

$$m \approx \left(\frac{nE(E^{(m+1)})}{(n!\alpha)^{1/n}} \right) - 1, \tag{11}$$

for the choice of reference-sample size, this guideline should be taken with some caution, because it depends on the process distribution. Solving guideline (11) for

m will have to be done using simple trial-and-error, or numerically, using a secant or quasi-Newton procedure (for more information on such procedures, see, e.g., Burden and Faires (1993)). However, it is interesting to note that Groeneboom (1988) and Stoumbos and Jones (2000) provide highly accurate closed-form approximations to $E(E^{(m+1)})$ for bivariate and trivariate normal and uniform distibutions that can be used to conveniently solve equation (11) for m, simply using a pocket calculator. They also provide highly accurate closed-form approximations to $Var(E^{(m+1)})$ that can be used to develop refinements to (11), based on the asymptotic normal distribution of the number of extreme points. Such a treatment, however, is beyond the scope of the present paper.

Table 4. Values for the expected number of extreme points $E(E^{(m+1)})$ and ARL at the expected minimum positive FAR for ERSs of $(m+1)$ trivariate uniform observations over the unit sphere

		ARL at Expected Minimum Positive FAR Computed Using Equation (10)				
$(m+1)$	$E(E^{(m+1)})$	$n=1$	$n=2$	$n=3$	$n=4$	$n=5$
600	103.9364	5.77	16.66	42.75	104.11	246.18
800	120.8950	6.62	21.89	64.39	179.76	487.24
1,000	135.8371	7.36	27.10	88.66	275.36	830.31
5,000	310.8146	16.09	129.39	925.11	6,278.35	41,368.92
10,000	441.9395	22.63	256.00	2,574.54	24,576.49	227,780.73
50,000	995.3273	50.23	1,261.76	28,170.84	597,018.27	12,284,335.19
100,000	1,409.9949	70.92	2,514.98	79,274.75	2,371,926.13	68,903,864.33
200,000	1,996.4244	100.18	5,017.93	223,418.35	9,442,342.49	387,451,025.71
300,000	2,446.4080	122.63	7,518.91	409,793.03	21,200,236.95	1,064,861,248.00
400,000	2,825.7619	141.55	10,018.87	630,319.54	37,641,679.51	2,182,495,564.35
500,000	3,159.9794	158.23	12,518.19	880,328.48	58,764,398.80	3,808,552,973.09
750,000	3,871.4664	193.73	18,764.70	1,615,640.75	132,042,688.55	10,477,558,082.17

6 Concluding Remarks

The SD chart is a nonparametric multivariate Shewhart-type control chart that simultaneously detects changes in the process location and scale and is easy to visualize and interpret. Cheng et al. (1997) stated that they were in the process of using this chart to develop and implement a unified threshold system of airline safety measures for the FAA. However, the use of the SD chart in a practical situation requires obtaining a reference sample whose size requirement has formerly been an unresolved issue.

This paper has demonstrated that the reference-sample size m required to design an SD chart with a reasonable FAR is highly dependent on the sample size n. The SD chart appears to have limited potential for practical applications that require individual observations ($n = 1$) due to the large reference-sample size requirements established in this paper. However, this paper has shown that the requirements on m drastically decrease when subgrouping is feasible, and in fact, become quite reasonable for practical situations, and especially those with $n \geq 3$.

This paper has also shown that in designing an SD chart with a reasonable FAR, the choice of reference-sample size m depends on the type of the underlying process distribution. General guidelines for the choice of m were given here, and a simple useful rule for chosing m was developed. This rule is given in equation (11) and should be used with some caution, because it depends on the process distribution, which in applications where the SD chart would be used, would be unknown.

Based on some preliminary empirical results, we feel that the requirements on the reference-sample size m should decrease significantly if EWMAs of the control statistic in (3) are plotted on a control chart. We also feel that the use of these EWMAs in the SD chart should even render the chart practical for certain applications that are limited to the use of individual observations. A similar investigation of the statistical properties of the simplicial-depth-based EWMA chart requires further nontrivial development of the mathematical theory. This chart will be considered in a future paper based on work currently underway.

Before concluding, it should be noted that Ajmani et al. (1997) considered a control chart for individual observations based on simplicial depth which seems to have been defined differently from the SD chart. Ajmani et al. (1997) compared the performance of this control chart to that of the T^2 chart for individual observations under deviations from multivariate normality.

Appendix

Efron (1965) derived integral expressions for the mean number of extreme points, $E(E^{(m)})$, in the convex hull of a random sample of m observations from bivariate and trivariate normal and uniform distributions. These expressions can be evaluated using numerical integration techniques, such as Gaussian quadrature (see, e.g., Burden and Faires (1993)), and are provided in this appendix.

First, consider a random sample of m observations from a multivariate normal distribution. Then, for $m \geq 3$ bivariate normal observations,

$$E(E^{(m)}) = 4\sqrt{\pi}\binom{m}{2}\int_{-\infty}^{\infty} \Phi^{m-2}(x)\phi^2(x)dx, \tag{A1}$$

and for $m \geq 4$ trivariate normal observations,

$$E(E^{(m)}) = 2 + 2\sqrt{3\pi}\binom{m}{3}\int_{-\infty}^{\infty} \Phi^{m-3}(x)\phi^3(x)dx, \tag{A2}$$

where $\phi(x)$ and $\Phi(x)$ respectively represent the univariate standard-normal density and cumulative distribution functions.

Next, consider a random sample of $m \geq 3$ observations from a continuous bivariate distribution which is uniform over the interior of the unit circle. The marginal density function of this distribution along the x-axis is given by

$$\lambda_2(x) = \begin{cases} \frac{2}{\pi}\sqrt{(1-x^2)}, & \text{for } -1 \leq x \leq 1 \\ 0, & \text{elsewhere,} \end{cases}$$

and its cumulative distribution function is given by

$$\Lambda_2(x) = \begin{cases} \frac{1}{2} + \frac{1}{\pi}\left(x\sqrt{(1-x^2)} + \sin^{-1}(x)\right), & \text{for } 0 \leq x \leq 1 \\ 1, & \text{for } 1 \leq x \\ 1 - \Lambda_2(-x), & \text{for } x < 0. \end{cases}$$

Then, for this bivariate uniform random sample,

$$E(E^{(m)}) = \frac{2\pi^2}{3}\binom{m}{2}\int_{-1}^{1} \Lambda_2^{m-2}(x)\lambda_2^3(x)dx. \tag{A3}$$

Finally, consider a random sample of $m \geq 4$ observations from a continuous bivariate distribution which is uniform over the interior of the unit sphere. The marginal density function of this distribution along the x-axis is given by

$$\lambda_3(x) = \begin{cases} \frac{3}{4}(1-x^2), & \text{for } -1 \leq x \leq 1 \\ 0, & \text{elsewhere,} \end{cases}$$

and its cumulative distribution function is given by

$$\Lambda_3(x) = \begin{cases} \frac{1}{2} + \frac{3}{4}x - \frac{1}{4}x^3, & \text{for } -1 \le x \le 1 \\ 0, & \text{for } x < -1 \\ 1, & \text{for } 1 < x. \end{cases}$$

Then, for this trivariate uniform random sample,

$$E(E^{(m)}) = 2 + \frac{35}{9}\binom{m}{3}\int_{-1}^{1} \Lambda_3^{m-3}(x)\lambda_3^4(x)dx. \tag{A4}$$

References

1. Ajmani, V.; Randles, R. H.; Vining, G. G.; and Woodall, W. H. (1997) "Robustness of Multivariate Control Charts". Proceedings of the Section on Quality and Productivity, 1997 Quality and Productivity Research Conference. American Statistical Association, Alexandria, VA, 152-157

2. Aldous, D. J.; Fristedt, B.; Griffin, P. S.; and Pruitt, W. E. (1991) "The Number of Extreme Points in the Convex Hull of a Random Sample". Journal of Applied Probability **28**, 287-304

3. Amin, R. W. and Reynolds, M. R., Jr. (1995) "Nonparametric Quality Control Charts Based on the Sign Statistic". Communications in Statistics—Theory and Methods **24**, 1597-1623

4. Amin, R. W. and Searcy, A. J. (1991) "A Nonparametric Exponentially Weighted Moving Average Control Scheme". Communications in Statistics—Simulation and Computation **20**, 1049-1072

5. Alt, F. B. and Smith, N. (1988) "Multivariate Process Control" in Handbook of Statistics, Volume 7, edited by P. R. Krishnaiah and C. R. Rao. Elsevier, Amsterdam, Netherlands, 333-351

6. Bakir, S. T. (1999) "Nonparametric Quality Control Charts: Executive Summaries". Preprint

7. Bakir, S. T. and Reynolds, M. R., Jr. (1979) "A Nonparametric Procedure for Process Control Based on Within-Group Ranking". Technometrics **21**, 175-183

8. Bhattacharya, P. K. and Frierson, D., Jr. (1981) "A Nonparametric Control Chart for Detecting Small Disorders". The Annals of Statistics **9**, 544-554

9. Bazaraa, M. S.; Sherali, H. D.; and Shetty, C. M. (1993) Nonlinear Programming, 2nd ed. John Wiley & Sons, New York, NY

10. Burden, R. L. and Faires, J. D. (1993) Numerical Analysis, 5th ed. PWS Publishing Company, Boston, MA

11. Cheng, A. Y.; Liu, R. Y.; and Luxhoj, J. T. (1997) "Thresholds for Safety Inspection Measurements Based on Control Charts". International Journal of Reliability, Quality and Safety Engineering **4**, 205-226

12. Coleman, D. E. (1997) "Individual Contributions" in "A Discussion on Statistically-Based Process Monitoring and Control" edited by D. C. Montgomery and W. H. Woodall. Journal of Quality Technology **29**, 148-149

13. Efron, B. (1965) "The Convex Hull of a Random Set of Points". Biometrika **52**, 331-343

14. Feller, W. (1971) An Introduction to Probability Theory and Its Applications, Vol. II, 2nd ed. John Wiley & Sons, New York, NY

15. Ghosh, B. K.; Reynolds, M. R., Jr.; and Hui, Y. V. (1981) "Shewhart $\overline{X}$ Charts with Estimated Process Variance". Communications in Statistics—Theory and Methods **A10**, 1797-1822

16. Groeneboom, P. (1988) "Limit Theorems for Convex Hulls". Probability Theory and Related Fields **79**, 327-368

17. Hackl, P. and Ledolter, J. (1991) "A Control Chart Based on Ranks". Journal of Quality Technology **23**, 117-124

18. Hackl, P. and Ledolter, J. (1992) "A New Nonparametric Quality Control Technique". Communications in Statistics—Simulation and Computation **21**, 423-443

19. Hayter, A. J. and Tsui, K.-L. (1994) "Identification and Quantification in Multivariate Quality Control Problems". Journal of Quality Technology **26**, 197-208

20. Hueter, I. (1994) "The Convex Hull of a Normal Sample". Advances in Applied Probability **26**, 855-875

21. Hueter, I. (1996) "Limit Theorems for the Convex Hull of Random Points in Higher Dimensions". Preprint

22. Hueter, I. (1999) "The Convex Hull of Samples from Self-Similar Distributions". Advances in Applied Probability. To appear

23. Hotelling, H. (1947) "Multivariate Quality Control" in Techniques of Statistical Analysis edited by C. Eisenhart, M. W. Hastay, and W. A. Wallis. McGraw-Hill, New York, NY

24. Janacek, G. J. and Meikle, S. E. (1997) "Control Charts Based on Medians". The Statistician **46,** 19-31

25. Kapatou, A. (1996) "Multivariate Nonparametric Control Charts Using Small Samples". Unpublished Ph.D. dissertation, Department of Statistics, Virginia Polytechnic Institute and State University, Blacksburg, VA

26. King, D. W. and Longnecker, M. T. (1990) "Computing the Distribution of the Wilcoxon Statistic with Applications to Process Control". Communications in Statistics—Simulation and Computation **19**, 381-399

27. Liu, R. Y. (1990) "On a Notion of Data Depth Based on Random Simplices". The Annals of Statistics **18**, 405-414

28. Liu, R. Y. (1995) "Control Charts for Multivariate Processes". Journal of the American Statistical Association **90**, 1380-1387

29. Liu, R. Y. and Singh, K. (1993) "A Quality Index Based on Data Depth and Multivariate Rank Tests". Journal of the American Statistical Association **88**, 252-260

30. Lowry, C. A. and Montgomery, D. C. (1995) "A Review of Multivariate Control Charts". IIE Transactions **27**, 800-810

31. Lu, C.-W. and Reynolds, M. R., Jr. (1999) "EWMA Control Charts for Monitoring the Mean of Autocorrelated Processes". Journal of Quality Technology **31**, 166-188

32. Mason, R. L.; Champ, C. W.; Tracy, N. D.; Wierda, S. J.; and Young, J. C. (1997) "Assessment of Multivariate Process Control Techniques". Journal of Quality Technology **29**, 140-162

33. Montgomery, D. C. (1996) Introduction to Statistical Quality control, 3rd ed. John Wiley & Sons, New York, NY

34. Pappanastos, E. A. and Adams, B. M. (1996) "Alternative Designs of the Hodges-Lehman Control Chart". Journal of Quality Technology **28**, 213-223

35. Park, C. and Reynolds, M. R., Jr. (1987) "Nonparametric Procedures for Monitoring a Location Parameter Based on Linear Placement Statistics". Sequential Analysis **6**, 303-323

36. Quesenberry, C. P. (1991) "SPC Q Charts for Start-up Processes and Short or Long Runs". Journal of Quality Technology **23**, 213-224

37. Quesenberry, C. P. (1993) "The Effect of Sample Size on Estimated Limits for $\overline{X}$ and X Control Charts". Journal of Quality Technology **25**, 237-247

38. Quesenberry, C. P. (1995) "On Properties of Q Charts for Variables". Journal of Quality Technology **27**, 184-203

39. Rigdon, S. E. (1995) "Preliminary Sample Size Requirements for Multivariate Control Charts". Proceedings of the 26th Annual Meeting of the Midwest Decision Sciences Institute, 240-243

40. Ryan, T. P. (1989) Statistical Methods for Quality Improvement, John Wiley & Sons, New York, NY

41. Stoumbos, Z. G. and Jones, L. A. (2000) "On the Properties and Design of Individuals Control Charts Based on Simplicial Depth". Nonlinear Studies **7**. To appear

42. Tukey, J. W. (1975) "Mathematics and the Picturing of Data". Proceedings of the International Congress of Mathematicians, Vancouver, **2**, 523-531

43. Wierda, S. J. (1994) "Multivariate Statistical Process Control—Recent Results and Directions for Future Research". Statistica Neerlandica **48**, 147-168

44. Willemain, T. R. and Runger, G. C. (1996) "Designing Control Charts Using an Empirical Reference Distribution". Journal of Quality Technology **28**, 31-38

Multivariate Process Monitoring for Nylon Fiber Production

Christina M. Mastrangelo[1], Joseph M. Porter[2] and Robert V. Baxley, Jr.[3]

[1] Department of Systems Engineering, University of Virginia, Charlottesville, VA 22903
[2] Proctor & Gamble Company, Cincinnati, OH 45202
[3] Solutia, Inc., Pensacola, FL 32560

Abstract. From the 1930s, literature on SPC schemes have been "captured" by the Shewhart paradigm of normality, independence and homogeneous variance. When in fact, the problems facing today's industries are more inconsistent than those faced by Dr. Shewhart in the 1930s. As a result of the advances in machine and sensor technology, process data can often be collected on-line. In this situation, the process observations that result from data collection activities will frequently not be independent, but autocorrelated. Autocorrelation has a significant impact on a statistical control chart: the process may not exhibit a state of statistical control when in fact, it is in control. As the prevalence of this type of data is expected to increase in industry, so does the need to control and monitor it. A second consequence of this technology shift is that many industries now operate in a data-rich environment. They have the ability to gather many observations on several quality characteristics in a short period of time. This gives rise to a situation probably not envisioned by Shewhart in the 1930s: too much process data and too little information about the process. This domain is ripe for the methods of data mining and multivariate quality control. This paper develops a multivariate statistical process monitoring scheme for a continuous process: the production of mono-filament nylon fibers. The data is abundant (over 80 process and quality variables) and has varying degrees of serial correlation. Various data mining methodologies are used to determine the driving variables which affect the response variable, spin breaks. The result is a substantially smaller set of variables which are then combined with two multivariate quality control methods. Preliminary results suggest that using a MEWMA control chart on the variables determined by a GMDH heuristic is a viable process monitoring tool in the presence of autocorrelated data.

1 Introduction

In the 1920's, Walter Shewhart visualized that the marriage of statistical methods and manufacturing processes would produce reliable and consistent quality products. Shewhart (1931) conceived the idea of statistical process control (SPC) and developed the well-known and appropriately named Shewhart control chart. However, from the 1930s to the 1990s, literature on SPC schemes have been "captured" by the Shewhart paradigm of normality, independence and homogeneous variance. When in fact, the problems facing today's industries are more inconsistent than those faced by Shewhart in the 1930s.

As a result of the advances in machine and sensor technology, process data can often be collected on-line. In this situation, the process observations that result from data collection activities will frequently not be serially independent, but autocorrelated. Autocorrelation has a significant impact on a control chart: the process may not exhibit a state of statistical control when in fact, it is in control. As the prevalence of this type of data is expected to increase in industry (Hahn 1989), so does the need to control and monitor it. Equivalently, literature has reflected this trend, and research in the area of SPC with autocorrelated data continues so that effective methods of handling correlated data are available. This type of data regularly occurs in the chemical and process industries, and is pervasive in computer-integrated manufacturing environments, clinical laboratory settings and in the majority of SPC applications across various manufacturing and service industries (Alwan 1991).

A second consequence of this technology shift is that many industries now operate in a data-rich environment (Runger and Willemain 1995). They have the ability to gather many observations on several quality characteristics in a short period of time. This gives rise to a situation probably not envisioned by Shewhart in the 1930s: too much process data and too little information about the process or product. This domain is ripe for the methods of multivariate quality control. Traditional control charting methods may be easily applied to process data. However, the major drawback to these charts is that they were developed to monitor one variable at a time. When a large number of process variables are measured, they are almost never independent but cross-correlated with each other. Univariate control charts do not expand to handle cross-correlated variables well (Kresta et al. 1991), nor do they capitalize on the correlative structure inherent in the data (Jackson 1980). Even when multivariate industrial processes are under formal control schemes, autocorrelated data may still be present. This predicament impacts multivariate control schemes because these schemes assume independent observation vectors.

While statistical process monitoring (SPM) methodologies have been incorporated into many manufacturing processes, few examples of a monitoring system using a large number of autocorrelated variables exist. In fact, most multivariate monitoring schemes discussed in the literature are theoretical and have not been incorporated into a production process. In addition, these SPM schemes deal with five or fewer input variables. n example of an environment characterized by a large number of variables with significant serial correlation may be found in mono-filament nylon fiber production. This paper studies the process of developing a multivariate SPM scheme for this environment.

The data comes from on-line sensors and off-line laboratory measurements at Solutia, Inc. The goal of the project is to monitor process variables so as to reduce the number of spin breaks. A spin break occurs when mono-filament nylon fibers break either during or shortly after they are extruded from a sieve-like mechanism known as a "pack". To return the production line to normal operation, the production process must be halted until the labor-intensive task of re-stringing the machinery is completed. Ideally, one solution to the spin-break problem is to prevent the breaks from occurring by monitoring the driving variables of the process.

1.1 The Nylon Production Process

To produce the mono-filament nylon fibers, Solutia employs a hybrid process consisting of batch and continuous processes as shown in Figure 1. The mixture of the chemicals that react to form the nylon is performed as a batch process. This mixture is then fed into a continuous production process that transforms the molten mixture into mono-filament nylon strands.

Once the molten polymer leaves the batch production process, it proceeds to the evaporator. The evaporator uses steam heat to reduce the concentration of water in the mix before it enters the reactor. This reduces the concentration of water in the molten nylon mix to approximately 80% of its previous concentration. The function of the reactor is to start the polymerization process. The nylon molecules join into segments in the reactor under elevated temperature and pressure. This joining of the nylon molecules increases the viscosity of the "fluid" flowing through the reactor on its way to the flasher. The flasher separates the liquid material from the vapor. From the flasher, the nylon proceeds to the finisher. The finisher is a large tank on its side half full with the nylon polymer. In the finisher, the polymer is transferred, by a screw drive, from the entrance point the finisher's exit. A sweep gas flows in the opposite direction (from exit point to entrance point) to remove the remaining water in the polymer. The removal of water from the nylon increases its viscosity. The polymer then leaves the finisher and flows through the transfer line to a pack.

The nylon polymer is extruded under pressure through pin size holes in the base of the pack. Circulating air around the semi-molten nylon fibers cools and solidifies them into strands. It is during this cooling stage that the fibers are most likely to break. Once the fibers have cooled to a manageable temperature, they are gathered into a bundle known as a "tow". This tow is then piddled into a heap and packaged for shipping.

The production line is peppered with sensors that collect data on various characteristics of the production process. The arrows in Figure 1 denote the placement of these on-line sensors. The on-line sensors collect data on process measurements such as temperature and pressure. In addition to on-line sensors, representative samples of the fibers are analyzed off-line in the laboratory to monitor product characteristics such as denier and the shape of the fiber.

The data used to construct and test the SPM methods consists of six data sets. These six data sets are from three different polymer lines and six different fiber-spinning machines. Every two spinning machines [i.e. 16,17; 22,23; 24,25] share one nylon polymer stream. This is shown in Figure 1 where the production line splits into two lines after the reactor. Each observation is the daily average of that variable. Each data set consists of 256 obervations on 54 variables. The data has also been preprocessed to replace missing values and two extreme temperature outliers.

To understand the levels of serial correlation present in the data, ARIMA models were estimated for each variable (see Box, Jenkins, and Reinsel 1994). Of the 54 variables, 30 are modeled as AR(1), and 10 as AR(2). Thus, 75% of the process variables exhibit low-order autoregressive behavior. Six variables are random noise, and 8 have miscellaneous non-stationary model forms. Approximately, 40% of the variables have high lag 1 autocorrelation (> 0.8). The response variable, unadjusted break frequency, is AR(2).

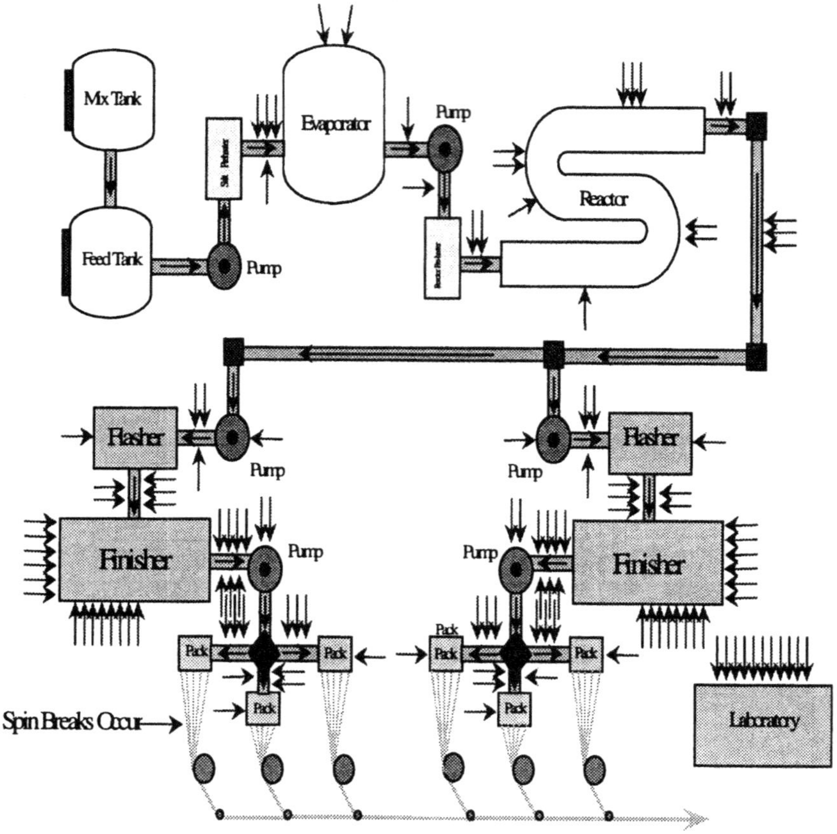

Fig. 1. Fiber Production Process

2 Variable Selection

Unaided, humans tend to perform poorly in extracting relevant information from data sets that contain large numbers of independent or interrelated variables. To compensate for this deficiency, considerable research has focused on the development of algorithms and heuristics that lend structure to high dimensional data sets. The implementation of one or more of these techniques is generically known as "data mining". Data mining strives to extract pertinent non-obvious information concerning the underlying structure of a process from the values of the attributes that describe that process. When the number of input and output variables are small, this is a simple task. In this application, pertinent non-obvious information includes determining the input, or process, variable(s) in which fluctuations contribute significantly to fluctuations in the output variable, or frequency of spin breaks. This suggests that a subset of the input variables drive the process's outcome. Two modeling techniques are used here: multiple regression and the Group Method of Data Handling (GMDH) .

Multiple regression is the cornerstone of many more complicated data mining techniques. It is a powerful induction tool that attempts to quantify the value of an output variable (a *response* or *dependent* variable) based on the summation of a series of weighted input variables (the *predictor* or *independent* variables). Because the pool of candidate regressors is large, two variable selection techniques were used to build the model: a directed search on t and the stepwise regression algorithm available in SAS Proc RSREG. In the model building process using stepwise selection, the default criteria values for entrance into the model and to remain in the model were used. We found the subset of variables did not change for a range of exit and entry values. See Montgomery and Peck (1982) and Neter et al. (1990) for additional information.

The Group Method of Data Handling (GMDH) induces a model by exploiting latent information pertaining to the structure and the interaction of the variables in a data set. By inducing a model solely from the data, an analyst theoretically does not have the opportunity to bias the model from his or her preconceived notions. This task of inducing solely from data is achieved by incorporating traditional statistical techniques, such as regression, with an evolutionary heuristic.

This evolutionary heuristic is critical to the GMDH methodology. The concept of evolving a mathematical model is similar to the evolutionary process of an organism. In this paradigm only the fittest mathematical models are allowed to survive and mate (Farlow 1984). The GMDH model creates a high order polynomial equation of the input variables that strives to mimic the fluctuations of the output variable. In essence, the model creates a polynomial estimation surface that by some metric best approximates the output data. One of the most appealing aspects of the GMDH is its ability to not only suggest the underlying structure of the data but also estimate the parameters of the model. This ability of the GMDH to "discover the structure or functional form of an empirical model" lends to its usefulness in determining the driving variables in a fiber production process (Barron 1984).

As in all model induction, the input and the output variables are known. Thus the implementation of the GMDH algorithm begins by designating the input variable(s) and one output variable. The raw data is then normalized, usually by a mean-sigma normalization, to create variables with a mean of zero and a variance of one. This normalization is performed to insure that one variable does not artificially skew the results due solely to the variable's magnitude or variance.

Once the input variables are normalized, the first "primeval" equations of the GMDH are calculated. In the methodology used here, the primeval model elements are composed of third-order polynomials that consist of all combinations of one, to, or three inputs (Brown and Elder, 1992):

$$y_{single} = w_0 + w_1 x_1 + w_2 x_1^2 + w_3 x_1^3 \tag{1}$$

$$y_{double} = \tag{2}$$
$$w_0 + w_1x_1 + w_2x_2 + w_3x_1^2 + w_4x_2^2 + w_5x_1x_2 + w_6x_1^3 + w_7x_2^3 + w_8x_2x_1^2 + w_9x_1x_2^2$$

$$y_{triple} = \tag{3}$$
$$w_0 + w_1x_1 + w_2x_2 + w_3x_3 + w_4x_1^2 + w_5x_2^2 + w_6x_2^2 + w_7x_1x_2 + w_8x_1x_3 + w_9x_2x_3$$
$$+ w_{10}x_1x_2x_3 + w_{11}x_1^3 + w_{12}x_2^3 + w_{13}x_3^3 + w_{14}x_2x_1^2 + w_{15}x_3x_1^2 +$$
$$w_{16}x_1x_2^2 + w_{17}x_3x_2^2 + w_{18}x_1x_3^2 + w_{19}x_2x_3^2$$

From this pool of equations only the "fittest", or those that best explain the fluctuations in the output variable, survive. The survival or the elimination of the current generation's equations are determined by a predefined "fitness rule". As in organic evolution, the survivors of the test of fitness become the parents of the next generation. For the next generation, the original input variables (x_1, x_2, x_3, ... x_{m-1}, x_m) are not used, but the predictions of the surviving models of the previous generation are.

This process iterates until the youngest generation of models fail to be superior to their parent generation in some measure of model fitness. This heuristic strives to find a compromise between the fit of the model and its complexity. In order to achieve an optimal compromise the objective of the heuristic employed is the minimization of predicted squared error (PSE). This is one of several model scoring techniques that relies on the an estimate of the underlying noise level in the data. Mallow's Cp and Akaike's Information Criterion are two other widely known and used techniques. The predicted squared error is defined as the sum of the fitted squared error (FSE) and a complexity penalty (KP): PSE = FSE + KP where

$$KP = CPM \times \left(\frac{2K}{N}\right) s_p^2 \tag{4}$$

CPM - complexity penalty multiplier; a user adjustable variable

K - total number of coefficients

N - number of data points in the training set

s_p^2 - an a-priori estimate of the true model error variance
- inversely proportional to N
(Model Quest Enterprise User's Manual, 1998)

The complexity penalty limits the size of the model. Without the complexity penalty, the model could theoretically grow to fit the noise of the data set. If this occurs the model is only applicable to that specific data set and looses its ability to generalize to other data. In other words, as the CPM increases, the model will be penalized more for complexity, which encourgages simpler models. If the CPM decreases, the model may become more complex and overfit the data. We used a default CPM value of 1.0. If after building a training model, the PSE of the output is much larger than the actual squared error, a better model could be built with a lower CPM. A lower bound on the CPM value is typically 0.01. Although manipulation of the CPM changed the order of the model, it did not significantly change the variables that the GMDH selected as inputs.

3 Process Monitoring Methods

In the specific area of multivariate statistical process monitoring with correlated data, relatively little work has appeared in the literature. Thus, this section generalizes to the broader realm of statistical monitoring with correlated data, then present recent work in multivariate quality control and, finally, the current work in multivariate quality control with correlated observation vectors.

3.1 SPC and Autocorrelated Data

An important assumption in the application of statistical control charts is that the observations generated by the monitored process are statistically independent or uncorrelated. In other words, a model for the observations from a process is $x_t = \mu + \varepsilon_t$ where μ is the process mean and ε_t is a series of independent and identically distributed (IID) random variables. It is well-known that control charts are robust to moderate departures from most underlying assumptions, such as normality in the case of variable control charts. Unfortunately, conventional control charts are very sensitive to correlated data (Montgomery and Friedman 1988, Harris and Ross 1991). Alwan (1992) studies the effects of autocorrelated data on control chart limit techniques and demonstrates that in the presence of even moderate levels of autocorrelation, an out-of-control point on the chart does not necessarily indicate a process change and many false alarms occur.

It is difficult to recognize a state of statistical control or determine departures from one if systematic, non-random patterns are present. When correlation is present, these non-random patterns are, in fact, common causes. Recently, several methods of applying control charts to autocorrelated data are found in the literature. Alwan and Roberts (1988), Montgomery and Friedman (1988), Wardell et al. (1992, 1994), to name a few, all present a similar approach to the problem of autocorrelated data: fit an appropriate time series model to the observations and apply control charts to the stream of residuals from this model. Since the residuals are IID random variables, a system change such as a process upset or assignable cause will increase the absolute magnitude of the residuals. Thus, plotting the residuals on a control chart will detect a process change. When the data is not modeled, non-random behavior is confounded with the common cause behavior of the data. Runger and Willemain (1995) present alternate approach: a model-free method that uses autocorrelated data directly.

One particular time series model, the ARIMA(0,1,1) or IMA(1,1) is an important case of the general ARIMA model because it is equivalent to the EWMA statistic. The EWMA control chart is often preferred over the Shewhart control chart because, in comparison, it is more sensitive and will detect small process changes. The versatility of the technique is well known, and the time series/SPC link makes it very powerful. Schmid (1995), VanBrackle and Reynolds (1997) and Lu and Reynolds (1998) study the run length properties of the EWMA and Shewhart charts.

However, the time series modeling approach may be awkward because control charts are often applied to several variables on many processes. Fitting and maintaining a time series model for each variable can easily become cumbersome. Montgomery and Mastrangelo (1991) and Alwan (1991) illustrate that a single control chart approach for correlated data is equivalent to a traditional two-chart approach; however, to use one control chart, the center line and control limits must be variable instead of fixed. Montgomery and Mastrangelo (1991) introduce a single control chart methodology for application to autocorrelated data. The Moving Centerline EWMA (MCEWMA) is a procedure that approximates the true underlying time series model with the familiar EWMA. Mastrangelo and Montgomery (1994) further characterize the method in terms of its run length and detection properties.

3.2 Multivariate Process Control

Multivariate SPC methods enable the simultaneous monitoring of several variables which are often related, or correlated. The majority of multivariate control chart procedures use Hotelling's T^2 statistic to test for out-of-control conditions (Alt 1985 , Lowry and Montgomery 1995, and Tracy, Young and Mason 1992, Sullivan and Woodall 1996). By using Hotelling's T^2 method, one has a single number criteria that coupled with the upper control limit produces a desired type I error and utilizes all of the available information with respect to means, variances and correlation between the variables.

Predominantly, multivariate control chart techniques make use of variations of the T^2 statistics that Hotelling developed. However, because this procedure signals when a statistically significant shift in the mean occurs, and it is based solely on the most recent observation, Hotelling's method is insensitive to small shifts in the mean vector. This situation gave rise to the exploration of applying more sensitive control chart techniques, such as the EWMA to multivariate data.

Lowry et al. (1992) present a multivariate EWMA (MEWMA):

$$z_t = \Lambda x_t + (I - \Lambda) z_{t-1} \tag{5}$$

where $\mathbf{x_t}'$ is a vector of the process variables at time t, Λ a diagonal matrix, I the identity matrix, and $\mathbf{z_t}$ the vector of the EWMA's. The MEWMA chart gives an out-of-control signal when

$$T^2 = z_i' \Sigma_z^{-1} z_i > h_4 \tag{6}$$

where $h_4 > 0$. Using simulation, they compare the ARL properties of four multivariate control chart procedures and find the MEWMA, using the exact covariance matrix, is an effective method in detecting shifts from the in-control mean vector when the process is initially out of control. Prabhu and Runger (1997) also study ARL performance and show that appropriate values for Λ depend on the number of variables and the shift size one wants to detect.

The primary advantage of a T^2 test is that it offers a single statistical test for a general shift in the mean vector of the process. However, the T^2 statistic is not immediately interpretable and once it signals, more work is need to diagnose which variable or variables have shifted. This interpretability problem increases the complexity and maintenance of a multivariate procedure and partially explains why most authors recommend use of other charts in addition to the T^2 chart (Mason et al. 1997, Kourti and MacGregor 1996, Mason et al, 1996, Runger and Alt 1996, Murphy 1987). The multivariate methods discussed in this section assumes that the random observation vectors are independent over time and have a known covariance structure. However, in many production processes, such as those found in the chemical industry, serially correlated observation vectors frequently occur.

3.3 Multivariate Process Control with Correlated Data

The literature in this area is not yet as extensive. Kramer and Schmid (1997) generalize the MEWMA to time series. Charnes (1995) uses simulation to study the type I error rate of several ARIMA models. Other work by Mastrangelo et al. (1996) have explored the use of principle component analysis (PCA) with correlated data, and they demonstrate that PCA, used in an exploratory manner, can detect a large process upset. Runger (1996) further shows that monitoring PCA residuals certain types of assignable causes. Nomikos and MacGregor (1995) use multi-way PCA to monitor process in batch manufacturing.

4 Results

This section analyzes of the performance of the T^2 and MEWMA control charts when applied to the Solutia data. All the results presented are for Spinning Machine 22. There is no predefined limit as to what is considered in-control or out-of-control in terms of the average number of spin-breaks per day. For comparison purposes, any level of spin breaks over 0.95 is considered to be out-of-control. The Machine 22 data set consists of seven instances when the average break frequency exceeds the 0.95 level. In order to determine the performance of the SPC methods, the charts analyze the level of correct detections as well as the number of false alarms. Throughout the analysis of the monitoring schemes, the following symbols are used:

Correct detection of OOC spin breaks: CC

False Alarm: FA

Failure to detect OOC spin breaks: O

In addition to using GMDH and multiple regression to determine candidate driving variables, stepwise regression (using SAS Proc RSREG) and the combined set of GMDH and multiple regression variables were also used as inputs to the control charts. Figure 2 is a scatter plot of the two conditions that are preferably minimized. The ideal variable selection and control chart method would plot at the origin. The line connecting two of the solutions acts a Pareto optimal frontier—the best solution of those explored. The control charts for these two methods are presented here.

Table 1 summarizes the key process variables found by multiple regression and GMDH modeling. The variables included in the multiple regression model are those with a significance level less than 0.05. Table 1 shows some common variahles; 'AVG_AGE' is a key variable and represents the average age of the pack. All of the variables are located at the latter end of the production line. Further analysis reveals that with the exception of the average-age variable, the key variables differ between spinning machines. Unique monitoring schemes would need to be developed for each machine. However, while the specific variables do differ, there are common function variables. For example, the finisher has multiple temperature variables and at least one is a key characteristic for each spinning machine.

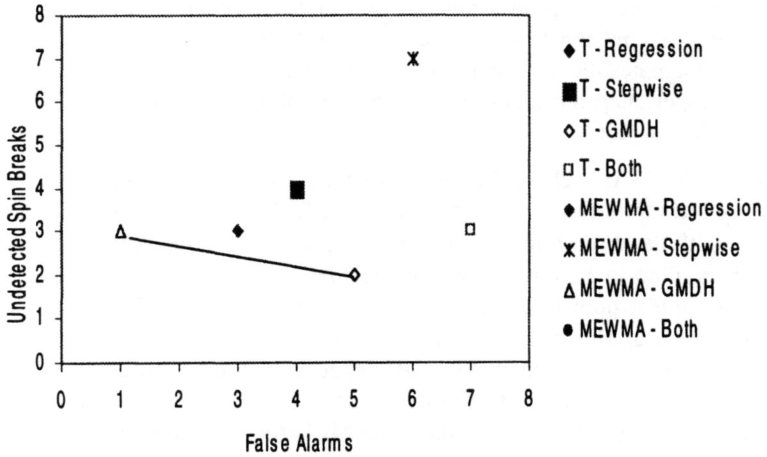

Fig. 2. Scatter plot of variable selection method/control chart

Table 1. Variables identified in selection process.

Variable Name	Description & Location	GMDH	Regression
AVG_AGE	Average age of pack.	X	X
T 010	Temperature, Flasher	X	X
SC 300	Pump Speed, Finisher	X	X
PI 392	Pressure, Pack	X	X
PC 273	Pressure Controller, Flasher	X	
PC 307	Pressure Controller, Finisher	X	
PI 267	Pressure, Flasher	X	
PI 385	Pressure, Manifold	X	
T 016	Temperature, Finisher	X	
TC 278	Temperature Controller, Flasher	X	
LB CDL	Lab Measurement		X
T 011	Temperature, Flasher		X
T 018	Temperature, Filter		X
T 023	Temperature, Manifold		X

Figure 3 consists of two time series plot: the upper plot shows the level of spin breaks over time and the lower is the T^2 control chart. A horizontal line in the upper plot designates the 0.95 spin break level. Ideally, every point above the 0.95 line on the upper chart has a corresponding point plotting above the upper control limit on the lower plot.

Overall, the MEWMA chart, Figure 4, displays much more gradually sloping monitoring lines than the T^2 chart because the MEWMA uses weighted historical data to smooth its observations. The weighting factor λ is set equal to 0.3 for all the variables. One interesting aspect of the MEWMA chart is that it caught the four out-of-control points between observation 151 and 175. The T^2 chart was only able to pick up the initial one or two out-of-control episodes in this series of four. This suggests the fundamental difference in the formulation of these two charts. The T^2 chart bases its determination of in-control versus out-of-control on each observation set individually, where the MEWMA also uses historical data to determine the state of the process.

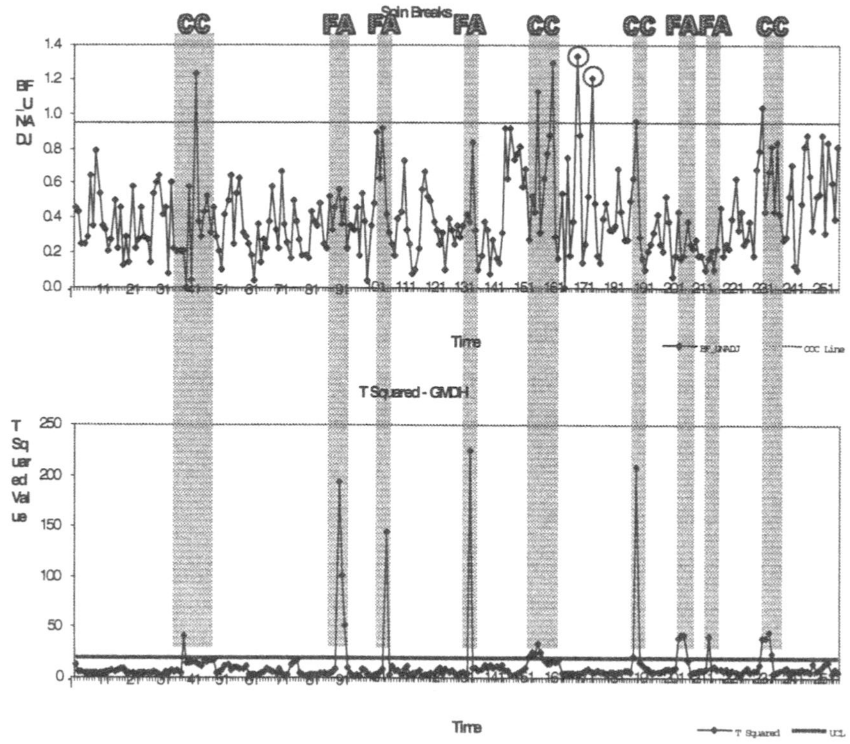

Fig. 3. T^2 Control Chart

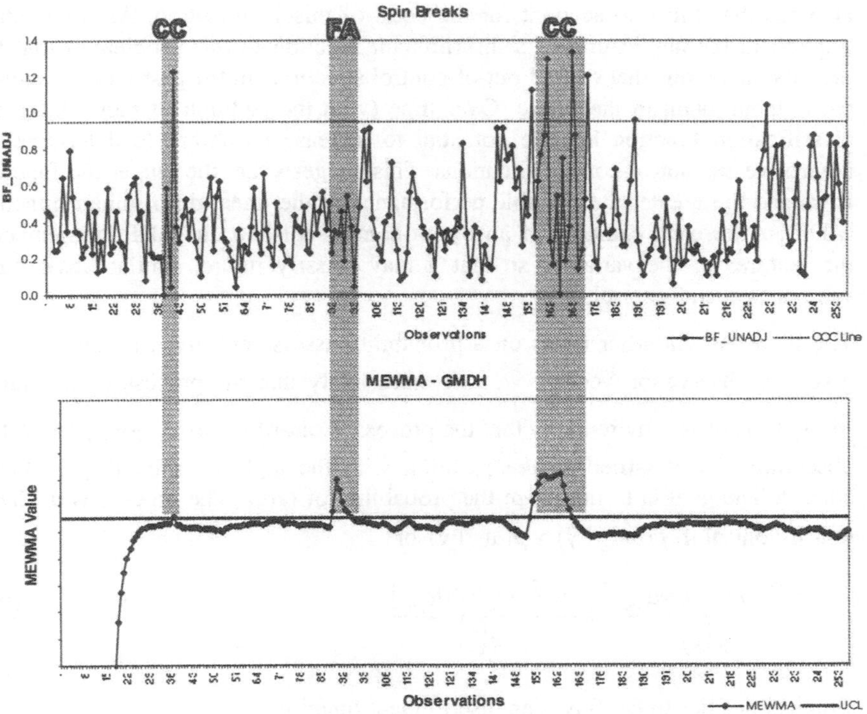

Fig. 4. MEWMA Control Chart

While the control charts were not completely effective in predicting spin breaks, they did "catch" the majority of them. The circles in Figure 3 imply that key process variables were not included in the model. On the other hand, the false alarms, in both figures, imply that the observation vector is a significant distance from the center of the process space, but the variables that pull the vector away do not tend to influence the occurance of spin breaks.

5 An Alternate Monitoring Scheme

An alternative way to view this problem is from a discriminant approach: classifying the process as being in an in-control or out-of-control state. Given the amount of historical process data, a multivariate Bayesian linear discriminant function could be trained to recognize a future out-of-control situation based on the current observation's similarity to past out-of-control episodes. The classifier also has the ability to account for the costs of misclassification. An assumption implicit in the application of a discriminant function to this problem is that the process deviations that caused out-of-control situations in the past will potentially cause them again in the future. Over time (with the addition of more data), the classification function has the potential to increase its ability to determine in-control versus out-of-control situations. This suggests that the longer the function is trained the greater the possible performance of the linear discriminant function (LDF). After each example of an out-of-control episode, the LDF "remembers" the features of the variables so that it may classify future, similar observation vectors as out-of-control.

The Bayesian classifier relies on a probability assessment given an observation: given an observation vector, $\vec{\hat{y}}$, is it more likely that the process is in-control, $p(\omega_{ic} \mid \vec{\hat{y}})$, or is it more likely that the process is out-of-control, $p(\omega_{ooc} \mid \vec{\hat{y}})$? The observation is classified as the partition with the highest probability of being correct. The goal is to minimize the probability of error. The process is declared out-of-control if $p(\omega_{ooc} \mid \vec{\hat{y}}) > p(\omega_{ic} \mid \vec{\hat{y}})$ or

$$\frac{p(\vec{\hat{y}} \mid \omega_{ooc})\Pr[\omega_{ooc}]}{p(\vec{\hat{y}})} > \frac{p(\vec{\hat{y}} \mid \omega_{ic})\Pr[\omega_{ic}]}{p(\vec{\hat{y}})}, \tag{7}$$

which simplifies to the Bayesian discriminant function:

$$\frac{P(\vec{\hat{y}} \mid \omega_{ooc})}{P(\vec{\hat{y}} \mid \omega_{ic})} > \frac{\Pr[\omega_{ic}]}{\Pr[\omega_{ooc}]}. \tag{8}$$

The "control limit" or decision line is influenced by the prior probabilities such that the decision line may move to the left or right (Therrien 1989). Multivariate normality is assumed as well as a common covariance matrix due to the limited amount of out-of-control data. Since the cost of classifying correctly is usually zero, the LDF becomes

$$\vec{m}_{ooc}^{T} K^{-1} \vec{m}_{ooc} - \vec{m}_{ic}^{T} K^{-1} \vec{m}_{ic} + (\vec{m}_{ic}^{T} - \vec{m}_{ooc}^{T}) K^{-1} \vec{\hat{y}} > 2 \ln \frac{\Pr[\omega_{ooc}]}{\Pr[\omega_{ic}]} \tag{9}$$

where K is the pooled variance for all groups, m is the mean of each category, $\vec{\hat{y}}$ is the point in question, and the right-hand side is the control limit. Figure 5 applies the LDF to the 1-period forecast errors of the key variables as determined by multiple regression. Since the method requires a sufficiently large training set of in-control and out-of-control states, only the last 86 observations are classified. None of the variable selection methods result in models that are able to detect the high spin break level at time period 173.

The process variables determined by GMDH modeling did not improve the classification rate. The LDF was also used with other data forms: the raw data and the residuals from ARIMA modeling. However the best results were obtained with the forecast errors and are presented here. An interesting extension to this work is to estimate and include the costs of misclassifiction.

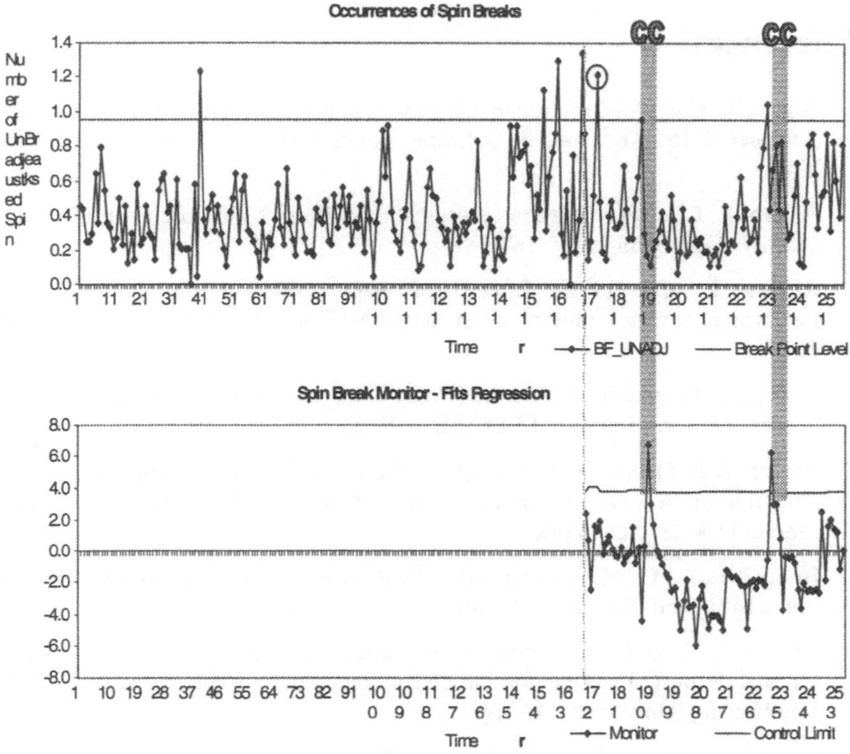

Fig. 5. Discriminant function of the ARIMA 1-period forecasts.

6 Summary

This work demonstrates the difficulty of developing multivariate schemes that monitor in-process variables believed to affect a key end-of-process variable, spin breaks. While the methods (and models) were not perfect in classifying all spin breaks, they did identify the majority of them. Additional analyses could be explored for this data. For example, different covariance estimation methods such as those presented in Sullivan and Woodall (1996); and additional variable selection methods such as classification trees. A recommended monitoring strategy for this process is a univariate chart monitoring package, and an MEWMA chart on the flasher and finisher variables. The GMDH may be used as a variable selection method for each spinning machine.

References

1. Alt, F.B. (1985) Multivariate Quality Control. Encyclopedia of Statistical Sciences **6** (S. Kotz and N. Johnson, eds.), 110-122. Wiley & Sons, New York

2. Alwan, L. C. (1991) Autocorrelation: Fixed Versus Variable Control Limits. Quality Engineering **4(2)**, 167-188

3. Alwan, L. C. (1992) Effects of Autocorrelation on Control Chart Performance. Communications in Statistics: Theory and Methods **21**, 1025-1049

4. Alwan, L. C. and H. V. Robert (1988) Time Series Modeling for Statistical Process Control. Journal of Business & Economic Statistics **6(1)**, 87-95

5. Barron, A.R. (1984) Predicted Square Error: A Criterion for Automatic Model Selection, in Self-Organizing Methods in Modeling: GMDH type Algorithms. Marcel Dekker, New York

6. Box, G. E. P., G. M. Jenkins and G.C. Reinsel (1994) Time Series Analysis, Forecasting and Control. 3rd Edition, Prentice Hall

7. Brown, D.E. and J.F. Elder (1996) Induction and Polynomial Networks, Internal Report. Institute for Parallel Computation, Department of Systems Engineering, University of Virginia

8. Charnes, J.M. (1995) Tests for Special Causes with Multivariate Autocorrelated Data. Computers in Operations Research **22(4)**, 443-453

9. Farlow, S.J. (1984) Self-Organizing Methods in Modeling: GMDH type Algorithms. Marcel Dekker, New York

10. Hahn, G.J. (1989) Statistics-Aided Manufacturing: A Look Into the Future. American Statistician **43**, 74-79

11. Harris, T. J. and W. H. Ross (1991) Statistical Process Control Procedures for Correlated Observations. The Canadian Journal of Chemical Engineering **69**, 48-57

12. Jackson, J.E. (1980) Principal Components and Factor Analysis: Part 1–Principal Components. Journal of Quality Technology **12**, 201-213

13. Kourti, T. and J.F. MacGregor (1996) Multivariate SPC Methods for Process and Production Monitoring. Journal Quality Technology **28**, 409-428

14. Kramer, H.G. and W. Schmid (1997) EWMA Charts for Multivariate Time Series. Sequential Anlaysis **16(2)**, 131-154

15. Kresta, J.V., MacGregor, J.F., and Marlin, T.E. (1991) Multivariate Statistical Monitoring of Process Operating Performance. Canadian Journal of Chemical Engineering **69**, 35-47

16. Lowry, C.A. and D.C. Montgomery (1995) A Review of Multivariate Control Charts. IIE Transactions **27**, 800-810

17. Lowry, C. A., W. H. Woodall, C. W. Champ and S. E. Rigdon (1992) A Multivariate Exponentially Weighted Moving Average Control Chart. Technometrics **34(1)**, 46-53

18. Lu, C. and M.R. Reynolds: Control Charts for Monitoring the Mean and Variance of Autocorrelated Processes, to appear in Journal of Quality Technology

19. Mason, R.L., C.W. Champ, N.D. Tracy, S.J. Wierda, and J.C. Young (1997) Assessment of Multivariate Process Control Techniques. Journal of Quality Technology **29(2)**, 140-143

20. Mason, R.L, N.D. Tracy, and J.C. Young (1996) Monitoring a Multivariate Step Process. Journal of Quality Technology **28(1)**, 39-50

21. Mastrangelo, C.M. and D.C. Montgomery (1995) Characterization of a Moving Centerline Exponentially Weighted Moving Average. Quality and Reliability Engineering International **11(2)**, 79-89

22. Mastrangelo, C.M., G.C. Runger, and D.C. Montgomery (1996) Statistical Process Monitoring with Principal Components. Quality and Reliability Engineering International **12(3)**

23. ModelQuest Enterprise User's Manual (1998) AbTech Corporation, Charlottesville, VA

24. Montgomery, D. C. and D. J. Friedman (1989) Statistical Process Control in a Computer-Integrated Manufacturing Environment, in Statistical Process Control in Automated Manufacturing, eds. J.B. Keats and N.F. Hubele. Marcel Dekker, New York

25. Montgomery, D.C. and C.M. Mastrangelo (1991) Some Statistical Process Control Methods for Autocorrelated Data. Journal of Quality Technology **23(3)**, 179-193

26. Montgomery, D.C. and E.A. Peck (1982) Introduction to Linear Regression Analysis. Wiley, New York

27. Murphy, B.J. (1987) Selecting Out of Control Variables with the T^2 Multivariate Quality Control Procedure. The Statistician **36**, 571-583

28. Neter J., W. Wasserman and M.H. Kutner (1990) Applied Staitistical Linear Models: Regression, Analysis of Variance and Experimental Design

29. Prabhu, S..S. and G.C. Runger (1997) Designing a Mulivariate EWMA Control Chart. Journal of Quality Technology **29(1)**, 8-15

30. Runger, G.C. (1996) Multivariate Statistical Process Control for Autocorrelated Processes. International Journal of Production Research **34(6)**, 1715-1724

31. Runger, G.C. and F.B. Alt (1996) Contributors to a Multivariate Statistical Process Control Chart Signal. Communications in Statistics—Theory and Methods **25(10)**, 2203-2213

32. Runger, G.C. and T.R. Willemain (1995) Model-Based and Model-Free Control of Autocorrelated Processes. Journal of Quality Technology **27**, 283-292

33. Schmid, W. (1995) On the Run Length of a Shewhart Chart for Correlated Data. Statistical Papers **36**, 111-130

34. Sullivan J.H. and W.H. Woodall (1996) A Comparison of Multivariate Control charts for Individual Observations. Journal of Quality Technology **28(4)**, 398

35. Therrien, C.W. (1989) Decision, Estimation and Classification. Wiley, New York

36. Tracy, N.D., J.C. Young, and R.L. Mason (1992) Multivariate Control Charts for Individual Observations. Journal of Quality Technology **24(2)**, 88-95

37. VanBrackle, L.N. and M.R. Reynolds (1997) EWMA and CUSUM Control Charts in the Persence of Correlattion, Communications in Statistics—Simulation **26(3)**, 979-1008

38. Wardell, D. G., H. Moskowitz and R. D. Plante (1992) Control Charts in the Presence of Data Correlation. Management Science **38(8)**, 1084-1105

39. Wardell, D.G., H. Moskowitz, and R.D. Plante (1994) Run-Length Distributions of Special-Cause Control Charts for Correlated Processes and Discussion. Technometrics **36(1)**, 3-27

The Management of SPC

Erwin M. Saniga[1] and Michael F. Pohlen

[1] University of Delaware, Department of Business Administration, Newark, DE 19716

Abstract. We analyze forty eight quality or productivity improvement projects we have managed from 1984-1998 in thirty four different organizations including large credit card banks, large chemical companies, medical service firms including hospitals, schools, manufacturing companies, and supermarkets. The type of projects ranged from the improvement of customer service to the improvement of product quality. Each project involved improving performance using standard SPC methodology.

Our analysis focuses on management factors that are associated with project design and implementation and on the strategic implications of the management of SPC throughout the firm.

We argue that while certain factors are found to be statistically significant the person implementing SPC must still be aware that the principles of sound decision support are a necessary condition for project and implementation success in particular. Moreover, our experience in a broad range of applications indicates that successful application requires specialized solutions tailored to specific problems even though the overall improvement *process* is structured.

I Introduction

In this paper we investigate the problem of the *management* of process improvement using SPC by studying 48 actual applications in a broad range of firms and over a broad range of functional applications. Our study involves investigating the relationships between design success and implementation success and some well known management variables identified in the project management and decision support literature.

Pinto and Slevin(1987,1988) and Pinto and Mantel(1990) have studied the process of project management, of which process improvement is a subset. These authors have found that support by management, accurate mission definition, clear communication between the project group and the rest of the organization, competence of the project manager in leadership, authority, technical expertise and interpersonal ability, and accurate monitoring and controlling of the project are important factors associated with project success.

There have also been an number of studies investigating the factors associated with project success in the field of decision support systems, which are systems that enhance decision making when decision making tasks are poorly structured or have some lack of structure to them. Using any of the common definitions it is easily argued that a process improvement system falls within the category of a DSS.

A number of studies have investigated the factors associated with DSS system success from the design and implementation perspective. For example, Sanders and Courtney(1985) found that top management support; user training and experience with DSS systems were variables associated with system success. Udo and Davis (1990) found that success in terms of benefits achieved is greater in larger organizations and those applications to unstructured tasks yields greater benefits than applications to routine tasks. Timeliness of output, the length of time of use and the range of alternatives allowed were other variables associated with systems benefits. Guimaraes, Igbaria and Lu(1992) find that important variables are user involvement, user training, top management support and task variety. Task variety is inversely related to user satisfaction; this finding implies that designers should limit the scope of the DSS. Task difficulty was found to be inversely related to user satisfaction, which implies that designers should not focus on unusually complex projects. Igbaria and Guimaraes(1994) concluded that user involvement in systems design is an important determinant of DSS success. Alavi and Joachimsthaler(1992) analyzed 144 findings from 33 studies and found that user related variables such as user involvement, training and experience are related to implementation. Igbaria and Guimaraes(1996) find that user understanding is an important factor in successful DSS design and implementation.

In this study we empirically investigate the relationship between some of these factors (and others) and project success. Our projects are all quality and productivity improvement projects done using traditional statistical process control (SPC) methods. In section II we discuss the actual projects. Section III contains the results of some data analysis on these projects and Section IV presents some strategic implications.

II The Projects and Data Collection

In 1983 the authors were co-designers of a curriculum in Information and Decision Support Systems to be offered by the University of Delaware. Six credits of that curriculum were allocated to a team based senior level project course where students were assigned to a local firm to design and implement a solution to a problem jointly determined by the authors and the corporate client. Students are assigned to the projects by preference and by project requirements. Since the initial projects in 1984 we have managed 158 projects, 48 of which were in the area of quality and productivity improvement. Each project is managed jointly by the authors in conjunction with the corporate client and is of fourteen-week duration. We estimate that the teams of students devote an average of one thousand person hours to each project.

The project sponsors represent a wide variety of firms in terms of industry as well as size. Some of the larger and better known firms are :Advanta Corporation, Atlantic Aviation Corporation, First USA bank, Hewlett Packard Company, MBNA Bank, Mellon Bank, Rodel, Inc., Tecot Electrical Supply Corporation, Unifirst Corporation and W.L. Gore and Associates, Inc.

In this study, we focus on the projects that we characterize as quality or productivity improvement projects. Figure I shows the functional area within a firm in which these projects took place. Note that forty-four of the forty-eight projects took place in the area of operations: process analysis, materials, forecasting and quality. Some examples of process analysis problems we solved include scheduling customer service representatives for several large credit card banks, scheduling preventive maintenance for a manufacturing firm, and improving payment processing productivity for a bank. Materials management projects included a service parts inventory system for a transportation company and making warehousing improvements for a manufacturer of high tech products. Forecasting projects included those for a large supermarket chain and for several large credit card banks. Traditional quality projects included service quality improvement for a chain of retirement homes, for a distribution firm, for a telemarketer and for traditional manufacturing firms. Product quality improvement projects were done for a number of manufacturing firms.

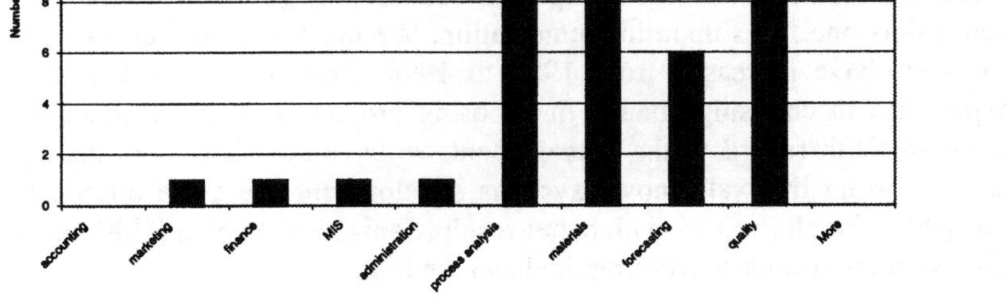

Fig. 1. Frequencies of productivity improvement projects by functional areas

As is obvious above we serve a variety of organizations and we solve problems in a number of different areas. Figure 2 shows that we also solve problems at all levels of management.

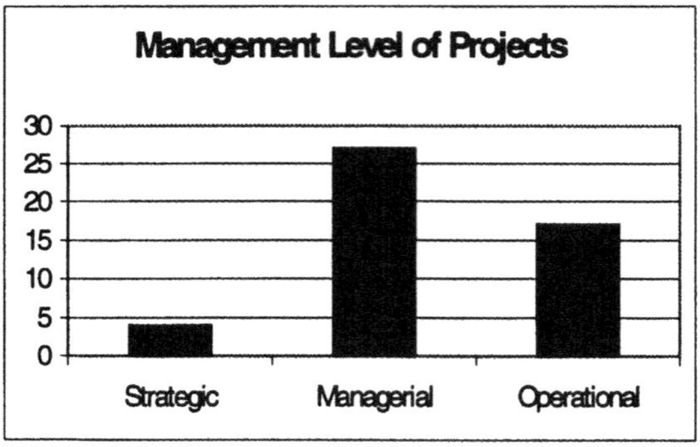

Fig. 2. Frequencies of projects by management level

When the curriculum was designed in 1983 a plan was made to collect and maintain data on project design success, project implementation success and a number of variables that were potentially related to those success variables. We define project design success or project success (PS) as a variable measuring the effectiveness of the solution to the organization's problem. Project implementation success (IS) is a variable measuring the degree to which the solution was successfully implemented. Both variables are measured subjectively using input from the client, the authors and the students. Figure 3 shows frequencies of project design success and Figure 4 shows a histogram of project implementation success. Note that the scores for project success are higher than for implementation success, a result that is not unexpected because design can be done without the constraints one faces upon implementation. We mention also that scores for both variables have increased from 1984 to 1998. This is perhaps because of our experience in choosing clients, in choosing projects and in the management of projects. With regard to the management we have switched from managing the projects using the well known systems development life cycle approach to one using SPC for all projects with better results Saniga and Pohlen (1996) present and discuss some results concerning the latter finding.

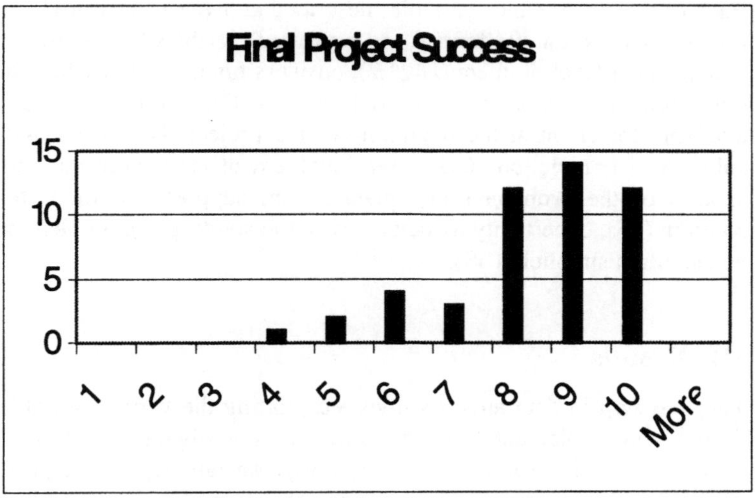

Fig. 3. Frequencies of project design success

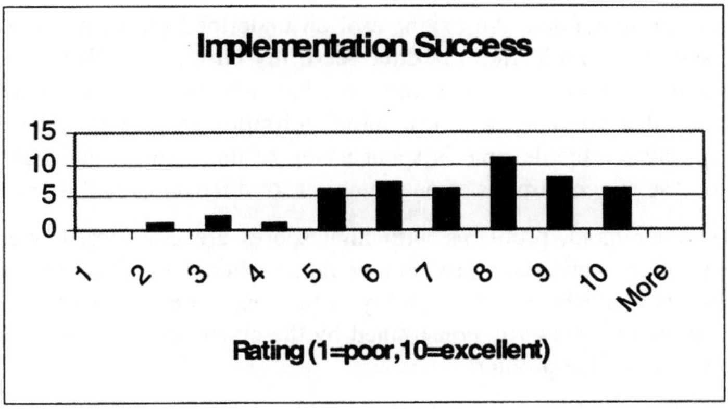

Fig. 4. Frequencies of project implementation success

Each of the performance variables described above was measured at the end of the project. In addition, we defined and measured variables for company size (S1, S2), management level of client (L1, L2), business function of problem, initial problem definition (ID), final problem definition (FD), effort ratio (E), information available for client at the beginning of the project (I), client's understanding of solution at completion (CU), user involvement (UI), user training in technical aspects of the problem (UT), management support (S), final structure of the solution (FS), uncertainty associated with the solution(U), problem difficulty(TD), and solution simplicity(SS).

III Results

The first step in data analysis involved plotting the various variables against the design and implementation variables and investigating the correlation matrix between all of the variables. Subsequently, we ran stepwise regressions of all of the independent variables versus the design and implementation success variables.

Initial problem definition is a variable measuring the degree as to which the problem is defined by the client and ranges from 1 to 10. This is the state of the problem definition at the onset of the project. We consider problems with low scores to be undefined. An example of an undefined problem would be where a client states: "I would like to better serve my customers" Here, the problem as defined is very broad with many possible sub problems to focus upon. We conjecture that problems with low initial definition are generally more difficult to solve because valuable time is spent trying to determine what problem to solve and we have the constraints of one semester, or three and a half months.

On the other hand, problems with high scores are considered overdefined. An example of an overdefined problem would be where a client wants the consultants to "buy a new machine so that quality of the final product would be better". Here, a solution is unnecessarily constrained by the client because the client is defining the solution, not the problem.

We hypothesize that initial definition can have an impact on project design and implementation success because of the issues raised above. Figure 5 shows a scatterplot of initial definition versus implementation success. Note that the form of the relationship we hypothesize is supported by these results. While we do not show the results here the same relationship appears when design success instead of implementation success is plotted.

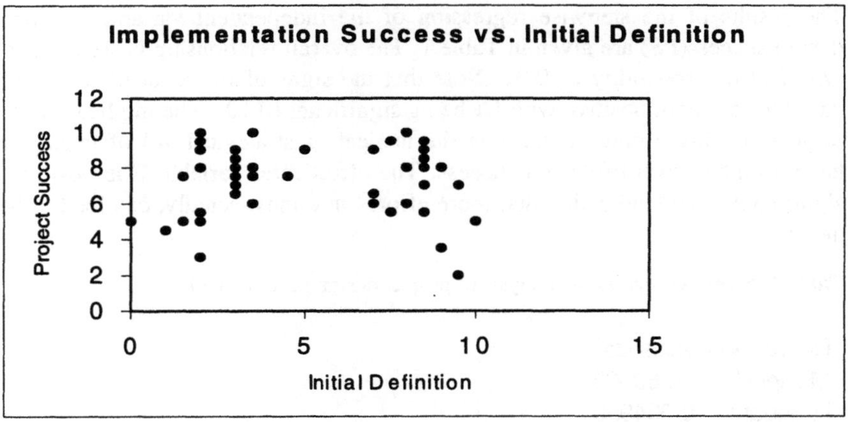

Fig. 5. Scatter diagram of implementation vs. initial definition

The same type of relationship appears for the variable measuring final problem definition, which is usually done at about midpoint in the project. These results appear in Figure 6. We conjecture that if the consultant can persuade the client to define the problem properly by focusing on appropriate sub problems for low definition projects and by removing the constraints of a pre determined solution for high definition projects there is a higher probability of implementation success. We conjecture that final definition scores of 5 are associated with project implementation success and lower or higher scores are associated with lower implementation scores. Thus, we define a new variable called revised final definition, or rfd where rfd= absolute value(5-fd).

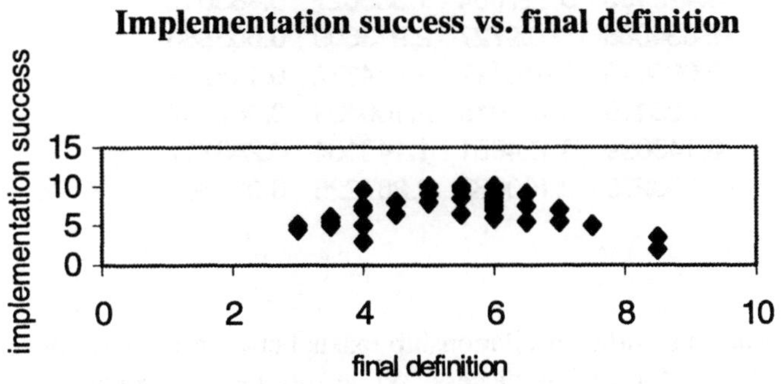

Fig. 6. Scatter diagram of implementation success vs. final definition

The results of the stepwise regression of the independent variables on project design success(PS) are given in Table 1. The overall relationship is significant at a type I error probability of 0.01. Note that the signs of the level of management variables are both positive with L1 being significant (0.10). The implication is that projects having a client contact in the tactical or strategic level of management have a higher level of design success. The effort ratio variable E is positive and significant (0.01) indicating that more effort , not unexpectedly, can lead to better design.

Table 1. Stepwise regression analysis of project design success (PS)

Regression Statistics	
Multiple R	0.83363
R Square	0.694939
Adjusted R	0.632362
Standard E	0.956567
Observatio	48

ANOVA

	df	SS	MS	F	Significance F
Regressior	8	81.29336	10.16167	11.1054	5.22E-08
Residual	39	35.68581	0.915021		
Total	47	116.9792			

	Coefficient:	Standard E	t Stat	P-value
Intercept	-0.505464	1.636618	-0.308847	0.759082
Level(L1)	0.563166	0.312758	1.800645	0.079496
Level(L2)	0.251744	0.543278	0.463379	0.645668
FD	0.172293	0.127094	1.355627	0.183017
E	2.834066	0.9727	2.913609	0.005888
I	-0.068016	0.046117	-1.474872	0.148275
UI	0.25419	0.121018	2.100439	0.042207
S	0.143099	0.124851	1.146164	0.258711
FS	0.230282	0.100682	2.287225	0.027691

A negative but insignificant relationship exists between information available for the client (I) and for design success. While this may be counterintuitive it also may support the well known phenomena of information overload identified by Ackoff (1967).

User involvement(UI) is significantly positively related (0.05) to design success. Arguments for UI and its impact are usually made for implementation results rather than for design results as noted in the literature cited in the first section. That it is significant here perhaps indicates that its effect is context dependent. More precisely, there are stages of solution in quality and productivity improvement problems that require user input such as the determination of rational subgroups, accurate data collection, identification of variables associated with special causes, etc. This user input in the general problem of DSS design may not be required as several of the authors cited in the first section have found.

The final structure of the solution (FS) is positive and significant (0.05). This indicates that the more structured the solution is in terms of its repeatability by the same or by different users the better the design. Of course, the final stage of most successful quality and productivity improvement projects involves providing the users with control charts to monitor a process, a tool that is completely structured. The results of the stepwise regression of the independent variables on implementation success (IS) are shown in Table 2. Note that the overall F is significant (0.01).

Table 2. Stepwise regression analysis of implementation success (IS)

SUMMARY OUTPUT

Regression Statistics		
Multiple R	0.933713	implementation
R Square	0.87182	
Adjusted R	0.845527	
Standard E	0.787975	
Observatio	48	

ANOVA

	df	SS	MS	F	Significance F
Regressior	8	164.7014	20.58768	33.15757	4.48E-15
Residual	39	24.21526	0.620904		
Total	47	188.9167			

	Coefficient	Standard E	t Stat	P-value
Intercept	-0.884915	1.459137	-0.606465	0.54772
rfd	-0.746803	0.250808	-2.977586	0.004974
Level(L1)	1.180536	0.25588	4.61364	4.20E-05
Level(L2)	0.011423	0.481725	0.023713	0.981203
FD	0.244937	0.136796	1.79052	0.081135
E	1.605524	0.806615	1.990445	0.053583
UI	0.237524	0.10115	2.348238	0.024027
S	0.364956	0.114302	3.192911	0.002784
TD	0.102579	0.07647	1.341419	0.187544

Implementation success, like design success, depends upon the client's level of management, with higher levels of management being associated with more successful implementations. This is not unexpected because higher levels of managers are more empowered allow necessary changes to be made.

The variables measuring final definition, FD and RFD are both significant (0.10) and (0.01) respectively. The conclusion that may be drawn by this is that an overdefined project may be better than an underdefined project in terms of implementation success. But the relationship between RFD and IS is stronger and supports the conjecture we make in the last section that proper definition is an important variable affecting implementation success. More precisely, a consultant should not underestimate the importance of properly defining the problem or project early in the course of the project because proper definition seems to be a necessary first step in progress success..

As with design success, implementation success depends upon effort ratio and user involvement, both being significant, (0.10) and (0.05) respectively.

The variable measuring management support is significant (0.01) and positive, a result that supports much of the previous literature on DSS design and implementation and on the performance of quality and productivity improvement projects.

It is interesting to note that a number of other variables we have investigated were omitted from the final stepwise regressions. A correlation analysis indicates a high degree of correlation between many of these variables such as solution simplicity, user training, uncertainty of solution, etc.

While we do not find these variables to be significant the complexity of the problem of design and implementation is such that we cannot state that they are of no importance in particular problems, only that they may not be important in general. In the next section we discuss this in more detail.

IV. Strategic Aspects of Results

Our results show that SPC may be applied in many areas of the firm apart from the production line. But successful implementation requires far more than a standardized approach such as those proposed in textbooks. In fact there are two other things to be considered in addition to the standard technical knowledge required for successful implementation of SPC throughout the whole firm. These two things are, first, the requirement that the solution designer recognize the fact that the methodology of decision support is an inherent part of SPC. The second is that while SPC provides a structured paradigm for project management, each particular application or solution requires specialized "expert" solutions instead of standardized "textbook" solutions for design and implementation to be successful. We address these in turn.

Knowledge of the methodology of decision support is an integral part of the successful application of SPC. Our findings quantitatively address many issues from the field of decision support such as user involvement, top management support, reduction of information overload, etc. It may be more important, though, to realize that it is not a consideration of just the factors that been found to be of statistical significance that is the most important finding. It is that successful implementation requires a consideration of all of the factors that researchers and practitioners have found to be of importance. Some studies that are of value in identifying these factors are Ackoff(1967), Alavi and Joachimsthaler(1992) and Guimaraes, Igbaria and Lu(1992).

A discussion of a number of things to consider when implementing SPC has been presented bv Saniga(1992). He argues that implementers must be aware of several important issues when solving quality or productivity problems. The first of these is that the implementer must develop an awareness of the amount of structure that can be designed into the solution, a decision that is usually but not always related to the level in the organization in which the decision exists, with top level decisions usually but not always being less structured than those at the operational level of the firm.

The second is that feedback must always be designed into any solution. Of course, one can argue that the use of a control chart in any standard production line application is a feedback device but we are stating here that the feedback must be made formal and of a degree of substance so that it could be considered a critical success factor for the firm. An example may illustrate this:

One of the problems in the study involved customer service improvement for one of the three largest credit card banks in the United States. While it is relatively straightforward to eliminate the special causes and monitor for common causes in the customer response given by service representatives it is much more important to ensure that this firm's response quality was the leader in the industry, because marketing research studies showed that the firm's competitive advantage came from their service quality. Thus standard control chart feedback, while excellent in terms of monitoring for special causes was insufficient in terms of providing feedback of variables critical to the success of the firm. These included providing measures that would allow the benchmarking of competitors, the estimation of the gap between expectations and actual service, and the estimation of the overall improvement in service over time.

A third consideration of the implementer is the determination of the role they must play in the solution process. Structured problems require a proactive role where the implementer leads in the solution process; unstructured problems are such that the designer assumes the role of evaluator of a solution.

The second thing we mention is that every problem we have been involved in required special "expert" solutions rather than standardized solutions to achieve the best results. There are at least two reasons for this.

The first is that our results show the possibility of extending SPC to almost any area of an organization. While SPC is an excellent paradigm to guide process improvement there is no need to argue that specialized skills are required when applying SPC in diverse functional areas such as marketing, finance, accounting, etc. While the steps of definition, measurement, improvement and monitoring are the same, the way in which these are done is different than that discussed in a standard textbook in quality, say. For example, the problem of measuring customer service quality is a substantive problem addressed by marketing researchers and requires specialized knowledge.

The second is that, in our considerable experience, every firm restricts possible solutions by the existence of their own organizational constraints. Thus the problem is not simply a matter of defining and measuring performance and subsequently improving and monitoring performance that needs to be done. It is the matter of convincing people that have not been introduced to productivity and quality improvement in areas such as marketing, finance, accounting, etc. that SPC is a strategic intervention and that it is required for competitive advantage. People like Deming have paved the road in this area and the popularization and demonstrated effectiveness of interventions like six sigma in the U.S. should help in this area. In many cases though it is the job of the implementer to come up with solutions that do improve performance but are not optimal because the organizational constraints cannot be eliminated and in some cases, even reduced.

V. Conclusions

The *management* of quality and productivity improvement projects strongly impacts design and implementation results. Those involved in this process should be aware of a number of factors we identify that are associated with project design and implementation success. They should also be aware that successful project management requires a thorough knowledge of the principles of SPC as well as the ability to seek out specialized or expert solutions to particular problems should the goal be application of SPC throughout the firm, an endeavor that will most certainly add to a firm's competitive advantage.

References

1. Ackoff, R. (1967) "Management Misinformation Systems", Management Science **14 (2)**

2. Alavi , M. and Joachimsthaler, E.A. (March 1992) "Revisiting DSS Implementation Research:A Meta-Analysis of the Literature and Suggestions for Rresearchers", MIS Quarterly

3. Guimaraes T. and Igbaria, M. and Lu, M: (1992) "The Determinants of DSS success: an integrated model", Decision Sciences **23 (2)**, 409-430

4. Igbaria, M. , and Guimaraes, T. (1996) "Exploring the Relationship Between EUC Problems and Successes."Information Resources Management Journal **9 (2)** 5-15

5. Igbaria, M. and Guimares, T. (1994) "Empirically testing the Outcomes of User Involvement in DSS Development." Omega **22 (2)** 157-172

6. Pinto, Jeffery K. and Slevin, Dennis P. (February 1987) "Critical Factors in Successful Project Implementation," IEEE Transactions of Engineering Management, **34 (1)** 22-27

7. Piinto, Jeffrey K. and Mantel, Samuel J. Jr. (November 1990) "The Causes of Project Failure," IEEE Transactions on Engineering Management., **37 (4)**, 269-275

8. Pinto, Jeffrey K. and Slevin, Dennis P. (June 19888 "Critical Success Factors Across the Project Life Cycle", Project Management Journal, **19 (4),** 67-75

9. Sanders, G. and Courtney, J. (1985) "A Field Study of Organizational Factors Influencing DSS Success" MIS Quarterly **9**, 77-93

10. Saniga, E. (1993) "Decision Support and Statistical Quality Control", International Journal of Quality and Reliability Management", **10 (2)**, 9-17

11. Saniga, E. and M. Pohlen (1996) "Performance Measurement, Improvement and Monitoring and DSS Success: Some Preliminary Empirical Results" NEDSI Conference Proceedings, 248-249.

12. Udo, G.J. and Davis, J.S. (1992) "Factors affecting decision support system benefits" Information & Management **23**

Part 3

Data Analysis and Process Capability Studies

Application of Statistical Causal Analysis to Process Analysis

Ken Nishina

Nagoya Institute of Technology, Gokiso-cho, Showa-ku, Nagoya 466-8555, Japan

Abstract. In this paper, statistical causal analysis is applied to the model building of manufacturing process mechanism in process analysis. Originally, the cause-and-effect diagrams, the relation diagrams and so on have been widely used as QC tools to clarify process mechanism. Although they can represent process mechanism as a causal model, the conventional process analysis, in which the multiple regression model has been mainly used, has lacked the viewpoint of causality. The aim of this paper is to identify some of the problems of the conventional process analysis and to illustrate that the statistical causal analysis of using the path analysis and the graphical modelling is very useful in process analysis. In this paper two real examples in which causal analysis is successfully applied to process analysis are presented.

1 Introduction

At first, a fictional but very instructive example is presented. An engineer of a chemical plant tried to make a prediction of the product yield per day (y) using the regression analysis. He collected the data, then discovered a very curious result in which the positive correlation between the yield and the sale amount of cola per day in the cafeterias of this plant (x_1) was strong. Therefore, he selected the sale amount of cola as an independent variable, and constructed an equation as follows:

$$y = \beta_0 + \beta_1 x_1 + e . \qquad (1)$$

Is his approach right? From the viewpoint of the statistical prediction, it is right. But he did not accept equation (1). Why? He was suspicious because there was not the causal relationship between the yield and the sale amount of cola. Is equation (1) available in another plant? Is equation (1) available next year? Then he tried to investigate why this correlation was strong. He explored a mechanism behind this correlation. The mechanism could be that there was a common cause behind the two variables. He found out that the common cause was temperature (x_2).

Then he reconstructed the following regression model:

$$y = \beta_0 + \beta_2 x_2 + e \,. \tag{2}$$

Equation (2) is more scientific than equation (1); therefore, the more universal prediction could be expected.

Next, he tackled process analysis for improvement of the yield. He was not satisfied with equation (2) because temperature was an uncontrollable factor. In this process some process parameters were adjusted according to the variation of temperature; moreover, some other factors were associated with temperature. Therefore, he tried to analyze the process mechanism related with temperature for process improvement.

As shown in the above example, the analysis of process mechanism is important in order to create more universal (not temporary) actions for process improvement. However, sometimes process mechanism is not clarified so distinctly, that is to say, it is a black box. In such a case, process mechanism can usually be considered as a set of some sequences of the causal relationship, which can be represented by the relation diagrams or the cause-and-effect diagrams. In the data analysis, conventionally and widely the multiple regression analysis has been used.

The aim of this paper is to indicate some problems of the conventional process analysis and to illustrate that the statistical causal analysis is useful for process analysis. In this paper two simple but instructive and real examples on process analysis using the statistical causal analysis are presented. In these examples the path analysis [see Wright (1934)] and the graphical modelling [see Whittaker (1990)] are taken as statistical methods for causal analysis.

2 Conventional process analysis and its problems

Process analysis is that the relationship between the results and the causes of a process is analyzed. Conventionally, the multiple regression analysis has been widely used. The main reasons for this are as follows:

(a) the multiple regression analysis is so popular that its software with many analytical functions is easily available and applicable,

(b) the retrospective data are available in the analysis; the data are not necessarily taken another time,

(c) the stepwise regression, the residual analysis and so on are applicable to the purposes of process analysis.

The purposes of process analysis depend on the quality of the prior knowledge.

They are as follows:

- confirmation of hypothesis
- identification of unknown causes
- selection of mainly active causes.

Regarding item (c) above, the residual analysis is helpful for the identification of unknown causes, and the stepwise regression is for the selection of mainly active causes.

As described above, the multiple regression analysis has been contributed to process analysis. However, the multiple regression models are not necessarily applicable to modelling the causal relationship because the relationships between the independent variables in the multiple regression models are not causality but correlation.

The aim of process analysis is not only to decrease the variation of process output but also to increase technical knowledge about the process. To realize this, "the formulation of the causal structure" should be added to the three purposes of process analysis above. In other words, a key point of process analysis is not only to identify the causes and how much they affect the variation of result but also what mechanism in process generates the variation of result. In tackling process analysis, we clarify process mechanism by using some traditional QC tools such as the cause-and–effect diagrams. It is just a causal model, which is considered to be a prior knowledge. However, in going ahead to the step of the statistical data analysis, we have mainly used the multiple regression models. This statistical data analysis lacks the viewpoint of causal analysis [see Tsubaki (1994)]. Box (1966) indicated the abuse of the regression analysis, also.

3 Statistical methods for causal analysis in process analysis

In this section, the path analysis and the graphical modelling are briefly explained.

The diagram in which the causal relationships between variables are represented by the directed graph (an arrow from a cause to a result) is called the path diagram. In the path analysis the causal relationships are formulated as the linear structural equations. Each equation of these structural equations is a multiple regression model, and generally its regression coefficients correspond to the path coefficients. However, it should be noted that there is some sort of distinction between the multiple regression models and the linear structural equations. In the multiple regression models all of the variables except a dependent variable are the exogenous variables. As described earlier, this means that the relationships between the independent variables are not causality but correlation. On the other hand, the linear structural equations represent the causality between the independent variables, which are the exogenous variables in the multiple regression models.

The graphical modelling is a form of multivariate analysis that represents the association structure (especially, the conditional independence) by using graphs. In the graph, nodes correspond to the variables and arcs to the association between two variables. When the association means a causal relationship, the arc is exchanged to an arrow which is drawn from a cause variable to a result variable.

In the graphical modelling the covariance selection [see Wermuth et al. (1977)] is carried out in a step-wise in order to reduce the model. This model reduction corresponds to deleting arcs. The covariance selection means that an element of the partial correlation matrix is set to the zero constraint just like the backward selection of the stepwise regression.

Now it should be noted that the path analysis is considered as a confirmatory method to test the hypothesis of causal model, which is portrayed in the path diagram. On the other hand, the graphical modelling is a type of exploratory method. This indicates a guideline on how to make proper use of the two methods.

The statistical causal analysis is based on the correlation matrix; however, the causal model can not be constructed without prior knowledge concerning the causal relationship because the correlation does not necessarily mean the causality. But the causality has not necessarily been clarified so definitely. Therefore, in the causal analysis our analytical viewpoint needs to go back and forth between the standpoint of hypothesis in the core engineering and the data analysis. Then this approach requires two kinds of approaches, the confirmatory approach and the exploratory one.

The path diagram and the graphical model are just the relation diagrams and can correspond to the cause-and-effect diagrams. Therefore, the causal analysis can be an applicable method for modelling the prior knowledge clarified by the widely used QC tools.

We recommend an outline of the gross algorithm of process analysis used for causal analysis as follows:

Step1: build a hypothesis from engineering studies using QC tools.

Step2: confirm this hypothesis using the path analysis.

Step3: explore another hypotheses using the graphical modelling if necessary.

Step4: build a new hypothesis from the interpretation of the results based on engineering studies.

Step5: repeat the algorithm from Step2 to Step4 taking additional data if necessary.

When a definete hypothesis can not be built, the first step may be to explore some hypotheses, which corresponds with Step3.

4 Real examples of causal analysis in process analysis

In this section two real examples, which the causal analysis can be successfully applied to, are presented. The first example is process analysis of a coating process of automobile bodies [Ito (1986)] using the path analysis. The second is a pulley manufacturing process [Nishina et al. (1997)] using the graphical modelling and the path analysis. These examples are real situations; however, in this paper only essential parts are explained. The author reanalyzes them.

4.1 Example 1: Taking an action to causal relationship between two variables (Coating process of automobile bodies)

There are many characteristics in the coating process of automobile bodies. In this paper only the transfer efficiency of coating (y) is considered as a final response variable and its improvement is undertaken. The process parameters are the distance of the spray gun (x_1), the diameter of the nozzle (x_2) and the booth temperature (x_3). The correlation matrix is as follows:

$$
\begin{array}{c} \\ R = \end{array}
\begin{array}{c}
\begin{array}{cccc} x_1 & x_2 & x_3 & y \end{array} \\
\begin{bmatrix}
1.00 & & & \\
-0.28 & 1.00 & & \\
0.52 & -0.13 & 1.00 & \\
-0.63 & -0.31 & -0.40 & 1.00
\end{bmatrix}
\end{array}
$$

At first, the result based on the multiple regression model is shown. The following regression equation is obtained by the stepwise regression,

$$
\begin{array}{l}
y = -0.78x_1 - 0.53x_2 + e_y \\
\quad (-7.43) \quad (-5.02)
\end{array}
\tag{3}
$$

where values in the parentheses are the t-value. Fig.1 shows the path diagram, which represents equation (3). The variable x_3 (temperature) is excluded by the stepwise regression. In Fig.1 the single-headed arrows and the double-headed arrows stand for the causality and the correlation, respectively. The coefficient of determination is not so high (0.655) because other parameters, especially ones of the conditions of coating material, are omitted in this analysis. From this result it is noted that the actions to be taken are the following: to decrease the variation of the distance of the spray gun and/or the diameter of the nozzle, or to make the robust design by controlling a variable which has an interaction with the distance of the spray gun and/or the diameter of the nozzle. However, in this analysis the causal relationship between the three variables is not considered. From the viewpoint of the coating engineering, the distance of the spray gun has been adjusted corresponding to the variation of temperature and sometimes the diameter of the nozzle has been adjusted to the distance of the spray gun also in order to maintain the transfer efficiency. Based on these engineering standpoints a path diagram can be portrayed in Fig.2 and formulated as the following linear structural equations. The path analysis is then carried out hypothesizing this path diagram.

$$x_1 = \beta_{13}x_3 + e_1$$

$$x_2 = \beta_{21}x_1 + e_2$$

$$y = \beta_{y1}x_1 + \beta_{y2}x_2 + e_y .$$

Fig.2 shows the result of the path analysis (by the maximum likelihood method). The index of fit is that χ^2=0.688, ϕ (degree of freedom)=2, P-value=0.709. In Fig.2 it should be noted that the path coefficient from x_1 to x_2 is negative (-0.28)□and both of the path coefficients from x_1 and x_2 to y are negative. This suggests that if the negative path coefficient from x_1 to x_2 is larger, it can reduce the variation of the transfer efficiency y. In other words, an alternative action is to standardize an adjustment procedure between the distance of the spray gun and the diameter of the nozzle in taking the conditions of courting materials into consideration.

This example illustrates that the causal relationship between the endogenous variables indicates an action for process improvement.

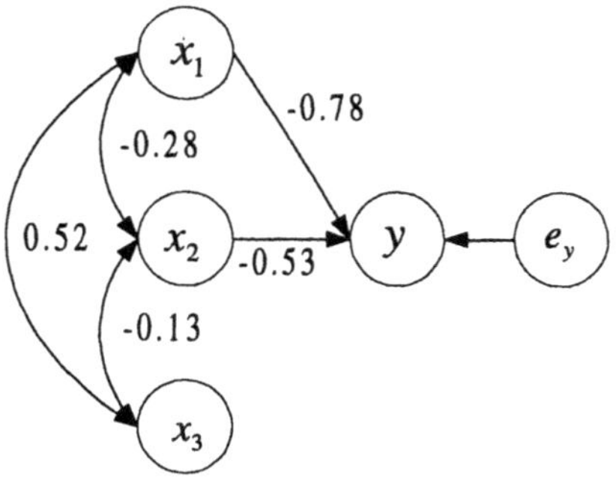

(Coefficient of determination=0.655)

Fig. 1. Path diagram of the multiple regression model (equation (3))

(Note 1: the single-headed arrows stand for causality and the double-headed arrows stand for correlation)

(Note 2: the numerical values on the single-headed arrows stand for standardized regression coeffients, and ones on the double-headed arrows stand for correlation coefficients)

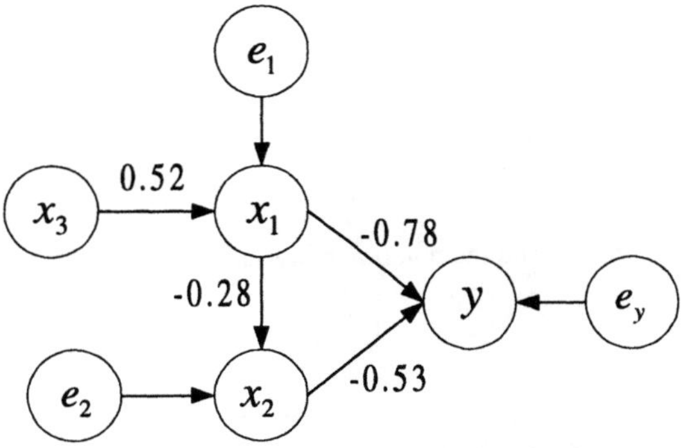

(χ^2=0.688, ϕ =2, P-value=0.709)

Fig. 2. Path diagram of the path analysis

4.2 Example 2: Increasing knowledge concerning process mechanism (Pulley manufacturing process)

In Example 2, it is researched to find the key element process in the processes and its mechanism for a final response. The final response is the ill-balance of the automotive generator's parts (hereafter, it is called pulley). This manufacturing process is composed of some element processes, which are connected in a series. It is shown in Fig.3. The symbols Δ shown in Fig.3 stand for the locations of taking data. However, some variables after the flanging process are omitted because of the convenience for essential explanation.

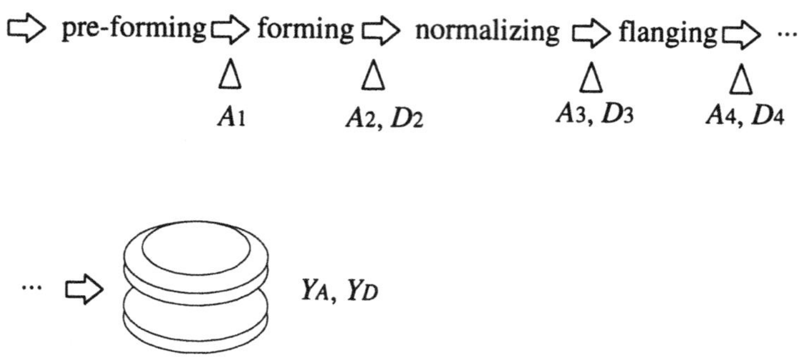

Fig. 3. Pulley manufacturing process and the notation of variables

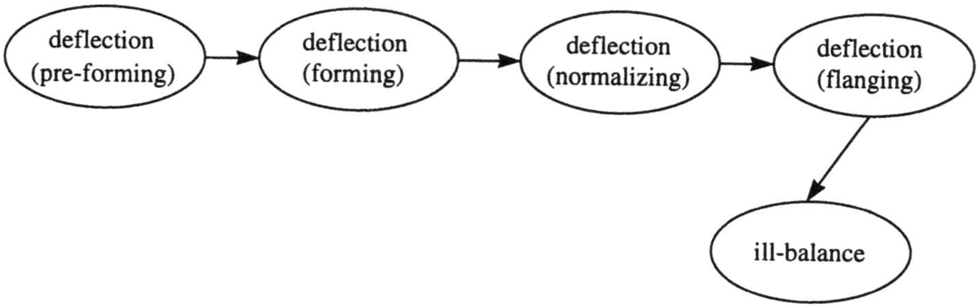

Fig. 4. Hypothesis H1

At first, it is considered to hypothesize that the outer diameter deflection, which is generated in the pre-forming process and/or the forming process, causes the ill-balance via the normalizing and the flanging processes (hereafter, this hypothesis is denoted by H1; see Fig.4). Therefore, the amount of outer diameter deflection and its direction are observed at each of the element processes (see Fig.3). The amount of outer diameter deflection based on the inner diameter is defined as its maximum amount. It is denoted by A. Its direction (denoted by D) is measured as an angle based on the direction of the maximum amount of outer diameter after pre-forming process. In Fig.3 the notations of these variables are shown also. Hereafter, the direction data and the amount data are analyzed separately. Their correlation matrices are as follows:

Amount:

$$R_A = \begin{array}{c} \begin{array}{ccccc} A_1 & A_2 & A_3 & A_4 & Y_A \end{array} \\ \begin{bmatrix} 1.00 & & & & \\ 0.32 & 1.00 & & & \\ 0.24 & 0.68 & 1.00 & & \\ 0.33 & 0.74 & 0.82 & 1.00 & \\ 0.29 & 0.73 & 0.62 & 0.61 & 1.00 \end{bmatrix} \end{array}$$

Direction:

$$R_D = \begin{array}{c} \begin{array}{cccc} D_2 & D_3 & D_4 & Y_D \end{array} \\ \begin{bmatrix} 1.00 & & & \\ 0.62 & 1.00 & & \\ 0.60 & 0.87 & 1.00 & \\ 0.84 & 0.61 & 0.53 & 1.00 \end{bmatrix} \end{array}.$$

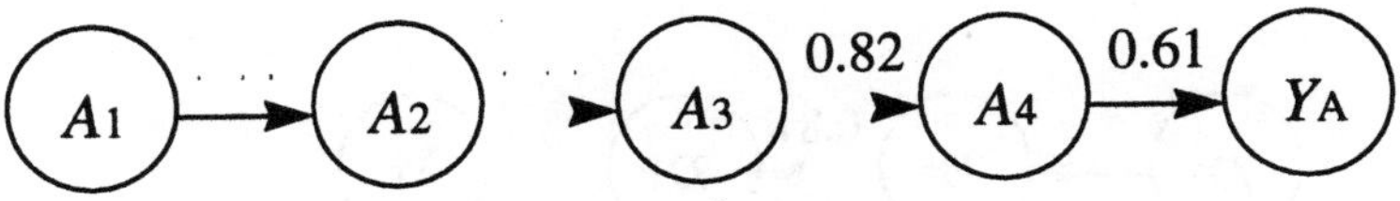

(χ^2=20.75, ϕ=6, P-value=0.002)

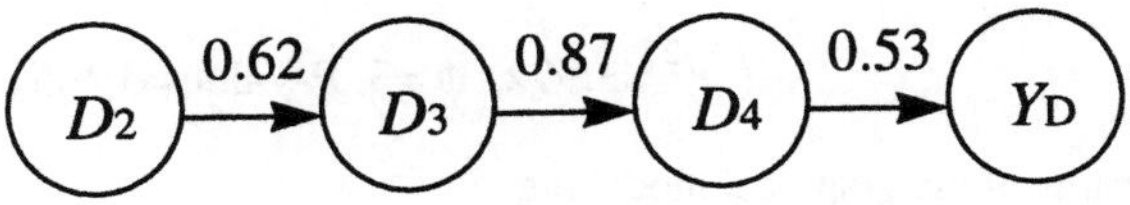

(χ^2=37.44, ϕ=3, P-value=0.000)

Fig. 5. Results of the path analysis based on the hypothesis H1

(Note: the error terms are omitted in Fig.5)

The path analysis (by the maximum likelihood method) is carried out based on the hypothesis H1. Their results are shown in Fig.5. Consequently the hypothesis H1 is rejected from the statistical viewpoint judging from the P-value (A: χ^2=20.75, ϕ=6, P-value=0.002; D: χ^2=37.44, ϕ=3, P-value=0.000). Therefore, another model should be explored.

It is undertaken to explore another model by using the graphical modelling (in this case the graphical modelling is equivalent to a sequence of the multiple regression analysis). The results are shown in Fig.6 (A: χ^2=2.664, ϕ=5, P-value=0.751; D: χ^2=3.972, ϕ=3, P-value=0.265). The numerical values in Fig.6 stand for partial correlation coefficients, which are derived after the covariance selection.

Some different hypotheses from the hypothesis H1 can be indicated from the results in Fig.6. They are interpreted from the engineering viewpoint as follows:

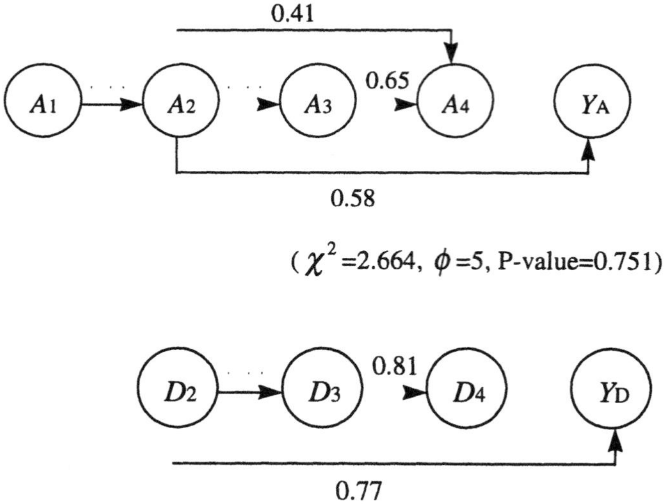

(χ^2=2.664, ϕ=5, P-value=0.751)

(χ^2=3.972, ϕ=3, P-value=0.265)

Fig. 6. Results of the graphical modelling

(Note: the numerical values stand for partial correlation coefficients)

At first it is noted that the deflection which comes from the pre-forming process does not cause the ill-balance so much as the deflection which comes from the forming does. Therefore, Fig.6 indicates that the key process is the forming process. Then the following hypothesis may be considered: the deflection, which comes from the forming process, causes the ill-balance via the normalizing and the flanging processes (the hypothesis H2; see Fig.7).

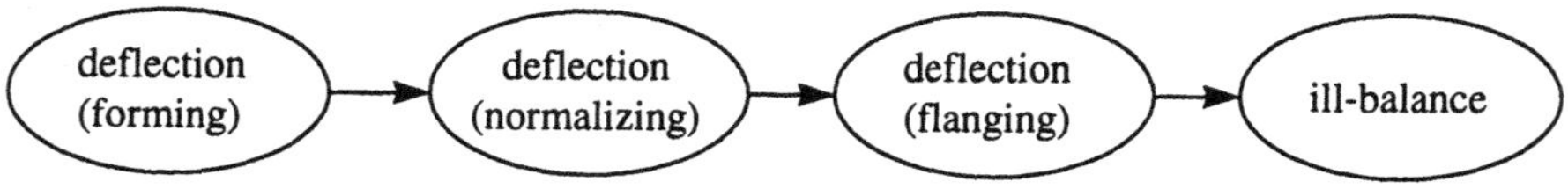

Fig. 7. Hypothesis H2

However, the results in Fig.6 do not support the hypothesis H2 because there is no arrow from A_4 and D_4 to Y_A and Y_D, respectively; but an arrow from A_2 and

D_2 to Y_A and Y_D, respectively. Truly, Fig.6 may indicate the following hypothesis (the hypothesis H3): what causes the ill-balance is not the deflection but a common cause between the deflection and the ill-balance. Provisionally this common cause is called the characteristic W (see Fig.8).

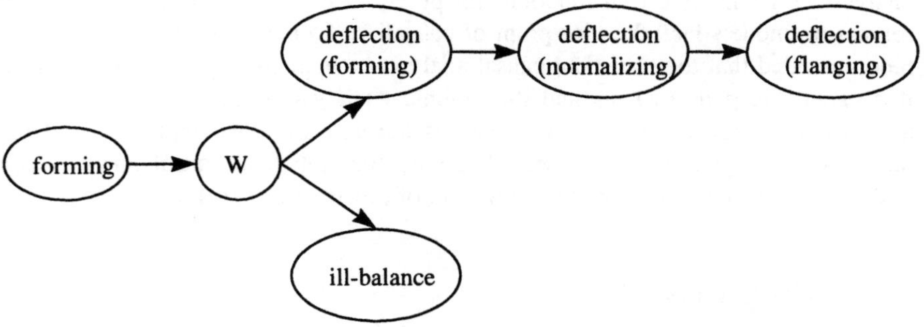

Fig. 8. Hypothesis H3

Based on engineering studies, the hypothesis H3 in Fig.8 can be explained as follows: the characteristic W can be regarded as the unevenness of wall thickness. The forming process generates the unevenness of wall thickness, and the unevenness causes the deflection, then the unevenness can shift upward or downward by the normalizing. If this mechanism is true, it is found out that the problem is the measurement location of the deflection. The measurement location is the vertical center of the pulley. Therefore, when the unevenness of wall thickness shifts upward or downward, the direction of the unevenness is still held but both the direction and the amount of the deflection at the vertical center changes. Consequently the unevenness of wall thickness, of which the direction has been held, causes the ill-balance.

In conclusion the hypothesis H3 is not rejected from the statistical standpoint and can be interpreted in the core engineering. The hypothesis H3 is new knowledge about the pulley manufacturing process. This example illustrates that our analytical viewpoint goes back and forth between the standpoint of the hypothesis in the core engineering and the data analysis by using the statistical causal analysis; therefore the causal analysis can increase knowledge concerning the process.

5. Concluding remarks

Originally, the statistical causal analysis has been developed in the social sciences. However, it is expected that the statistical causal analysis can be applied to the engineering field also. The conventional QC offers some tools for clarifying the causality in process mechanism. For example, the cause-and-effect diagrams and the relation diagrams; therefore, the statistical causal analysis should be applied to the SQC approaches combining the conventional QC tools.

Conventionally, the multiple regression analysis has been widely used in process analysis. It has many useful functions for process analysis. However, the multiple regression models lack the viewpoint of causality. To make up for this lack, it is recommended that the statistical causal analysis be applied to process analysis. In this paper, the path analysis and the graphical analysis were applied to process analysis, in which the former was used as the comfirmatory approach, and the latter as the exploratory approach. Applying the statistical causal analysis can realize the improvement from the viewpoint of process mechanism.

Acknowledgments

The author would like to thank Mr. Norio Irikura of DENSO Co. and Mr. Tetsuji Ito of Toyota Auto Body Co. for providing useful information, and the Technometrics research group of Japanese Society for Quality Control for discussing the causal analysis.

References

1. Box, G.E.P (1966) "Use and abuse of regression". Technometrics, **8 (4)**, 625-629
2. Ito,T. (1986) Multivariate data analysis in Industry, Part 2, Chapter 1, JUSE (in Japanese)
3. Nishina, K., Fujiwara,H. and Irikura,N. (1997) "Process analysis of parts manufacturing using graphical modelling". Journal of the Japanese Society for Quality Control **27**, 534-543 (in Japanese)
4. Tsubaki,H.(1994) "From regression analysis to causal analysis", Standardization and Quality Control, **47 (5)**, 111-116 (in Japanese)
5. Wermuth,N. and Scheidt,E. (1977) "Fitting a covariance selection and path analysis". J. Amer. Statist. Assoc. **75**, 88-92
6. Whittaker, J. (1990) Graphical Models in Applied Multivariate Statistics. John Wiley, Chichester
7. Wright,S. (1934) "The method of path coefficients". Ann. Math. Statist. **5**, 161-215

Detecting Changes in the Mean from Censored Lifetime Data

Stefan Steiner[1] and R. Jock MacKay[2]

[1] Dept. of Statistics and Actuarial Science, University of Waterloo, Canada
[2] Dept. of Statistics and Actuarial Science, University of Waterloo, Canada

Abstract. In many industrial and medical applications observations are censored either due to inherent limitations or cost/time considerations. For example, with many products their lifetimes are sufficiently long that it is infeasible to test all products until failure even using accelerated testing. As a result, often a limited stress test is performed and only a proportion of the true failure times are observed. In such situations, it may be desirable to monitor the process quality using repeated lifetesting on samples of the process output. However, with highly censored observations a direct application of traditional monitoring procedures is not appropriate. In this article, Shewhart type control charts based on the conditional expected value weight are developed for monitoring processes where the censoring occurs at a fixed level. An example is provided to illustrate the application of this methodology.

1 Introduction

In industrial applications censored observations are collected for process monitoring purposes. As an example, consider the manufacture of metal electrical boxes used to cover transformers in residential neighbourhoods. These boxes are outside and are thus subjected to ravages of nature, and it is undesirable and dangerous if they rust. During the production process the boxes are painted to inhibit rust formation. It is desired to monitor the painting process to ensure that the rust resistant capabilities of the paint process does not deteriorate. Of course, even under poor regular conditions the boxes do not rust quickly and it would typically take years at a minimum before rust is visible. To speed up the rusting process an accelerated salt fog endurance test is performed. From each test box a panel is cut, scratched in a prescribed manner, and then put in a 30° Celsius salt spray chamber. Test units in the chamber are checked daily for rust, and stay in the chamber until either rust appears or a maximum of 20 days. The maximum is necessary to limit the cost of testing, and to allow more rapid feedback. In this way, censored observations of the time to rust under the accelerated test conditions are obtained. We may monitor the paint process by analyzing the observed censored data. Since our goal to detect deterioration in process, the accelerated nature of the test may be ignored so long as the testing is performed in the same manner each time, assuming the relationship between accelerated and normal conditions does not change.

Similar situations that result in censored life time data occur in medical applications. In addition, analogous situations may occur with failure strength data. For simplicity, in this article we refer to the variable of interest as a lifetime although it may just as well be a strength. We assume that the distribution of the failure times can be modeled with a Weibull distribution. We consider the Weibull distribution because it is very flexible and a popular choice in life testing (Nelson, 1986).

Monitoring the process quality in situations where there is a substantial amount of censoring can be challenging. However, if the censoring proportion is not large, say less than 25%, monitoring the process quality using a direct application of an $\overline{X}$ control chart with probability limits on the observed lifetime is reasonable. This is the approach followed by Schneider et al. (1995). Also, when the censoring proportion is very high, say greater than 95%, it is feasible to use a traditional np chart where only the number of censored observations is recorded (Montgomery, 1991). In this article, we explore a number of possible monitoring procedures that are superior to the either of these methods in detecting mean shifts when the censoring proportion lies between 25-95%. These proposed methods take into account both the censoring and the fact that the underlying process output distribution is Weibull. The proposed charts include two likelihood based approached and three approaches based on the conditional expected value.

To set notation, let W be a random variable whose distribution represents the failure times. Assuming a Weibull distribution, the probability density and survivor function of W, $f(w;\alpha,\beta)$ and $S(w;\alpha,\beta)$ respectively, are given by (1), where α and β are the scale and shape parameters of the Weibull respectively.

$$
\begin{aligned}
f(w;\alpha,\beta) &= \frac{\beta}{\alpha^\beta} w^{\beta-1} \exp\left(-\left[\frac{w}{\alpha}\right]^\beta\right) \\
S(w;\alpha,\beta) &= \exp\left(-\left[\frac{w}{\alpha}\right]^\beta\right)
\end{aligned}
\tag{1}
$$

With this parameterization of the Weibull, the mean and variance of W, μ and σ^2 , are given as

$$
\begin{aligned}
\mu &= \alpha\Gamma\left(1+\frac{1}{\beta}\right) \\
\sigma^2 &= \alpha^2\left\{\Gamma\left(1+\frac{2}{\beta}\right) - \Gamma^2\left(1+\frac{1}{\beta}\right)\right\}
\end{aligned}
$$

where $\Gamma(x)$ is the Gamma function (Abramowitz and Stegun, 1972). The standard form of the Weibull is given by $W^* = W/\alpha$ and has probability density function $f(w; 1, \beta)$. Note that the standardized Weibull still depends on β. The Weibull distribution is commonly used to model lifetimes because it is very flexible. The shape parameter, β, determines whether the failure rate increases with time ($\beta > 1$) or decreases with time ($\beta < 1$). When $\beta = 1$ the distribution of W is exponential.

In this article, we assume the censoring is in the right tail as is typical for lifetime data. However, similar monitoring procedures may be derived for other censoring patterns. Denote the fixed censoring time as C. Then, from (1), the probability of censoring, denoted p_c, equals

$$p_c = \exp\left(-(C/\alpha)^{\beta}\right) \tag{2}$$

It is well known (Lawless, 1982) that the log-likelihood of a sample of n observations from a Weibull distribution censored at the fixed level C is

$$\log L_W(\alpha, \beta) = (n - r) \log S(C; \alpha, \beta) + \sum_{u \notin D} \log f(w_i; \alpha, \beta) = \tag{3}$$

$$(n-r)\left[-\left(\tfrac{C}{\alpha}\right)^{\beta}\right] + \sum_{i \notin D} \left\{\log(\beta) - \log(\alpha) + (\beta - 1)[\log(w_i) - \log(\alpha)] - \left(\tfrac{w_i}{\alpha}\right)^{\beta}\right\}$$

where w_i equals an observed failure time, D represent the set of all observations that were not censored, n equals the sample size, and r equals the number of uncensored observations. We shall design monitoring procedures to detect decreases in the mean lifetime caused by decreases in α, since process changes that cause changes to the shape parameter β seem less likely. Notice that this implies that decreases in the mean lifetime are accompanied by decreases in the process variability. In addition, with a high censoring rate, detecting decreases in the mean lifetime is possible, but increases will be very difficult to detect, since such changes would lead to more censoring. As a result, the proposed monitoring procedures are all one-sided. It is well known that the Weibull distribution is related to the extreme value and exponential distributions through simple transformations. Since these other distributions will be used later we introduce them now. Letting $V = \beta \log(W/\alpha), V$ has a standardized minimum extreme value distribution with location and scale parameters η and ω equal to zero and unity respectively. The probability density and survivor function of the general extreme value distribution are

$$f(v; \eta, \omega = \frac{1}{\omega} e^{(v-\eta/\omega} \exp\left(-e^{(v-\eta)/\omega}\right)$$
$$S(v; \eta, \omega) = \exp(-e^{(v-\eta)/\omega}) \tag{4}$$

The mean and variance of V are $\eta + \gamma\omega$ and $\omega^2\pi^2/6$ respectively, where $\gamma = -.577$, the derivative of the Gamma function evaluated at unity, or minus Euler's constant.

Letting $X = (W/\alpha)^\beta, X$ has a standardized exponential distribution with mean equal to unity. The probability density and survivor function of the general exponential distribution are given by (5). The mean and variance of X are θ and θ^2 respectively.

$$\begin{aligned} f(x;\theta) &= \exp(-x/\theta)/\theta \\ S(x;\theta) &= \exp(-x/\theta) \end{aligned} \tag{5}$$

This article is organized in the following manner. In Section 2, four distinct methods for monitoring the process mean from samples that contain censored observations from an underlying Weibull distribution for failure times are discussed. The methods compared include maximum likelihood and score based approaches, and two methods based on conditional expected value (CEV) weights. For each method the appropriate test statistic given in terms of the original Weibull scale is given. In addition, the determination of an appropriate level for the control limit in each case is discussed, and design figures are given where possible. The control limits are set to give a false alarm rate of .0027. The choice of .0027 comes from the false alarm rate of the standard Shewhart $\overline{X}$ chart. Section 3 compares the monitoring procedures in terms of their power to detect decreases in the process mean when the monitored output is censored at different rates. The best choice of procedure depends on the application, but we recommend the CEV extreme value approach as the best compromise choice. Finally, Section 4 illustrates the use the proposed CEV extreme value weight control charting procedure in the example described above.

2 Monitoring Procedures

The goal of the monitoring procedure is to quickly detect decreases in mean lifetime. To accomplish this the first step is to estimate the in-control process performance to provide a benchmark for comparison. In this initial step, called the implementation phase, the process parameters are estimated from a fairly large initial sample of the process output, a control chart is built, and we verify that the initial sample comes from an in-control process (Montgomery, 1991). With an initial censored Weibull sample, we may derive maximum likelihood estimates (MLEs) for the underlying Weibull parameters using the likelihood function given by (3). Computational methods for determining the MLEs from (3) are well documented, and have well known properties. See Lawless (1982). Of course, with censored observations the Weibull parameters are less well estimated than when all failure times are observed. The amount of lost information can be evaluated by comparing the expected (Fisher) information available in censored and uncensored samples (Escobar and Meeker, 1986). As a result, the usual recommendation of 20 subgroups of size 5, or 100 units total, must be increased. Using Fisher information appropriate sample sizes for any amount of censoring can be established. In the rest of the article we assume that initial Weibull parameters, denoted α_0 and β_0, have been determined from an in-control sample of sufficient size that the sampling variability can be ignored. This assumption corresponds to that made in the usual application of an $\overline{X}$ chart.

Now we turn attention to the ongoing monitoring procedure. An ideal procedure is able to detect decreases in the mean quickly, has an interpretable test statistic in the units of the original problem, and is easy to design. Note also that the procedure should work well with the small samples typical used in most process monitoring applications. We consider four distinct approaches: the Weibull CEV, MLE, exponential CEV (or score), and the extreme value CEV approaches.

2.1 Weibull CEV Method

Traditional $\overline{X}$ control charts use the sample average as test statistic, and set control limits at the estimated process mean plus or minus three standard errors $(\sigma/\sqrt{n})$. Thus, a naive application of a control chart in our application would ignore the censoring and set a lower control limit based on the three standard deviation unit rule. This is a poor choice in our application since the position of the control limit does not take into account the substantial skewness in the distribution of the sample averages caused by the underlying Weibull distribution and the censoring of the observations. The resulting naive control chart would have a much lower false alarm rate than desired, and thus also much lower power. This problem can be alleviated by taking into account the distribution of the sample average and using probability limits to set the control limit. However, we can do better still by not ignoring the censoring. One suggestion is to replace all censored observations with their conditional expected value (CEV) based on the initial parameter estimates. In other words, rather than work with the original sample that contains many observations that are censored at C, we use a sample of CEV weights where the censored observations are replaced by some weight larger than C.

It can be shown (Lawless, 1982) that assuming a Weibull distribution the conditional expected value of a observation censored at C, evaluated at α_0 and β_0, is

$$CE(W) = E(W \mid W \geq C) = \frac{\alpha_0 \Gamma^*(z_C, 1 + 1/\beta_0)}{\exp(-z_C)} \tag{6}$$

where $z_C = (C/\alpha_0)^{\beta_0}$, and $\Gamma^*(x, a) = \int_{y=0}^{x} y^{a-1} \exp(-y) dy$ is the incomplete Gamma function (Abramowitz and Stegun, 1972). The resulting sample of Weibull CEV weights is defined by

$$s(W) = \begin{cases} w & \text{if } w \leq C (\text{not censored}) \\ CE(W) & \text{if } w > C (\text{censored}) \end{cases}, \tag{7}$$

The conditional expected value (CEV) weight is a logical choice because then the expected sample average equals the process mean. Denoting the sample of n CEV weights as $s_1(W), s_2(W), \ldots, s_n(W)$, the test statistic is $\sum_{i=1}^{n} s_i(W)/n$, the sample average of the $s(W)$ weights. To distinguish this procedure from other procedures discussed later we call this approach the Weibull CEV method.

An appropriate lower control limit, one that gives .0027 probability of a false alarm, can be obtained through simulation for any values of α_0 and β_0. We may design the chart for the standardized Weibull to eliminate the dependence on α_0, but unfortunately, the design is complicated by a dependence on the shape parameter β_0 that can not be removed. As a result, no design figures are given for the Weibull CEV approach since a different figure would be needed for each value of β_0.

2.2 MLE Approach

Another way to monitor for decreases in the mean is to use a likelihood based approach. We are interested in detecting decreases in the mean caused by decreases in α. Based on the likelihood function (3), the maximum likelihood estimate (MLE) for α, denoted $\hat{\alpha}$, given β_0 is easily obtained. To avoid possible division by zero (when all the observations in the sample are censored) we will use $1/\hat{\alpha}$ as test statistic. We get $\frac{1}{\hat{\alpha}} = \left[\frac{1}{r}\sum_{i=1}^{r} t_i + (n-r)C\right]^{-1/\beta_0}$. Using the notation introduced in the previous section the resulting sample of MLE based weights is defined by

$$s(M) = \begin{cases} w \text{ if } w \leq C \text{ (not censored)} \\ C \text{ if } w > C \text{ (censored)} \end{cases}, \tag{8}$$

Based on (8), the appropriate test statistic from each sample of size n for the MLE approach is $(r/\sum_{i=1}^{n} s_i(M))^{1/\beta_0}$.

Since the test statistic is $1/\hat{\alpha}$, decreases in α will be manifest as increases in the test statistic. As a result, the MLE approach requires an upper control limit (UCL). The position of the appropriate UCL for the MLE based control chart depends on the sample size, and the in-control probability censored. Figure 1 gives the appropriate UCL derived using simulation to have a theoretical false alarm rate of .0027, and assuming $\alpha_0 = \beta_0 = 1$.

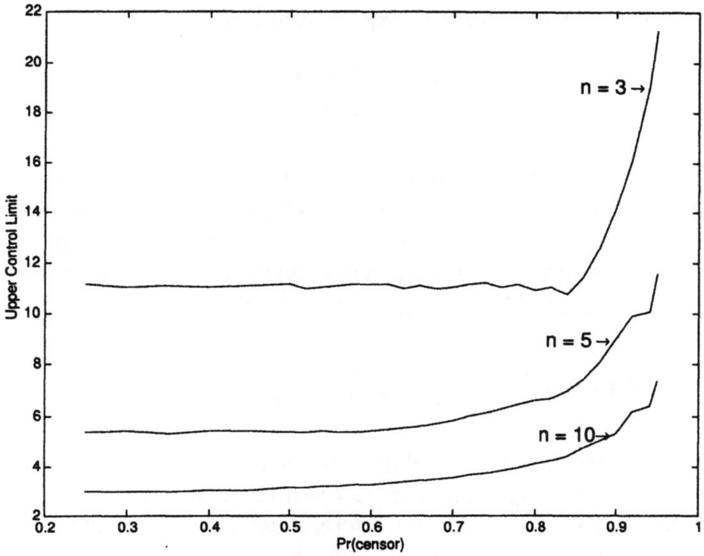

Fig. 1. Standardized Upper Control Limit, ucl_0, for the MLE chart

Through the connection between the Weibull and the exponential distributions, the UCL appropriate in any given example problem equals $(\alpha_0 ucl_0)^{-1/\beta_0}$, where α_0 and β_0 are the actual in-control process parameter, and ucl_0 is the standardized upper control limit given by Figure 1.

2.3 Score and Exponential CEV Method

Another likelihood based approach is to use α-scores. The α-score is defined as the derivative of the log likelihood function with respect to α evaluated at the parameter values of the stable (in-control) process. Due to their derivation, the α-scores form the basis of optimal tests for small changes in the process mean caused by changes in α. From (3), α-scores equal

$$\frac{\partial \log L}{\partial \alpha}\Big|_{\substack{\alpha=\alpha_0 \\ \beta=\beta_0}} = \begin{cases} -\frac{\beta_0}{\alpha_0} + \frac{\beta_0}{\alpha_0}\left(\frac{w}{\alpha_0}\right)^{\beta_0} & \text{if } w \leq C \text{ (not censored)} \\ \frac{\beta_0}{\alpha_0}\left(\frac{C}{\alpha_0}\right) & \text{if } w > C \text{ (censored)} \end{cases} . \tag{9}$$

The corresponding test statistic is given by the sample average of the α-scores. One general disadvantage of using scores is that they do not usually have a physical interpretation. However, in this case, the α-scores are a simple linear translation of the conditional expected value weights derived on the exponential scale. To see this we determine the CEV weight for censored observations on the standardized exponential distribution scale. Based on the translating the censoring time to the exponential scale, we get

$$CE(X) = E\left(X \mid X \geq (C/\alpha_0)^{\beta_0}\right) = (C/\alpha_0)^{\beta_0}) + 1 \tag{10}$$

Thus the CEV weight on the exponential scale is

$$\begin{cases} x = (w/\alpha_0)^{\beta_0} & \text{if } w \le C \quad \text{(not censored)} \\ CE(X) & \text{if } w > C \quad \text{(censored)} \end{cases} .$$

Comparing the α-scores and the exponential CEV weight we see that one is a linear function of the other. This means that the CEV exponential method is optimal in the same sense as the α-score approach. Translating back to the Weibull scale, a sample average on the exponential scale is equivalent to using the test statistic $\alpha_0 \left(\sum_{i=1}^{n}(s_i(X)/\alpha_0)^{\beta_0}/n\right)^{1/\beta_0}$ where $s(X)$ is given by (11).

$$s(X) = \begin{cases} w & \text{if } w \le C \quad \text{(not censored)} \\ \alpha_0 CE(X)^{1/\beta_0} & \text{if } w > C \quad \text{(censored)} \end{cases} . \tag{11}$$

The appropriate lower control limit derived using simulation for the exponential CEV weight is given in Figure 2 assuming $\alpha_0 = \beta_0 = 1$. The irregular parts of the figures are due to the discreteness inherent in the problem. The appropriate control limit given α_0 and β_0 is $(\alpha_0 lcl_0^X)^{1/\beta_0}$, where lcl_0^X is read from Figure 2.

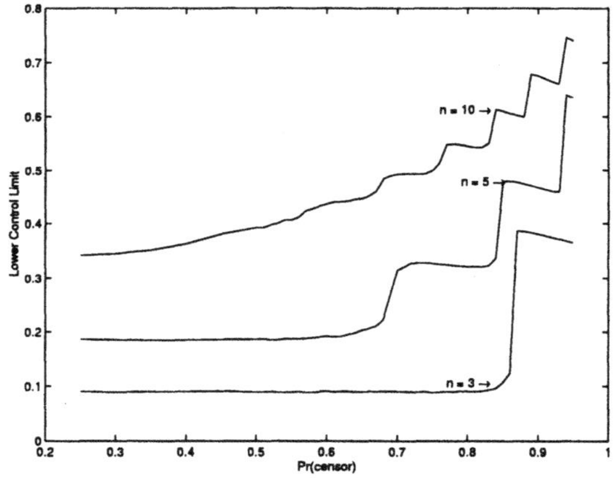

Fig. 2. Standardized Lower Control Limit, lcl_0^X, for the CEV Exponential chart

2.4 Extreme Value CEV Method

The final approach we shall consider is the CEV extreme value method. With this approach the censored observations are replaced by their conditional expected values derived on the extreme value scale. Working with the standardized extreme value distribution, the conditional expected value for a censored observation is

$$CE(V) = E\left(V \mid V \geq \beta_0 \log(C/\alpha_0)\right)$$
$$= \int_{x=\beta_0 \log(C/\alpha_0)}^{\infty} xe^x \exp(-e^x)dx/S\left(\beta_0 \log(C/\alpha_0)\right) \quad (12)$$

where S is the survivor function of the standardized extreme value distribution given by (4). The value of $CE(V)$ can be determined through numerical integration. Given $CE(V)$ from (12) we get the weights given by (13) on the original Weibull scale.

$$s(V) = \begin{cases} w & \text{if } w \leq C \quad \text{(not censored)} \\ \alpha_0 \exp(CE(V)/\beta_0) & \text{if } w > C \quad \text{(censored)} \end{cases}, \quad (13)$$

$\left(\sum_{i=1}^{r} \beta_0 \log(w_i/\alpha) + (n-r)CE(V)\right)/n$, the sample average on the extreme value scale, corresponds to $\left(\prod_{i=1}^{n} s_i(V)\right)^{1/n}$, or the geometric sample average, on the original Weibull scale. Figure 3 gives the appropriate lower control limit for the extreme value CEV approach derived through simulation to give a false alarm rate of .0027, and assuming $\alpha_0 = \beta_0 = 1$. The appropriate lower control limit for any values of α_0 and β_0 is $\alpha_0 \exp(lcl_0^V/\beta_0)$, where lcl_0^V is read from Figure 3.

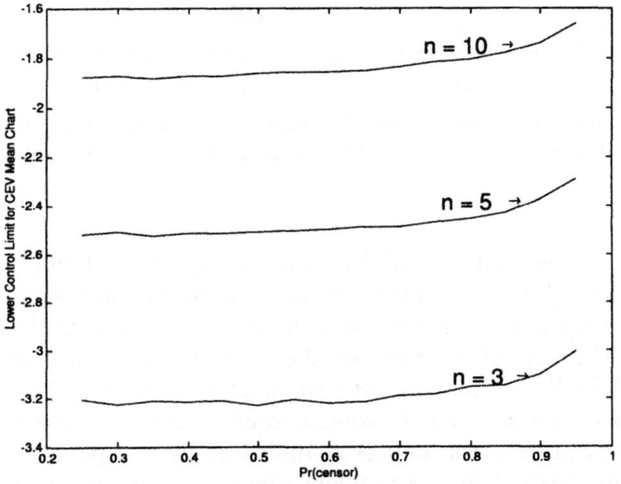

Fig. 3. Standardized Lower Control Limit, lcl_0^V, for the CEV Extreme chart

2.5 Summary of Methods

As shown, the four distinct methods all correspond to replacing the censored observations with a different value. The MLE approach uses C and adjusted the denominator, while each of the other methods uses a CEV weight derived on a different scale. It is interesting to compare the different replacement values when they are all translated back to the original Weibull scale. Figure 4 shows the resulting replacement value plotted against the censoring time C. Note the scales on the plots in Figure 4 are different.

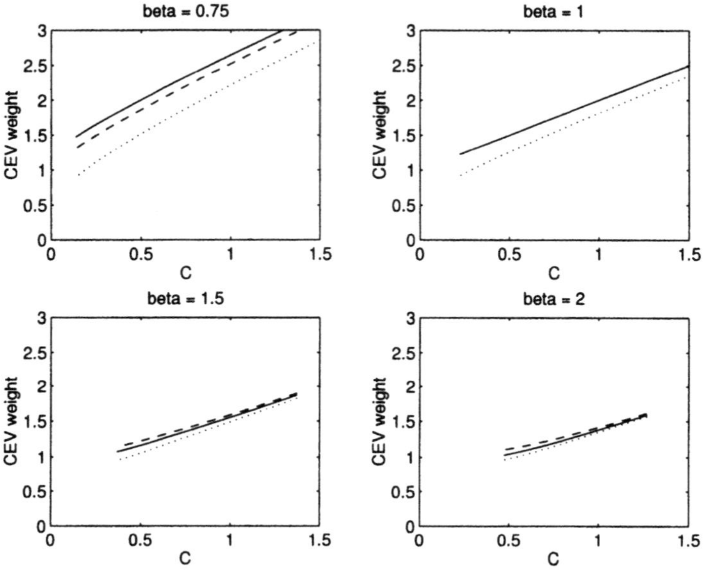

Fig. 4. Conditional Expected Value Weights as a Function of the Censoring Time solid line - Weibull CEV, dotted line - exponential CEV, dashed line - extreme value CEV

Figure 4 shows that the CEV weights are all larger than C, as expected. However, the values are similar. In fact, the major difference between the approaches lies more in the test statistic used to monitor for decreases in the mean, and the ease of determining the control limit. The test statistic for the Weibull CEV method is the easiest, since we use the (arithmetic) sample average, as is typically for $\overline{X}$ control charts. However, the determination of an appropriate control limit is complicated by a dependence on β_0 that cannot be removed. This is a problem, since it means that it is not possible to give one figure from which the appropriate control limit can be determined for any situation. On the other hand, the test statistics for the exponential CEV and MLE approaches are not simple, since they are not interpretable in terms of lifetimes and thus may be difficult for production personnel to understand. The design of control limits for the exponential CEV and MLE approaches, on the other hand, is easier than for the Weibull CEV approach, and generally applicable design figures are given in previous sections. From this perspective the best approach is the extreme value CEV method. The extreme value CEV control limits can also be determined independent of β_0, and the test statistic is the geometric sample average. This test statistic is not as familiar to production personnel as the sample average, but it does retain interpretability. Of course, in each case, we could simply use the sample average as test statistic with what every replacement value was suggested. However, as seen with the Weibull CEV method, then the design of control limits can not be divorced from the value of β_0. Table 1 summarizes the comparison of the four methods in terms of the replacement value, or censored weight, test statistic and ease of design.

Table 1. Method Summary and Comparison

Method	Censored Weight	Test Statistic	Design
Weibull CEV	$CE(W)$	$\sum s_i(W)/n$	depends on β_0
Extreme CEV	$\alpha_0 \exp(CE(V)/\beta_0)$	$(\prod s_i(V))^{1/n}$	Figure 3
Exponential CEV	$\alpha_0(CE(X)^{1/\beta_0}$	$\alpha_0(\sum_{i=1}^{n}(s_i(X)/\alpha_0)^{\beta_0}/n)^{1/\beta_0}$	Figure 2
MLE	C	$(r/\sum_{i=1}^{n} s_i(M))^{1/\beta_0}$	Figure 1

3 Power Comparison of the Control Charts

In this section, we compare the power of the four control charts, namely the Weibull CEV, exponential CEV (alpha score), MLE, and the extreme value CEV methods, for detecting mean shifts in the process. Since the rate of censoring plays a role, we compare the power of the charts with in-control censoring rates of 50% and 90%. Figure 5 gives results for subgroups of size 5 or 10, where control limits are determined through simulation to have false alarm rates equal to .0027. For all approaches other than the Weibull CEV method, the control limits can be determined from the design figures given in Section 2. For comparison purposes the performance in the uncensored case is given with a dashed line in each plot. The Weibull CEV, exponential CEV and MLE approaches are all virtually indistinguishable in term of power. The extreme value CEV approach has good power, but is less powerful than the other methods. The results are given in terms of standard deviation shifts in the extreme value location parameter because then the results can be generalized as explained below. Decreases in the extreme value mean correspond to decreases in on the Weibull scale.

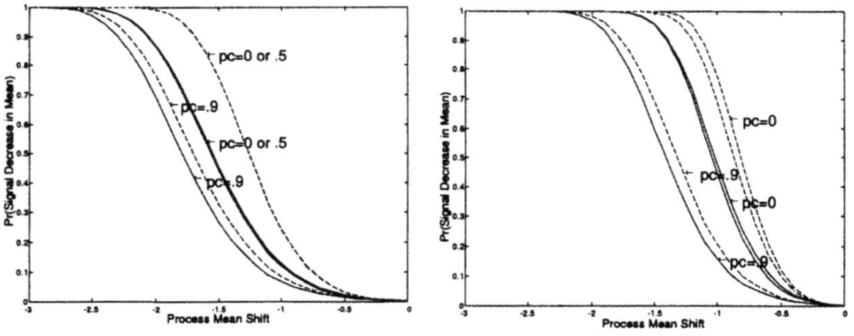

Fig. 5. Power to Detect Shifts in the Mean of the Extreme Value Distribution, $\alpha_0 = \beta_0 = 1$, dashed line - CEV Weibull/ CEV exponential/ MLE method, solid line - Extreme value method

From Figure 5 we see that for moderate censoring proportions, such as 50% censoring, there is almost no loss in power to detect process mean decreases. The loss in power becomes more pronounced when the censoring rate approaches 90%. One reason why we can detect decreases in the mean quickly even with high censoring rates is that as the mean decreases the censoring rate also decreases, and thus each sample contains more information. Clearly, based on these results, there is a tradeoff between information content of the subgroup and the data collection costs. In many applications, the censoring proportion is under our control through the censoring level C. Setting it so that there are few censored observations provides the most information, but will be the most expensive. The optimal tradeoff point depends on the testing costs and the consequences of false alarms and/or missing process changes.

Figure 5 can be used to determine the approximate power to detect a decrease in alpha from α_0 to α_1 given β_0 for any of the methods. The power of the Weibull CEV approach actually depends on β_0, but the difference is small for β_0 in the usual range .5 to 2. The horizontal axis gives the mean shift in terms of the standard deviation units of the location parameter of the extreme value distribution. Using the relationship between the Weibull and extreme value distributions, the number of standard deviation units on the extreme value scale is given by $\beta_0\sqrt{6}\log(\alpha_0/\alpha_1)/\pi$. Thus, for example, if $\alpha_0 = 2.5$ and $\beta_0 = 1.25$ a decrease to $\alpha_1 = .5$ corresponds to a 1.57 standard deviation shift in the location parameter of the extreme value distribution. Thus, from Figure 5, our chance of detecting this shift with a single sample using the exponential CEV method with samples of size 5 and an in-control censoring rate of 50% is around 80%.

The control charts derived in this article were designed to detect decreases in mean lifetime caused by decreases to the scale parameter of the Weibull. However, it is interesting to explore the power of the various methods when decreases in the mean lifetime are due to changes in the mean holding the variance constant. This corresponds to simultaneous changes to both the scale and shape parameters of the Weibull. As explained in Section 1, this kind of process change is not as compelling since it implies the shape of the distribution changes. With simultaneous changes to α and β it is difficult to give general results because the performance will depend on β_0. For illustration purposes Figure 6 compares the performance of the monitoring procedures when $\beta_0 = 1$. Based on this limited comparison, the Weibull CEV, exponential CEV and MLE approaches again give very similar power. Now, however, the extreme value CEV approach is the most powerful. It is not clear why with the Weibull CEV, exponential CEV and MLE approaches we do better with 90% than with no censoring.

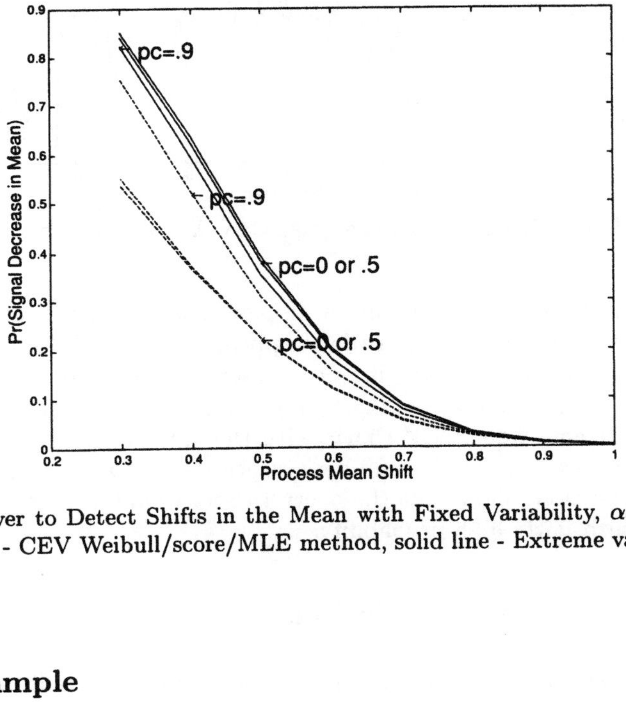

Fig. 6. Power to Detect Shifts in the Mean with Fixed Variability, $\alpha_0 = \beta_0 = 1$, dashed line - CEV Weibull/score/MLE method, solid line - Extreme value method

4 Example

In the rust test example described in the introduction, an initial sample of 100 subgroups of size 3 was selected from historical monitoring records. The censoring time was 20 days. In the data there was a 70% censoring rate. Using the MLE procedure described in Lawless, 1982, we estimate $\alpha_0 = 48.04$ and $\beta_0 = 1.51$. It was also verified that even though our data are discrete a Weibull distribution was a reasonably good fit to the data. Based on the analysis presented in this article, the extreme value CEV method was used. With a censoring time equal to 20 and α_0 and β_0 we get $CE(V) = -.023$ from (12). Translating this into the original Weibull scale means that all censored observations were assigned a weight equal to 47.3. Since the salt spray chamber has a maximum capacity of 60 units a monitoring procedure based on subgroups of size 3 was utilized. Thus, with the estimated in-control Weibull parameters the standardized lower control limit for the extreme value CEV $\overline{X}$ chart is -3.2. This value was determined through simulation, but may be approximated from Figure 3. Scaling the control limits by the in-control parameter estimates gives a lower control limit of 5.9 for the extreme value CEV $\overline{X}$ chart. The resulting CEV $\overline{X}$ chart for the example data is given by the first 100 subgroups in Figure 7.

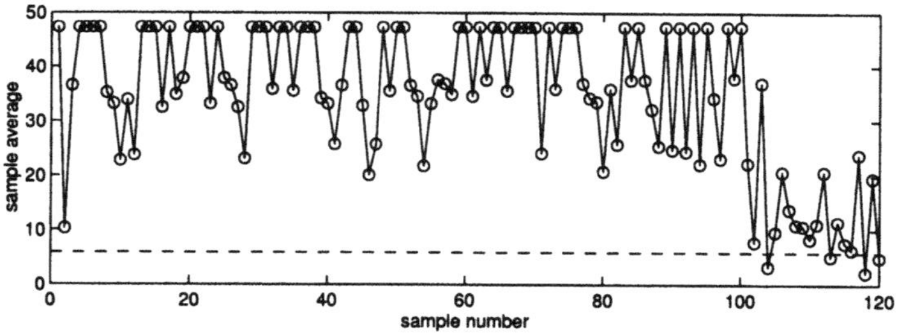

Fig. 7. Example Extreme Value CEV $\overline{X}$ charts with $n = 3$

Figure 7 shows that in the initial implementation there were no out-of-control points. Thus, the initial data appears to come from an in-control process, and we should have obtained reasonably accurate parameter estimates. As a result, we may continue to monitor the process for deterioration using the extreme value CEV charts with the given control limit. The final 20 subgroups are simulated assuming a process shift to $\alpha_1 = 12$. These simulated observations show that the control chart would signals a decrease in the mean with fairly high probability.

5 Summary and Conclusions

In applications where observed data may be censored, traditional process monitoring approaches, such as $\overline{X}$ charts, have undesirable properties such as large false alarm rates or low power. In this article, adapted control charting procedures to monitor the process mean applicable when observations are censored at a fixed levels are proposed. The monitoring procedure is derived assuming the process has an underlying Weibull distribution. A number of possible procedures are compared, including procedures based on maximum likelihood, score and conditional expected value. A chart based on the idea of replacing all observations by their conditional expected values (CEV) weights calculated on the extreme value scale is recommended. The procedure is illustrated with an example.

Acknowledgments

This research was supported, in part, by the Natural Sciences and Engineering Research Council of Canada, and General Motors of Canada.

References

1. Abramowitz, M., Stegun, I.A. (1972) Handbook of Mathematical Functions. Dover, New York
2. Escobar, L.A., Meeker, W.Q., Jr. (1986) Elements of the Fisher Information Matrix for the Smallest Extreme Value Distribution and Censored Data. Applied Statistics **35**, 80-86
3. Lawless, J.F. (1982) Statistical Models and Methods for Lifetime Data. John Wiley and Sons, New York
4. Montgomery, D.C. (1991) Introduction to Statistical Quality Control. Second Edition, John Wiley and Sons, New York
5. Nelson, W. (1990) Accelerated Testing. John Wiley and Sons, New York
6. Ryan, T.P. (1989) Statistical Methods for Quality Improvement. John Wiley and Sons, New York
7. Schneider, H., Kasperski, W.J., Ledford, T., Kravshaar, W. (1995-1996) Control Charts for Skewed and Censored Data. Quality Engineering **8**(2) 263-274

A Graphical Method to Control Process Capability

Kerstin Vännman
Division of Quality Technology & Statistics, Luleå University of Technology
SE-971 87 Luleå, Sweden

Abstract. We suggest that one should focus on the process parameters (μ, σ) instead of the capability index alone by using process capability plots, both when defining what is meant by a capable process and when deciding whether a process can be considered capable or not. First we introduce the so called the (δ, γ)-plot, to define a capability region. The (δ, γ)-plot is based on a capability index but focuses on the parameters (μ, σ). We have earlier defined a class of capability indices, containing the indices C_p, C_{pk}, C_{pm}, and C_{pmk}. By varying the parameters of the class various indices as well as various capability regions with different properties are obtained.

Under the assumption of normality we suggest two process capability plots to be used when interpreting the results of the estimated indices in the above mentioned class for deciding whether a process can be considered capable or not. One of these plots is the so called (δ^*, γ^*)-plot, which is based on the estimated index and is a generalization of the (δ, γ)-plot. The other is the so called confidence rectangle plot. This plot is based on a confidence region for (δ, γ), which is plotted in the (δ, γ)-plot. These plots are compared and discussed from different aspects. An advantage with using a process capability plot, compared to using the capability index alone, is that we will instantly get visual information, simultaneously about the location and spread of the studied characteristics, as well as information about the capability of the process. When the process is non-capable, the plots are helpful when trying to understand if it is the variability, the deviation from target, or both that need to be reduced to improve the capability. In this way the proposed graphical methods give a clear direction for quality improvement.

1 Introduction

Process capability indices have gained considerable popularity in industry, and today process capability studies seem to be facing a new boom. See, e.g., Deleryd (1998b). According to Kotz & Lovelace (1998) today calculation of capability index values for key processes has become as standard as SPC was during the infancy of the modern quality assurance movement. They claim that process capability indices are here to stay.

When focusing on the capability of a process there are mainly two characteristics of importance, the location and the spread in relation to the specified requirements of the process. The closer the process output is to the target value and the smaller the process spread, the more capable is the process. Capability indices combine these characteristics of a capable process and transform them into a single number reflecting the capability of the process. The larger the value of a process capability index, the more capable the process.

Capability indices have received much interest in the statistical literature during recent years. For thorough discussions of different capability indices and their properties see, e.g., Chan, Cheng & Spiring (1988), Gunter (1989), Boyles (1991), Pearn, Kotz & Johnson (1992), Kotz & Johnson (1993), Pignatiello & Ramberg (1993), Vännman & Kotz (1995a, b), Vännman (1995, 1997a, b, 1998), Carlsson (1995), Spiring (1997), and Kotz & Lovelace (1998).

The two most widely used capability indices in industry today are

$$C_p = \frac{USL - LSL}{6\sigma} \quad \text{and} \quad C_{pk} = \frac{\min(USL - \mu, \mu - LSL)}{3\sigma}, \tag{1}$$

where $[LSL, USL]$ is the specification interval, μ is the process mean, and σ is the process standard deviation under stationary controlled conditions. The process parameters μ and σ in (1) are usually unknown and are estimated by appropriate sample statistics.

When capability indices were introduced the motivation for the denominator 6σ in C_p seems to be associated with the probability of non-conformance, i.e., the probability of obtaining a value outside the specification limits. According to today's modern quality improvement theories the focus should not be on the probability of non-conformance only. It is also very important to use target values and to keep the process on target. These theories are based, among other things, on Taguchi's quality philosophy, in which reduction of variation from the target value is the guiding principle, see Taguchi (1986). According to this philosophy the specification limits, in connection with manufacturing, must not be interpreted as permission to be anywhere within the tolerance interval. Instead target values should be used and attention should focus on meeting the target instead of meeting the tolerances. See, e.g., Bergman & Klefsjö (1994) and Phadke (1989). Hence if μ is far away from the target T the process should not be considered capable even if σ is so small that the probability of non-conformance is small. Examples illustrating these ideas are found, e.g., in Phadke (1989) and Sullivan (1984).

Boyles (1991) showed that C_{pk} becomes arbitrarily large as σ approaches 0 and this characteristic makes C_{pk} unsuitable as a measure of process centering. A large value of C_{pk} does not in itself tell anything about the distance between μ and T. The same, of course, is true for C_p. A capability index which takes into account that the process mean, μ, may differ from the target value, T, is C_{pm} (Chan, Cheng & Spiring, 1988), where

$$C_{pm} = \frac{USL - LSL}{6\sqrt{\sigma^2 + (\mu - T)^2}} . \tag{2}$$

In order to gain sensitivity with regard to departures of the process mean from the target value, and at the same time generalize the indices above, Vännman (1995) defined a family of capability indices, depending on two non-negative parameters, u and v, as

$$C_p(u, v) = \frac{d - u|\mu - M|}{3\sqrt{\sigma^2 + v(\mu - T)^2}} , \tag{3}$$

where $d = (USL - LSL)/2$, i.e., half the length of the specification interval, $M = (USL + LSL)/2$, i.e., the mid-point of the specification interval. The indices in (1) and (2) are obtained by setting $u = 0$ or 1 and $v = 0$ or 1 in (3), i.e., $C_p(0, 0) = C_p$, $C_p(1, 0) = C_{pk}$, $C_p(0, 1) = C_{pm}$. The family in (3) also includes the index C_{pmk} introduced by Pearn, Kotz & Johnson (1992). We get $C_p(1, 1) = C_{pmk}$.

In this paper we will focus on the case when the specification interval is defined in advance and, furthermore, is two-sided with the target value T at the mid-point M of the specification interval, which is quite common in practical situations. Henceforth we will assume that $T = M$. In this case Vännman (1995, 1997a) and Vännman & Kotz (1995a, b) have studied the index in (3) and derived statistical properties of the estimated index as well as suggested suitable choices of the parameters u and v. They propose the indices $C_p(0,3)$ and $C_p(0,4)$ for consideration of the practitioners in those situations where the normality of the process is not severely violated and the target value coincides with the midpoint of the specification interval. These indices are superior to both C_{pm} and C_{pk} by being more sensitive to departures of the process mean from the target value and possessing smaller relative biases and mean square errors.

Process capability indices combine information about closeness to target and process spread and express the capability by a single number. For management it is convenient with a single-number summary of what is happening on the production floor, which simplifies the supervision of production activities. However, it may in some cases also be held as one of the major drawbacks of the indices. If for instance the process is found non-capable, the operator is interested in knowing whether this non-capability is caused by the fact that the process output is off target, or that the process spread is too large, or if the result is a combination of these two factors. This information cannot instantly be found from the index value.

In order to circumvent this shortage of process capability indices we suggest two graphical methods to be used when interpreting the results of the estimated capability indices. When the process is non-capable, these methods are helpful when trying to understand if it is the variability, the deviation from target, or both that need to be reduced to improve the capability. In this way the proposed graphical method gives a clear direction for quality improvement.

The results obtained in this paper can easily be generalized to asymmetric tolerances or one-sided tolerances using the results in Vännman (1997b, 1998).

2 Capability Regions Defined by (δ, γ)-Plots

When using process capability indices it is common to require that in order to consider a process as capable, the process capability index has to exceed a certain value k. Some commonly used values are $k = 1.0$, $k = 4/3$, $k = 5/3$, or $k = 2.0$. Here we will use the index $C_p(u, v)$ in (3) and define the studied process to be capable if $C_p(u, v) > k$, for given values of u, v, and k. In order to better understand what restrictions this definition imposes on the process parameters (μ, σ) we suggest the use of the so called (δ, γ)-plot, cf. Figure 1. It is simply a contour plot of $C_p(u, v) = k$, called a capability level curve, as a function of δ and γ, where

$$\delta = \frac{\mu - T}{d} \quad \text{and} \quad \gamma = \frac{\sigma}{d}\,. \tag{4}$$

It is obtained by rewriting the index in (3) as a function of δ and γ, solving the equation $C_p(u,v) = k$ with respect to γ, and then plotting γ as a function of δ. We easily find the following expressions

$$\gamma = \sqrt{\frac{(1-u|\delta|)^2}{9k^2} - v\delta^2}, \quad |\delta| \bullet \frac{1}{u+3k\sqrt{v}}. \tag{5}$$

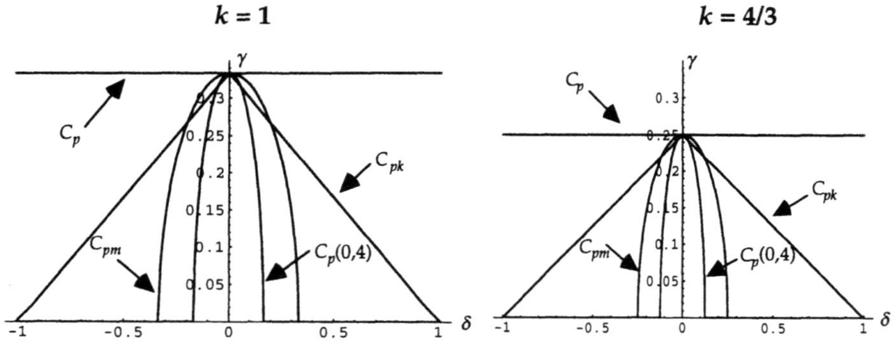

Fig. 1. The (δ, γ)-plots show the capability level curves for the process capability indices C_p, C_{pk}, C_{pm}, and $C_p(0,4)$ for $k = 1$ and $k = 4/3$. C_p corresponds to the straight line, C_{pk} to the two lines forming a triangle together with the δ-axis, C_{pm} to the semi-circle, and $C_p(0,4)$ to the semi-ellipse.

Values of the process parameters μ and σ which give (δ, γ)-values inside the region bounded by the capability level curve $C_p(u, v) = k$ and the δ-axis will give raise to a $C_p(u, v)$-value larger than k, i.e., a capable process. We can call this region of the (δ, γ)-plot the capability region. Furthermore, values of μ and σ which give (δ, γ)-values outside this region will give a $C_p(u, v)$-value smaller than k, i.e., a non-capable process. Figure 1 shows some examples of capability regions when $k = 1$ and $k = 4/3$. We see clearly the different requirements with regard to closeness to target and small spread for a capable process using the different indices. Among the process capability indices in Figure 1, the indices C_{pm} and $C_p(0,4)$ are to be preferred, since they take into account the closeness of the process output to the target value, which C_p and C_{pk} do not.

The reason for making the contour plot as a function of (δ, γ) instead of a function of (μ, σ) is to obtain a plot which is invariable irrespective of the values of the specification limits. If the process is on target, i.e., $\mu = T$, we have $\delta = 0$. If $\mu = LSL$ we have $\delta = -1$, and if $\mu = USL$ we have $\delta = 1$. Furthermore, with this scaling δ is a measure of how far away from the target value the process mean is relative to d, where d is half the length of the specification interval, while γ measures how large the process standard deviation σ is relative to d. This is a convenient interpretation of closeness to target as well as spread.

We suggest that the (δ, γ)-plot be used as a complement to the capability index when deciding how to define a capable process and what capability index to use. With the (δ, γ)-plot we use the concept of a process capability index but focus on the process parameters (μ, σ). We believe this to be important since then, already at the outset, we focus on closeness to target as well as small spread. By using a plot we enhance the two concepts closeness to target and small spread in a more efficient way than by using an index only. For instance, if we choose $u = 0$, $v = 1$, i.e., C_{pm}, to define the capability region of the (δ, γ)-plot we obtain the region bounded by a semi-circle and the δ-axis. See Figure 1. This means, for a capable process, that the maximum allowed deviation of μ from T relative to d, which occurs when σ is 0, is equal to the maximum allowed variability of σ relative to d, which occurs when $\mu = T$. We can say that we assign the same weights on a deviation of μ from T as on a σ-deviation. If a deviation of μ from T relative to d is considered more serious than a σ-deviation relative to d, then we should choose an index with $v > 1$, e.g., $C_p(0,4)$, which has been shown to be an index, among the indices in the $C_p(u, v)$-family, sensitive to departures of the process mean from the target value and possessing good statistical properties. The index $C_p(0,4)$ gives rise to a capability region in the (δ, γ)-plot bounded by a semi-ellipse and the δ-axis. See Figure 1.

The idea behind the (δ, γ)-plot is not new. Gabel (1990) presents a process performance chart, which is a plot similar to the (δ, γ)-plot but with a different scaling and for C_{pk} only. Boyles (1991) introduces the (μ, σ)-plot as an effective graphical method for theoretically comparing and contrasting different capability indices. This is simply a contour plot of one or more indices as functions of the process parameters (μ, σ) in the region $LSL \bullet \mu \bullet USL$.

In most situations the process parameters μ and σ are unknown and need to be estimated. Hence the estimated indices have to be treated as random variables and their distributions have to be taken into account. In such situations the (δ, γ)-plot cannot be used directly. It needs to be adjusted before it can be utilized for deciding whether a process is capable or not. In the next section we will develop a decision rule, based on sample statistics and a process capability plot similar to the (δ, γ)-plot, to be used for deciding whether a process can be considered capable or not.

3 Interpreting Capability Indices Using the (δ^*, γ^*)-Plot

In this paper we treat the case when the studied characteristic of the process is normally distributed. Let $X_1, X_2, \dots, X_n$ be a random sample from a normal distribution with mean μ and variance σ^2 measuring the studied characteristic. We will assume that a capability region has been defined by (5) for given values of u, v, and k, i.e., the studied process is defined to be capable if $C_p(u,v) > k$.

To obtain an appropriate decision rule it seems reasonable to consider a hypothesis test with the null hypothesis $H_0 : C_p(u,v) \bullet k$ and the alternative hypothesis $H_1 : C_p(u,v) > k$. As a test statistic we will use the estimator $C_p^*(u,v)$ of $C_p(u,v)$ defined as

$$C_p^*(u,v) = \frac{d - u|\mu^* - T|}{3\sqrt{\sigma^{2*} + v(\mu^* - T)^2}} \quad (6)$$

where the mean μ is estimated by the sample mean and the variance σ^2 is estimated by its maximum likelihood estimator, i.e.,

$$\mu^* = \bar{X} = \frac{1}{n}\sum_{i=1}^{n} X_i \quad \text{and} \quad \sigma^{2*} = \frac{1}{n}\sum_{i=1}^{n}(X_i - \bar{X})^2 . \quad (7)$$

The null hypothesis will be rejected whenever $C_p^*(u,v) > c_\alpha$, where the constant c_α is determined so that the significance level of the test is α. The decision rule to be used is then that, for given values of α and n, the process will be considered capable if $C_p^*(u,v) > c_\alpha$ and non-capable if $C_p^*(u,v) \bullet c_\alpha$.

Now we can illustrate the decision rule in a plot corresponding to the (δ, γ)-plot, described in the previous section, to obtain an estimated capability region. To do so we introduce the notation

$$\delta^* = \frac{\mu^* - T}{d} = \frac{\bar{X} - T}{d} \quad \text{and} \quad \gamma^* = \frac{\sqrt{\sigma^{2*}}}{d} = \frac{\sigma^*}{d} \tag{8}$$

for the estimators of δ and γ, respectively. We will then make a contour plot of $C_p^*(u, v)$, when $C_p^*(u, v) = c_\alpha$, as a function of δ^* and γ^* . This plot will be called a (δ^* , γ^*)-plot. As in the previous section we easily find the following function to make the (δ^* , γ^*)-plot

$$\gamma^* = \sqrt{\frac{(1 - u|\delta|)^2}{9c_\alpha^2} - v\delta^2}, \quad |\delta| \bullet \frac{1}{u + 3c_\alpha\sqrt{v}} \quad . \tag{9}$$

We can now use the (δ^* , γ^*)-plot to define an estimated capability region in a way corresponding to using the (δ, γ)-plot. We consider a process capable if (δ^* , γ^*) falls inside the estimated capability region. Otherwise it will be considered non-capable. Hence, instead of calculating the estimated capability index in (6) and compare it with c_α, we calculate (δ^* , γ^*) in (8) and use a graphical method to make the decision. See an example in Section 4.

In order to find the critical value c_α we have to calculate $P(C_p^*(u, v) > c_\alpha)$. Hence we need the distribution of $C_p^*(u, v)$, which has been derived by Vännman & Kotz (1995a) and in a simplified form by Vännman (1997a). Let F_ξ denote the cumulative distribution function of ξ, where ξ is distributed according to a central χ^2 -distribution with $n - 1$ degrees of freedom, and let f_η denote the probability density function of η, where η is distributed according to the standardized normal distribution $N(0 \, , \, 1)$. From Vännman (1997a) we have that when $x > 0$ then, for $u \bullet 0$ or $v \bullet 0$, the cumulative distribution function of $C_p^*(u, v)$ can be written as

$$F_{C_p^*(u,v)}(x) = 1 - \int_{\frac{-D}{u+3x\sqrt{v}}}^{\frac{D}{u+3x\sqrt{v}}} F_\xi\left(\frac{(D-u|t|)^2}{9x^2} - vt^2\right) f_\eta(t-g)\, dt, \qquad x > 0\ , (10)$$

where $D = \frac{d\sqrt{n}}{\sigma}$ and $g = \frac{(\mu - T)\sqrt{n}}{\sigma}$.

For the complete cumulative distribution function of $C_p^*(u,v)$, including the case x • 0, see Vännman (1997a). When deriving the critical value c_α we are interested in $x > 0$ only.

We see from (10) that the distribution of $C_p^*(u,v)$ depends on μ and σ but not solely through $C_p(u,v)$. Hence for a given value of $C_p(u,v)$ the probability $P(C_p^*(u,v) > c_\alpha)$ is a function of the unknown process parameters μ and σ. To calculate this probability we need given values of both μ and σ.

Since our null hypothesis $H_0 : C_p(u,v)$ • k is composite, for a significance level α, the critical value c_α is defined as the unique smallest x for which $\alpha(x)$ • α, where

$$\alpha(x) = \sup\{P(C_p^*(u,v) > x) : (\mu, \sigma) \text{ such that } C_p(u,v) \bullet k\}. \qquad (11)$$

See, e.g., Bickel & Doksum (1977), p. 170. To obtain $\alpha(x)$ we have to investigate $P(C_p^*(u,v) > x)$ for all values of (μ, σ) for which $C_p(u,v)$ • k, i.e., for all values of (μ, σ) outside the capability region defined by the (δ, γ)-plot, cf. Figure 1. Vännman & Kotz (1995a) have shown that when μ moves away from T (δ moves away from 0) in either direction, then $P(C_p^*(u,v) > x)$ decreases, for any given values of x and σ. Furthermore, from (10) we easily find that when $\mu = T$ (i.e., $\delta = 0$) then $P(C_p^*(u,v) > x)$ is a decreasing function of σ, for any given value of x. Hence it is reasonable to assume that the supremum in (11) will be found along the capability level curve in (δ, γ)-plot defined by $C_p(u,v) = k$. Numerical investigations support this assumption. See Vännman (1997c). Hence we assume that we can reduce the null hypothesis to $H_0 : C_p(u,v) = k$. However, this is still a composite hypothesis. To derive c_α analytically does not seem feasible, instead it has been derived numerically using Mathematica (Wolfram, 1991) for given values of n and α. In the following we will show some examples for the case $k = 1$, $n = 80$, and $\alpha = 0.05$.

Consider first the case when the capability region has been defined using $u = 0$ and $v = 1$, i.e., the index C_{pm} . See Figure 1. Numerical investigations, using (10), yield the critical value $c_{0.05} = 1.151$. Hence the process will be considered capable if $C^*_{pm} = C^*_p(0, 1) > 1.151$ and non-capable if C^*_{pm} • 1.151. The corresponding estimated capability region is plotted in Figure 2 and the process will be considered capable if the observed value of (δ^* , γ^*), defined in (8), falls inside the region defined by the semi-circle (continuous curve). Otherwise it will be considered non-capable. In Figure 2 b) we have also plotted the capability level curve corresponding to $C_{pm} = 1$ together with the estimated capability curve corresponding to $C^*_{pm} = 1.151$ to enhance the difference between the capability region and the estimated capability region. The estimated capability region will always be smaller than the corresponding capability region.

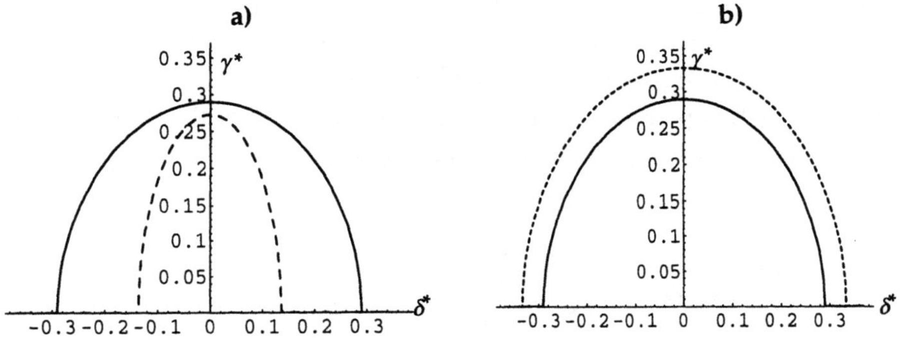

Fig. 2. a) The estimated capability region bounded by the semi-circle defined by $C^*_{pm} = c_{0.05} = 1.151$ (continuous curve) and the δ^* -axis, when $n = 80$. The estimated capability re5gion bounded by the semi-ellipse defined by $C^*_p(0, 4) = c_{0.05} = 1.225$ (dashed curve) and the δ^* -axis, when $n = 80$. b) The capability level curve corresponding to $C_{pm} = 1$ (dashed curve) together with the estimated capability curve corresponding to $C^*_{pm} = 1.151$ (continuous curve).

Using (10) together with (5) with $k = 1$ we can express the probability that the process will be considered capable, i.e., $P(C^*_{pm} > 1.151)$, as a function of δ and plot it as in Figure 3 a). We can see how the probability changes when the parameters δ and γ move along the curve defined by $C_{pm} = C_p(0,1) = 1$. We see that when $u = 0$ and $v = 1$ we get the maximum of $P(C^*_{pm} > 1.151)$ when $\delta = 0$, i.e., when the process is on target. When δ moves away from 0 then $P(C^*_{pm} > 1.151)$ decreases. The reason for the fact that the probability $P(C^*_{pm} > 1.151)$ changes quite a lot when moving along the curve $C_{pm} = 1$ is that the shape of the distribution of $C^*_p(u,v)$ may change considerably when the process is on target and when it is off target. See Vännman & Kotz (1995) and Vännman (1997c).

We now consider the case when the capability region has been defined using $u = 0$ and $v = 4$. Numerical investigations, using (10), yield the critical value $c_{0.05} = 1.225$. Hence the process will be considered capable if $C^*_p(0,4) > 1.225$ and non-capable if $C^*_p(0,4) \bullet 1.225$. The corresponding estimated capability region is plotted in Figure 2 a) and the process will be considered capable if the observed value of (δ^*, γ^*) falls inside the region defined by the semi-ellipse (dashed curve). Otherwise it will be considered non-capable. The probability that the process will be considered capable, i.e., $P(C^*_p(0,4) > 1.225)$, when δ and γ move along the curve defined by $C_p(0,4) = 1$, is plotted in Figure 3 b). Here we see that the maximum no longer occurs when the process is on target. Instead it is close to $\delta = \pm 0.12$.

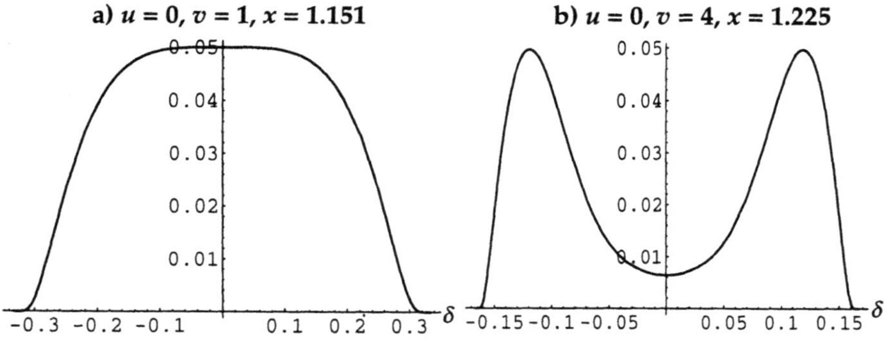

Fig. 3. The probability that the process is considered capable, i.e., $P(C^*_p(u,v) > x)$, given that $C_p(u,v) = 1$, for $n = 80$.

Using (10) the power function can be calculated for each test based on $C_p^*(u,v)$. In this way a suitable sample size can be determined to ensure that the test will detect, with large probability, if the process gives rise to values of δ and γ in the capability region. See Deleryd & Vännman (1999) for some examples of power functions.

The maximum likelihood estimator in (7) was used above as an estimator of σ^2, when estimating $C_p(u,v)$. If instead the sample variance s^2 is used as an estimator of σ^2, when estimating $C_p(u,v)$, we can carry out the same reasoning as above using this second estimator. We only have to utilize a simple relationship between the two estimators of $C_p(u,v)$, see Vännman (1995), to derive the cumulative distribution function of this second estimator of $C_p(u,v)$ using (10).

4 A Practical Example

To illustrate the ideas presented above we will consider a practical example described by Pearn & Chen (1997). They have collected some sample data from rubber edge production in a factory. The upper and lower specification limits, *USL* and *LSL*, respectively, for this particular model of rubber edge are set to 8.94 and 8.46 (in grams). The target value for the rubber edge weight is set to the mid-point between the two specifications limits, i.e., $T = 8.70$. Hence we have $d = 0.24$. They define the process as capable if $C_{pk} > 1$ and use the significance level $\alpha = 0.05$. The collected sample data, a total of 80 observations, are given in Pearn & Chen (1997). From these observations we get the observed values of μ^* and σ^* to be 8.6234 and 0.0519, respectively.

We will here consider the two cases when the process is defined as capable if $C_{pm} > 1$ or if $C_p(0,4) > 1$. For illustrations of the case using C_{pk} see Vännman (1997c). From Section 3 we have that $c_{0.05} = 1.151$ when using C_{pm} and $c_{0.05} = 1.225$ when using $C_p(0,4)$. See also Figures 2 and 3.

Based on the sample data we then calculate the observed values of δ^* and γ^* in (8) and find them to be –0.319 and 0.216, respectively. Then the point with coordinates (–0.319, 0.216) is plotted, together with the two estimated capability regions, in the (δ^*, γ^*)-plot. The result is found in Figure 4 a).

We can see that the point (–0.319, 0.216) is outside the estimated capability regions in both cases and hence the process cannot be considered capable using either of the two definitions. This is the same conclusion as was reached by Pearn & Chen (1997) using C_{pk} alone.

What is obvious from the graphical method used here is that the non-capability to a large extent is due to the large deviation of the process mean from the target value. The estimated deviation is more than 30% of d and the process mean is smaller than the target value. Furthermore, the spread is not very small. The estimated standard deviation is about 20% of d. We get this information about both the deviation of the process mean from the target value and the process spread instantly by looking at the (δ^*, γ^*)-plot.

When it was found that the quality condition of the process was unsatisfactory then the process was adjusted, and a second sample of 80 observations was made. The second sample is given in Pearn & Chen (1997). From these observations we get the observed values of μ^* and σ^* to be 8.6628 and 0.0516, respectively. Based on the second sample we then calculate the observed values of δ^* and γ^* in (8) and find them to be –0.155 and 0.215, respectively.

Then the point with coordinates (–0.155, 0.215) is plotted, together with the two estimated capability regions, in the (δ^*, γ^*)-plot. The result is found in Figure 4 b). We can see that the point (–0.155, 0.215) is now inside the estimated capability region for C_{pm} and hence the process can be considered capable according to the definition of a capable process based on $C_{pm} > 1$. This is the same conclusion as was reached by Pearn & Chen (1997) using C_{pk} alone. However, if the process is defined as capable according to the definition $C_p(0,4) > 1$ it will not be considered capable. Hence, although after adjustments of the process, the large deviation of the process mean from the target value has been reduced to about 15% of d the deviation is still too large, according to the definition based on $C_p(0,4) > 1$, to make the process capable.

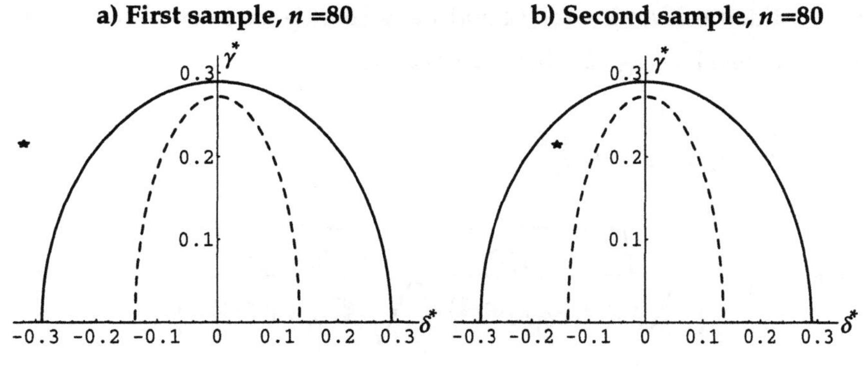

Fig. 4. The estimated capability region bounded by the semi-circle defined by $C^*_{pm} = 1.151$ (continuous curve) and the δ^*-axis. The estimated capability region bounded by the semi-ellipse defined by $C^*_p(0, 4) = 1.225$ (dashed curve) and the δ^*-axis. In a) the observed value (–0.319, 0.216) of (δ^*, γ^*) from the first sample is plotted. In b) the observed value (–0.155, 0.215) of (δ^*, γ^*) from the second sample is plotted.

5 Interpreting Capability Using Confidence Regions

In this section we will discuss an alternative method to be used instead of the (δ^*, γ^*)-plot to decide whether a process can be considered capable or not. This method can be used for any definition of the capability region in the (δ, γ)-plot, not only for regions defined by (5).

The method is based on the definition of a capable process using a capability region in a (δ, γ)-plot. We then calculate a confidence region for (δ, γ), with confidence level $1 - \alpha_0$, and plot this region in the (δ, γ)-plot. The decision rule to be used is the following. For given values of α and n, the process will be considered capable if the whole confidence region for (δ, γ) is inside the capability region in the (δ, γ)-plot. The value of α_0 has to be determined so that the probability that the process is considered capable, i.e., that the confidence region is inside the capability region, given that $C_p(u, v) = k$, equals the significance level α.

Under the assumption of a random sample from a normal distribution we can calculate a confidence rectangle for (μ, σ) as follows. See, e.g., Bickel & Doksum (1977), pp. 162–163. Let I_μ denote a confidence interval for μ

with confidence level $1 - \alpha_0/2$ and I_σ a confidence interval for σ with confidence level $1 - \alpha_0/2$. Then we have

$$I_\mu = \bar{X} \pm t_{1-\alpha_0/4}(n-1)\frac{s}{\sqrt{n}}$$

$$\text{and} \quad I_\sigma = \left[\sqrt{\frac{(n-1)s^2}{\chi^2_{1-\alpha_0/4}(n-1)}}, \sqrt{\frac{(n-1)s^2}{\chi^2_{\alpha_0/4}(n-1)}}\right], \tag{12}$$

where $\bar{X}$ is the sample mean, s^2 is the sample variance, $t_\nu(n-1)$ is the νth quantile from the t-distribution, and $\chi^2_\nu(n-1)$ is the νth quantile from the χ^2-distribution, both with $n - 1$ degress of freedom. Then the rectangle $I_\mu \infty I_\sigma$ is a confidence rectangle for (μ, σ) with confidence level at least $1 - \alpha_0$ according to the Bon-Ferroni inequality. From (4) and (12) we then obtain that $I_\delta \infty I_\gamma$ is a confidence rectangle for (δ, γ) with confidence level at least $1 - \alpha_0$, where

$$I_\delta = \frac{\bar{X} - T}{d} \pm t_{1-\alpha_0/4}(n-1)\frac{s}{d\sqrt{n}} = \delta^* \pm t_{1-\alpha_0/4}(n-1)\frac{\gamma^*}{\sqrt{n-1}}$$

$$\text{and} \quad I_\gamma = \frac{1}{d} \cdot I_\sigma = \left[\sqrt{\frac{n\gamma^{*2}}{\chi^2_{1-\alpha_0/4}(n-1)}}, \sqrt{\frac{n\gamma^{*2}}{\chi^2_{\alpha_0/4}(n-1)}}\right]. \tag{13}$$

Since $1 - \alpha_0$ denotes the probability that the confidence region contains (δ, γ) we cannot expect that the probability is α_0 that the whole confidence region for (δ, γ) is inside the capability region when $C_p(u,v) = k$. Hence we have to determine α_0 so that the probability to consider the process capable, i.e., the probability that the whole confidence region for (δ, γ) is inside the capability region, when $C_p(u,v) = k$, is equal to a pre-determined significance level α. This has been done by Deleryd & Vännman (1998) for the case $u = 0$ and $v = 1$, i.e., for C_{pm}, when $\alpha = 0.05$.

It was found through a simulation study, since so far there are no theoretical results available, that a 76% confidence level for the confidence rectangle (13) corresponds to $\alpha = 0.05$ in this case. Furthermore, this result seems to be independent of the values of n and k.

The probability of considering the process to be capable will vary when the parameters δ and γ move along the curve defined by $C_{pm} = 1$. Using the confidence rectangle plot this probability has its maximum for $\delta = \pm 1/3$. Using the (δ^*, γ^*)-plot, the corresponding probability, has its maximum for $\delta = 0$, see Figure 3 a).

When we consider the case when the process is defined as capable if $C_p(0,4) > 1$, a simulation study, in analogy with the one in Deleryd & Vännman (1998), shows that an 80% confidence level for the confidence rectangle (13) corresponds to $\alpha = 0.05$ in this case. This result seems to be independent of the values of n and k. Furthermore, the probability of considering the process to be capable will vary when the parameters δ and γ move along the curve defined by $C_p(0,4) = 1$. It has its maximum for $\delta = \pm 1/3$. Compare with Figure 3 b) for the corresponding probability, using the (δ^*, γ^*)-plot.

We will now apply this method to the practical example from Pearn & Chen (1997) described in Section 4, where $n = 80$ and $\alpha = 0.05$. When the process is defined as capable when $C_{pm} > 1$ we calculate the confidence rectangle (13) with confidence level 76%. Using the quantiles $t_{0.94}(79) = 1.57177$, $\chi^2_{0.06}(79) = 60.4634$, and $\chi^2_{0.94}(79) = 99.4240$, and the observed values of μ^* and σ^* from Section 4 we obtain the following results. See also Figure 5.

For sample 1: $I_\delta = [-0.358, -0.280]$ and $I_\gamma = [0.193, 0.249]$.

For sample 2: $I_\delta = [-0.193, -0.117]$ and $I_\gamma = [0.192, 0.248]$.

When the process is defined as capable when $C_p(0,4) > 1$ we calculate the confidence rectangle with confidence level 80%. Using the quantiles $t_{0.95}(79) = 1.66437$, $\chi^2_{0.05}(79) = 59.5223$, and $\chi^2_{0.95}(79) = 100.749$, and the observed values of μ^* and σ^* from Section 4 we obtain the following observed confidence intervals using (13):

For sample 1: $I_\delta = [-0.360, -0.278]$ and $I_\gamma = [0.192, 0.251]$.

For sample 2: $I_\delta = [-0.195, -0.115]$ and $I_\gamma = [0.191, 0.250]$.

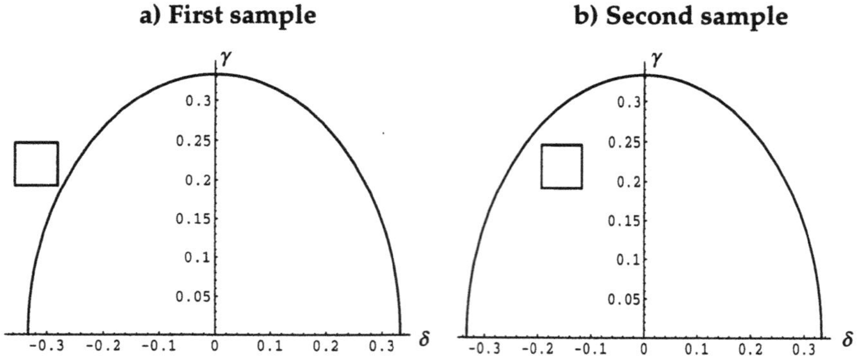

Fig. 5. The capability region bounded by the semi-circle defined by $C_{pm} = 1$ and the δ -axis. In a) the observed 76% confidence rectangle of (δ, γ) from the first sample is plotted. In b) the observed 76% confidence rectangle of (δ, γ) from the second sample is plotted.

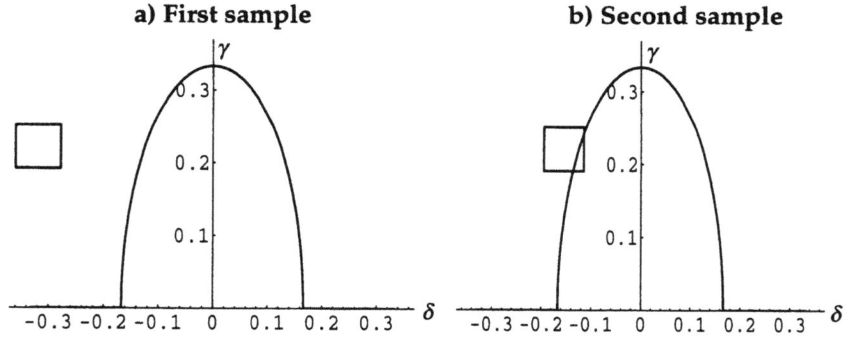

Fig. 6. The capability region bounded by the semi-ellipse defined by $C_p(0,4) = 1$ and the δ -axis. In a) the observed 80% confidence rectangle of (δ, γ) from the first sample is plotted. In b) the observed 80% confidence rectangle of (δ, γ) from the second sample is plotted.

In Figures 5 and 6 the confidence rectangles for the two cases have been plotted in the (δ, γ)-plot. We see that the conclusions are the same as in Section 4, when using the (δ^*, γ^*)-plot.

6 Discussion and Comparison of the Two Methods

One quite common problem in industry today is that practitioners interpret estimates of process capability indices as true values, see Deleryd (1998a). This leads to quite a few erroneous decisions. We believe that the studied capability plots show the uncertainty associated with these estimates more clearly than do estimated capability indices alone. In the confidence rectangle plot the uncertainty is shown by the rectangle, and in the (δ^*, γ^*)-plot by plotting the capability level curve together with the estimated capability level curve as in Figure 2b). Furthermore, the visual impact of a plot is more efficient than numbers like estimates or confidence limits.

From Figures 4 and 6 we can also clearly see the advantage of using process capability plots compared to using process capability indices alone. If we had studied sample 1 by using C_{pm} only we would have received the estimate $C_{pm}^* = 0.865$ together with a 95% confidence interval [0.761, 0.957], indicating that the process is non-capable, but not how to adjust the process in order to improve it. If using process capability plots, as in Figures 4 and 6, we see instantly that the process is non-capable, mainly because the process is off-target, but also that the process spread is not very small. The major conclusion is that process capability indices and process capability plots used together will help the practitioner to make proper decisions when improving the capability of processes.

The studied process capability plots are invariable irrespective of the values of the specification limits. Hence we can monitor, in the same plot, several characteristics of a process and at the same time retain the information on the location and spread of the process. In this way more information is obtained on how to improve the process compared to using traditional capability indices alone.

If we study the same single characteristic over a longer period of time we can plot the estimated indices from different time periods in one process capability plot and follow the change over time in a lucid way. When focusing on the time aspect and with the sample size n constant it seems that the (δ^*, γ^*)-plot will give the most clear picture. However, when focusing on the time aspect and with n non-constant, it seems that the confidence rectangle plot is to be preferred. For more details and discussions about the two methods from a practitioners point of view see Deleryd & Vännman (1998).

In Deleryd & Vännman (1998) the two capability plots discussed here have also been compared from a theoretical point of view based on power comparisons. The two plots are compared when using C_{pm} as an index and studying the null hypothesis $H_0 : C_{pm}$ • 4/3 and the alternative hypothesis $H_1 : C_{pm} > 4/3$. The significance level is 0.05 for each plot. The sample sizes $n = 30, 50, 80$ are studied and the power is calculated for each one of $C_{pm} = 5/3$, 1.8, and 2.0. The results from this comparison show that for values of δ not far from 0, i.e., when the process output is close to target, the (δ^*, γ^*)-plot gives the largest power. But for values of δ close to 0.25, implying γ-values close to 0, the confidence rectangle plot gives the largest power. For an example of the power functions see Figure 7.

In practical situations it is most common first to try to get the process on target and then to work successively with the reduction of the process spread. The reason for this particular order is that, in the long run, the number of defective products will be minimised compared to starting with minimising the process spread while ignoring the centering of the process. When following this practical procedure the process will be centred, in a relatively short time, and hence, based on the results above, the (δ^*, γ^*)-plot would be more efficient to use. For more details, see Deleryd & Vännman (1999).

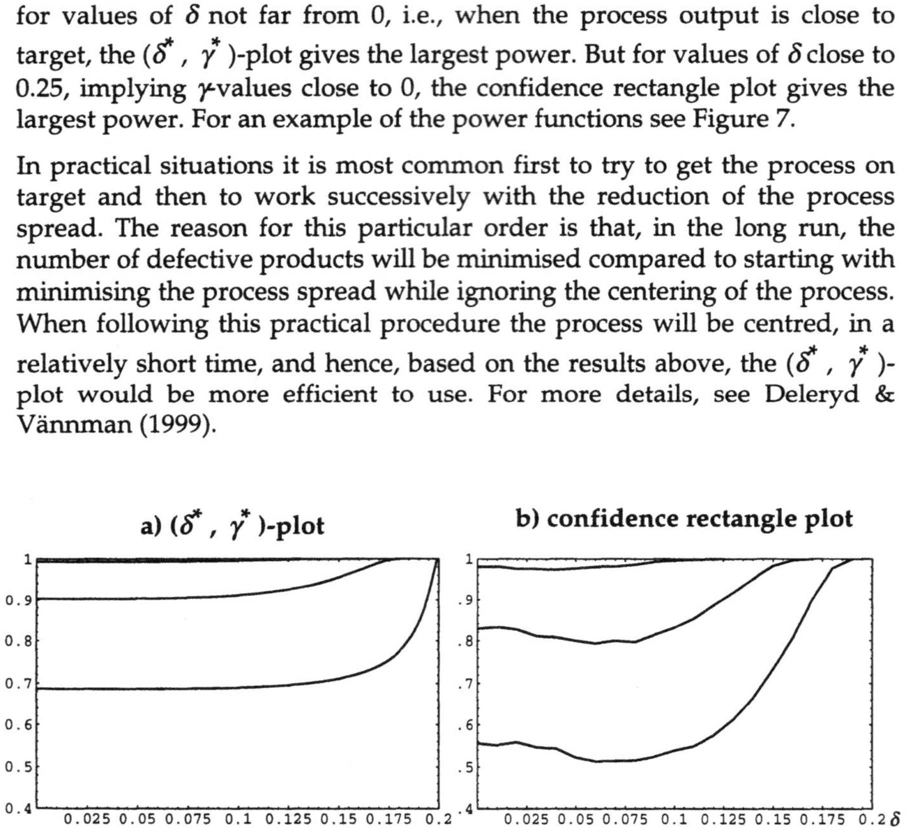

Fig. 7. The power function, i.e., the probability of considering the process capable for a given value of C_{pm}, when $n = 50$, using a) the (δ^*, γ^*)-plot and b) the confidence rectangle plot, plotted as a function of δ. In each figure, the top curve corresponds to $C_{pm} = 2.0$, the middle curve to $C_{pm} = 1.8$, and the bottom curve to $C_{pm} = 5/3$.

However, the most important conclusion is to use any of the plots due to the advantages already discussed, compared to using the capability index alone.

7 Concluding Remarks

From an improvement point of view, more information on how to improve the process is received when using process capability plots compared to using process capability indices alone. By using the studied process capability plots we focus on the two parameters of importance when controlling and improving a process, closeness to target and process spread. The recommendation is that practitioners should use process capability plots as a complement to process capability indices. The use of process capability indices admits them a new role when used together with process capability plots.

Using the suggested process capability plots, compared to using the capability index alone, we will, from one single plot, instantly get visual information simultaneously about the location and spread of the studied characteristic, as well as information about the capability of the process. If the process is considered non-capable, we can see immediately and clearly whether it is the spread, the deviation from target, or both that need to be reduced to improve the capability. In this way the proposed plots give clear directions for quality improvement. If the process is considered capable we can also get information whether the studied characteristic is close to target, has small spread, or both. The capability indices were introduced to focus on the process variability and closeness to target and relate these quantities to the specification interval and the target value. We believe that the process capability plots discussed here will do this in a more efficient way than the capability index alone.

8 Acknowledgement

The author is grateful for the financial support provided by the Volvo Research Foundation, project number 96:08.

References

1. Bergman, B. & Klefsjö, B. (1994) *Quality from Customer Needs to Customer Satisfaction*. McGraw-Hill, New York and Studentlitteratur, Lund

2. Bickel, P. J. & Doksum, K. A. (1977) *Mathematical Statistics. Basic Ideas and Selected Topics*. Holden-Day, Inc., Oakland, California

3. Boyles, R. A. (1991) The Taguchi Capability Index. *Journal of Quality Technology* **23**, 17–26. Corrigenda in *Journal of Quality Technology*, 23, 382–383

4. Carlsson, O. (1995) On the Use of the C_{pk} Index for Detecting Instability of a Production Process. *Economic Quality Control* **10**, 39-50

5. Chan, L. K., Cheng, S. W. & Spiring, F. A. (1988) A New Measure of Process Capability: Cpm. *Journal of Quality Technology* **20**, 162–175

6. Gunter, B. H. (1989) The Use and Abuse of C_{pk} : Parts 1–4. *Quality Progress*, **22**; January, 72–73; March, 108–109; May, 79–80; July, 86–87

7. Deleryd, M. (1998a) On the Gap Between Theory and Practice of Process Capability Studies. *International Journal of Quality & Reliability Management* **15**, 178–191

8. Deleryd, M. (1998b) *Enhancing the Industial Use of Process Capability Studies*. Doctoral Thesis 1998:19, Luleå University of Technology, SE-971 87 Luleå, Swede

9. Deleryd, M. & Vännman, K. (1999) Process Capability Plots—A Quality Improvement Tool. *Quality and Reliability Engineering International* **15**, 1–15

10. Kotz, S. & Johnson, N. L. (1993) *Process Capability Indices*. Chapman & Hall, London

11. Kotz, S. & Lovelace, C. R. (1998) *Introduction to Process Capability Indices: Theory and Practice*. Arnold, London

12. Pearn, W. L. & Chen, K. S. (1997) A Practical Implementation of the Process Capability Index C_{pk}. *Quality Engineering* **9**, 721–737

13. Pearn, W. L., Kotz, S. & Johnson, N. L. (1992) Distributional and Inferential Properties of Process Capability Indices. *Journal of Quality Technology* **24**, 216–231

14. Phadke, M. S. (1989) *Quality Engineering Using Robust Design*. Prentice Hall, Englewood Cliffs, New Jersey 07632

15. Pignatiello, Jr, J. J. & Ramberg, J. S. (1993) Process Capability Indices: Just Say "No!". *ASQC Quality Congress Transactions 1993*, Boston, 92–104

16. Spiring, F. A. (1997) A Unifying Approach to Process Capability Indices, *Journal of Quality Technology* **29**, 49–58

17. Sullivan, L. P. (1984) Reducing Variability: A New Approach to Quality. *Quality Progress* **17(7)**, 15–21

18. Taguchi, G. (1986) *Introduction to Quality Engineering*. Asian Productivity Organization, Tokyo, Japan

19. Vännman, K. (1995) A Unified Approach to Capability Indices. *Statistica Sinica* **5**, 805-820

20. Vännman, K. (1997a) Distribution and Moments in Simplified Form for a General Class of Capability Indices. *Communications in Statistics—Theory and Methods* **26**, 159–179

21. Vännman, K. (1997b) A General Class of Capability Indices in the Case of Asymmetric Tolerances. *Communications in Statistics—Theory and Methods* **26**, 2049–2072

22. Vännman, K. (1997c) *Interpreting Process Capability Using Graphical Methods*. Research Report, Division of Quality Technology & Statistics, 1997:7

23. Vännman, K. (1998) Families of Capability Indices for One-sided Specification Limits. *Statistics* **31**, 43-66

24. Vännman, K. & Kotz, S. (1995a) A Superstructure of Capability Indices—Distributional Properties and Implications. *Scandinavian Journal of Statistics* **22**, 477–491

25. Vännman, K. & Kotz, S. (1995b) A Superstructure of Capability Indices—Asymptotics and Its Implications. *International Journal of Reliability, Quality, and Safety Engineering* **2**, 343–360

26. Wolfram, S. (1991) *Mathematica. A System for Doing Mathematics by Computer*. Wolfram Research Inc., 100 Trade Center Drive, Champaign, Illinois 61820-7237, USA, and Addison-Wesley Publishing Company, Redwood City, California

Confidence Limits for the Process Capability Index C_{pk} for Autocorrelated Quality Characteristics

ERIK WALLGREN

University of Örebro, ESA/Department of Statistics, S-701 82 Örebro

Abstract. A generalization of the C_{pk}-index, C_{pkr}, is introduced to cover situations when the data are generated from an AR(1)-process. Different approximations of the sampling distributions for an estimator are studied, and Monte Carlo simulations are performed to compare empirical coverage rates to nominal confidence levels for lower confidence levels for the C_{pkr}-index.

1 Introduction

The process capability index C_{pk}, introduced by Kane 1986, is probably the most used capability index in process industry. It relates the absolute deviation of the actual process mean from specification limits to the actual process variability and the index is defined as

$$C_{pk} = \min(USL - \mu, \mu - LSL)/3\sigma = (d - |\mu - T|)/3\sigma \tag{1.1}$$

where *USL* and *LSL* are upper and lower specification limits and μ and σ are the mean and standard deviation of a quality characteristic of the process. Finally, T is the midpoint and d the half-width of the specification interval, *(USL-LSL)/2*.

Normally μ and σ are unknown and have to be estimated. With the traditional estimators $\overline{X}$ and *S* for μ and σ a natural estimator for C_{pk} becomes

$$\hat{C}_{pk} = (d - |\overline{X} - T|)/3S \tag{1.2}$$

A common assumption when studying this capability index, and other indices, is that data can be considered as observations on identically, independently and normally distributed random variables, IIDN. In practice, however, some of these conditions are not fulfilled, and the assumption of independence is sometimes violated.

The aim of this paper is to derive appropriate lower confidence limits for C_{pk} when the data are observations on identically and normally, but not independently, distributed random variables. As a model for the dependency a 1.st order autoregressive model, AR(1), is used. Two different approaches are used to develop approximate confidence limits. Finally, in an extensive Monte-Carlo study the derived confidence limits will be compared to corresponding limits when the presence of dependence is ignored.

2 Moments and Distribution of $\hat{C}_{pk}$

The distribution of $\hat{C}_{pk}$ can be recognized by writing (1.2) as

$$\hat{C}_{pk} = \min\left[\left(USL - \overline{X}\right)/3S, \left(\overline{X} - LSL\right)/3S\right] \tag{2.1}$$

Here each of the ratios is a scaled non-central t distributed random variable and hence $\hat{C}_{pk}$ is distributed as a bivariate non-central t variable. The first two moments for this variable are presented by Chou and Owen (1989), and in a different form, by Zhang et al (1990). In the same paper by Zhang et al is also shown that the estimator for C_{pk} is asymptotically unbiased and, by Monte Carlo simulations, the distribution of $\hat{C}_{pk}$ is weakly positively skewed. For practical use, however the moments for this variable are too complicated and in the report by Zhang et al., approximate moments are derived. They also show that, unless the sample size is very small, a normal approximation is appropriate,

With these results, an approximate $(1-\alpha)\,\%$ confidence interval for C_{pk} is

$$\hat{C}_{pk} \pm z_{1-\alpha/2} \cdot \left(\hat{V}\left(\hat{C}_{pk}\right)\right)^{1/2} \tag{2.2}$$

where $\hat{V}\left(\hat{C}_{pk}\right)$ is an estimate of the approximate variance of $\hat{C}_{pk}$ and $z_{1-\alpha/2}$ is a percentile in a standard normal distribution.

However, even the approximate variance $V(\hat{C}_{pk})$, derived by Zhang et al., is, for practical applications, too complicated. In an earlier report Heavlin (1988) suggests the variance estimator

$$\hat{V}\left(\hat{C}_{pk}\right) \approx (n-1)/(9n(n-3)) + \hat{C}_{pk}^2(1+6/(n-1))/(2(n-3)) \tag{2.3}$$

For large samples Bissel (1990) suggests the reduced version:

$$\hat{V}\left(\hat{C}_{pk}\right) \approx 1/9n + \hat{C}_{pk}^2/(2(n-1)) \tag{2.4}$$

which Kushler and Hurley (1992) reduce further to

$$\hat{V}\left(\hat{C}_{pk}\right) \approx \hat{C}_{pk}^2/(2(n-1)) \tag{2.5}$$

In a large simulation study, Kushler and Hurley (1992) compares the confidence intervals based on the three variance estimates (2.3), (2.4) and (2.5), (and three more) with respect to empirical coverage rates and to nominal confidence levels. They find that all the intervals studied are conservative when μ is close to T. However, when μ differs from T, the approximations show different variance estimations and Kushler and Hurley conclude that, for a general use, the variance estimation suggested by Bissel is to be recommended. In the following sections of this report, the approximate variance (2.4) will be used.

3 Generalization of C_{pk} for Autocorrelated Variables

Suppose that the observations on a quality characteristic follow an AR(1)-process, i.e.

$$X_t - \mu = \rho(X_{t-1} - \mu) + \varepsilon_t \tag{3.1}$$

where μ is the actual process mean and ρ is the 1.st order autocorrelation coefficient. Assume that ε_t is the noise of the process where $\varepsilon_t \sim IIN(0,\sigma^2)$, $t = 1,\ldots,n$.

It is well known that $V(X_t) = \sigma^2/(1-\rho^2)$, asymptotically, and with this variance, the C_{pk}-index for a quality characteristic generated by an AR(1) process can be defined as,

$$C_{pkr} = \min(USL - \mu, \mu - LSL)/3\sigma_r \tag{3.2}$$

where $\sigma_r^2 = \sigma^2/(1-\rho^2)$ and here the extra '*r*' indicates the presence of an autocorrelation.

C_{pk} and C_{pkr} are compared graphically in Figure 3.1 in a ($(\mu - T), \sigma$) coordinate system. Let C_{pk}=c. Then the C_{pkr}-index can be represented by the straight lines $\sigma = d/3c \pm (\mu - T)/3c$ and if C_{pkr}=c, the lines change to

$$\sigma = d(1-\rho^2)^{1/2}/3c \pm (\mu - T)(1-\rho^2)^{1/2}/3c$$

Further, as a reference, the figure includes the index $C_p = (USL - LSL)/6\sigma$, where $C_p = c$ is represented by the line $\sigma = d/3c$. It can be noted that the presence of an autocorrelation can be interpreted as adding an extra component of variability, and to maintain a constant value of C_{pkr}, when $|\rho|$ increases, the variance of the noise has to decrease.

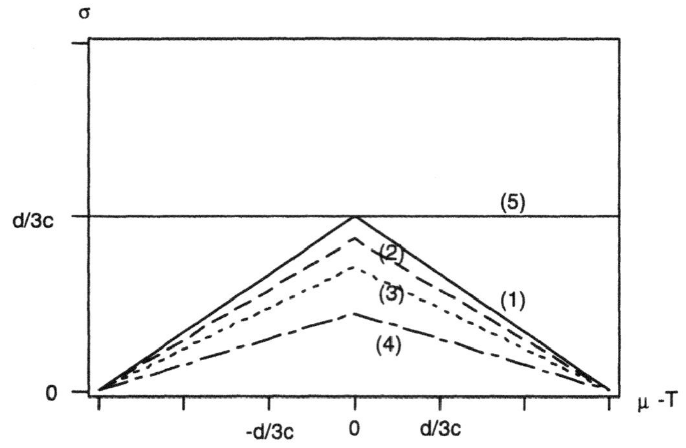

Fig. 3.1. Contours for C_{pkr} =c, when (1):ρ=0, (2): ρ=.5, (3) ρ =.7 and (4): ρ =.9. As a reference, C_p =c, (5) is included in the figure.

4 Estimation of the C_{pkr} -index

4.1 Moments of $\hat{C}_{pkr}$

4.1.1 General Remarks

The C_{pkr} -index includes three parameters, μ, σ and ρ which all have to be estimated. The ML-estimators for the parameters are discussed in Wallgren (1996), and to avoid situations when the sample autocorrelation $|\hat{\rho}| > 1$ a modified estimator for ρ was introduced which also is used in this paper. The alternative estimator was studied by simulations and it was found that the coverage rate for a lower confidence limit was almost exactly the same as for the ML-estimator.

The estimators are

$$\hat{\mu}_t = \sum_{t=1}^{n} X_t / n \tag{4.1}$$

$$\hat{\rho} = (\sum_{t=1}^{n} X_t X_{t-1} - n\hat{\mu}_t \hat{\mu}_{t-1}) / (\sum_{t=1}^{n} X_{t-1}^2 - n\hat{\mu}_{t-1}^2) \tag{4.2}$$

$$\hat{\sigma}^2 = (1 - \hat{\rho}^2) \sum_{t=1}^{n} (X_t - \hat{\mu}_t)^2 / n \tag{4.3}$$

Let the sample mean, in (4.1), be denoted $\overline{X}$ and let S^2 denote $\sum_{t=1}^{n} (X_t - \hat{\mu}_t)^2 / n$. Then, from (4.3), $S^2 = \hat{\sigma}^2 / (1 - \hat{\rho}^2)$ which emphasizes the fact that S^2 is an estimator of the total process variance, σ_r^2.

When $\overline{X}$ and S^2 are calculated from n consecutive observations on an AR(1)-process the moments can be found by simple, but tedious, calculations:

$$E(\overline{X}) = \mu \tag{4.4}$$

$$V(\overline{X}) = \kappa_1 \sigma_r^2 / n \tag{4.5}$$

$$E(S^2) \approx \sigma_r^2 (1 - \kappa_1 / n) \tag{4.6}$$

and

$$V(S^2) \approx 2\kappa_2 \sigma_r^4 / n \tag{4.7}$$

where $\kappa_1 = (1 + \rho)/(1 - \rho)$ and $\kappa_2 = (1 + \rho^2)/(1 - \rho^2)$. For $V(S^2)$, see Priestley, (1981). Zhang (1998) gives expressions for $E(S^2)$, $V(\overline{X})$ and $V(S^2)$ for a discrete stationary Gaussion process.

An estimator for the C_{pkr} -index is then

$$\hat{C}_{pkr} = \min\left(USL - \overline{X}, \overline{X} - LSL\right)/3S \tag{4.8}$$

which is identical to the traditional estimator for C_{pk}.

To find the variance for $\hat{C}_{pkr}$ it is essential that (4.1)-(4.3) are pairwise uncorrelated, at least asymptotically. This is proved by Zhang (1998) for $\overline{X}$ and S^2. Since $S^2 = \hat{\sigma}^2/(1-\hat{\rho}^2)$ and $\hat{\rho} = \text{cov}(X_t, X_{t-1})/S^2$ all three are uncorrelated if $\overline{X}$ and $\text{cov}(X_t, X_{t-1})$ are uncorrelated. A proof is given in APPENDIX A.

4.1.2 The Bissel Approach

Two alternative approaches are made and the first one is based on the variance estimator (2.4), suggested by Bissel.

The C_{pkr} -index can be written as

$$C_{pkr} = \left(1-\rho^2\right)^{1/2} \cdot C_{pk} \tag{4.9}$$

and a natural estimator for C_{pkr} is then

$$\hat{C}_{pkr} = \left(1-\hat{\rho}^2\right)^{1/2} \cdot \hat{C}_{pk} \tag{4.10}$$

To find the moments for the square root term, moments for $\hat{\rho}$ can be used. Approximate moments for the square root function of ρ can be found by Taylor linearization:

Denote $g(\hat{\rho}) = \left(1-\hat{\rho}^2\right)^{1/2}$. Then

$$E\left(g(\hat{\rho})\right) \approx \left(1-\hat{\rho}^2\right)^{1/2} \tag{4.11}$$

and

$$V(g(\hat{\rho})) \approx (g'(\rho))^2 \cdot V(\hat{\rho}) = \rho^2 / n \tag{4.12}$$

Under the condition that $\hat{\rho}$ is uncorrelated to $\hat{\mu}$ and $\hat{\sigma}$, the mean and the variance for $\hat{C}_{pkr}$ can be approximated with

$$E(\hat{C}_{pkr}) \approx E\left((1-\hat{\rho}^2)^{1/2}\right) \cdot E(\hat{C}_{pk}) \tag{4.13}$$

$$V(\hat{C}_{pkr}) \approx \left[E\left((1-\hat{\rho}^2)^{1/2}\right)\right]^2 \cdot V(\hat{C}_{pk}) + V\left((1-\hat{\rho}^2)^{1/2}\right) \cdot \left[E(\hat{C}_{pk})\right]^2 + V\left((1-\hat{\rho}^2)^{1/2}\right) \cdot V(\hat{C}_{pk}) \tag{4.14}$$

Insertion of the approximate variances (2.4) and (4.12) gives

$$V(\hat{C}_{pkr}) = C_{pk}^2 \left[(n + \rho^2(n-1))/(2n(n-1))\right] + \left((1-\rho^2) + \rho^2/n\right)/9n \tag{4.15}$$

and, after an insertion of (4.10), and for large samples, this can be approximated by,

$$V(\hat{C}_{pkr}) = C_{pkr}^2 \cdot \kappa_2 / (2(n-1)) + (1-\rho^2)/9n \tag{4.16}$$

4.1.3 Taylor Linearization

This approach is based on a bivariate Taylor linearization of the C_{pkr}-index.

If $\overline{X}$ is normally distributed $|\overline{X} - T|$ can be written

$$\left|\overline{X}-T\right| = +\left[\left(\left(\overline{X}-T\right)/\sigma\left(\overline{X}\right)\right)^2 \cdot \sigma^2\left(\overline{X}\right)\right]^{1/2} = +\sigma\left(\overline{X}\right)\cdot\left(\chi^2(1,\lambda)\right)^{1/2} \tag{4.17}$$

where $\chi^2(1,\lambda)$ is a non-central χ^2-variable with 1 df and the noncentrality parameter $\lambda = \left((\mu - T)/\sigma\left(\overline{X}\right)\right)^2$. Then an estimator for the C_{pkr}-index becomes

$$\hat{C}_{pkr} = \left[d - \sigma\left(\overline{X}\right)\cdot\left(\chi^2(1,\lambda)\right)^{1/2}\right]/3S \tag{4.18}$$

i.e. a function, here denoted *f*, of the two independent variables S^2 and $\chi^2(1,\lambda)$ whose first two moments are (4.6), (4.7), and $(1+\lambda)$ and $2(1+2\lambda)$ respectively.

The first two approximate moments for $\hat{C}_{pkr}$ can then be found by a Taylor linerarization.

$$E\left(\hat{C}_{pkr}\right) \approx \left[d - \sigma\left(\overline{X} \cdot (1+\lambda)^{1/2}\right)\right] / \left[9\sigma_r^2(1-\kappa_1)/n\right]^{1/2} \tag{4.19}$$

For large samples $\sigma^2\left(\overline{X}\right) \approx 0$ and $(1-\kappa_1/n)^{1/2} \approx 1$. Then

$$E\left(\hat{C}_{pkr}\right) \approx C_{pkr} \tag{4.20}$$

To simplify the notations, let χ^2 denote $\chi^2(1,\lambda)$. Then the variance becomes

$$V\left(\hat{C}_{pkr}\right) \approx V\left(\chi^2\right)\cdot\left[f'_{\chi^2}\left(E\left(\chi^2\right), E\left(S^2\right)\right)\right]^2 + V\left(S^2\right)\cdot\left[f'_{S^2}\left(E\left(\chi^2\right), E\left(S^2\right)\right)\right]^2 \tag{4.21}$$

After insertion of the moments, and some algebra, the expression for the variance can be reduced to

$$V\left(\hat{C}_{pkr}\right) \approx C_{pkr}^2 \cdot \kappa_2 / \left[2n(1-\kappa_1/n)^3\right] + \kappa_1 / \left[9n(1-\kappa_1/n)\right] \cdot (1+2\lambda)/(2+2\lambda) \quad (4.22)$$

which for large sample sizes further reduces to

$$V\left(\hat{C}_{pkr}\right) \approx C_{pkr}^2 \cdot \kappa_2 / 2n + \kappa_1 / 9n \cdot (1+2\lambda)/(2+2\lambda) \quad (4.23)$$

Finally, the last ratio, $(1+2\lambda)/(2+2\lambda)$, obviously belong to the interval (.5,1) and for large samples the ratio is close to 1 which reduces the variance

$$V\left(\hat{C}_{pkr}\right) \approx C_{pkr}^2 \cdot \kappa_2 / 2n + \kappa_1 / 9n \quad (4.24)$$

4.2 Sample Size

Traditionally the sample sizes in quality control work are small. A sample size of n=5 is common in control charts for means etc. But capability indices requires larger samples and even a sample size of n=40 leaves substantial uncertainty, Kotz, Johnson (1993). However, when measurements are autocorrelated, even such sample sizes are too small. Box, Jenkins, Reinsel (1994), recommends at least 50 observations in estimating an autocorrelation function. In the Monte Carlo simulations outlined in the following sections sample sizes smaller than n=50 are therefore not used.

4.3 A Sampling Distribution of $\hat{C}_{pkr}$

An explicite expression for a probability density function for $\hat{C}_{pkr}$ can probably not be found, except for, as mentioned in section 2, the case when $\rho = 0$. In this special case, the distribution is found to be slightly positively skewed and it can for large sample sizes be approximated by a normal distribution.

Approximate sampling distributions for $\hat{C}_{pkr}$ are given in the below Figures 4.1 and 4.2. The distributions are simulated for four different parameter settings, see

Table 4.1, and here the parameters μ and σ are chosen, together with *T=0* and *d=3*, to make $C_{pkr} = 1$ when ρ =0 and $C_{pkr} = \sqrt{(1-\rho^2)}$ when ρ ≠0. (The combination $\mu = 0$ and $\sigma = 1$ is in section 5 referred to as Case A and the combination $\mu = 1.5$ and $\sigma = .5$ as Case B). In Figure 4.1, n=50 and in Figure 4.2, n=200. For both figures panels a) and b) show results for Case A and panels c) and d) for Case B. For each combination 10 000 samples were simulated.

Table 4.1. Sets of parameters for sampling distributions of $\hat{C}_{pkr}$

μ	σ	ρ
0	1	0
0	1	.5
1.5	.5	0
1.5	.5	.5

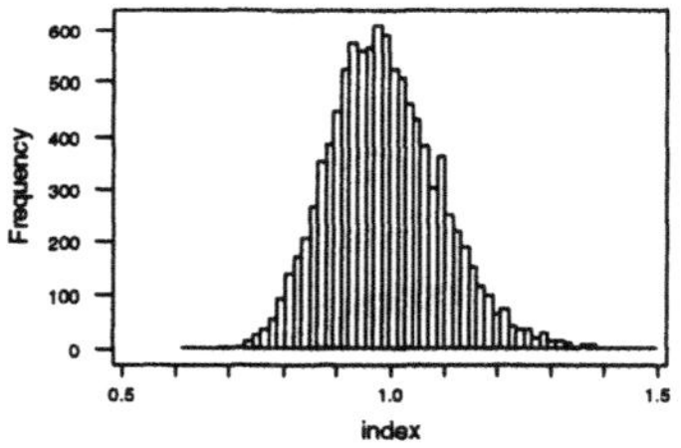

(a) μ =0, σ =1 and ρ =0. $C_{pkr} = 1$

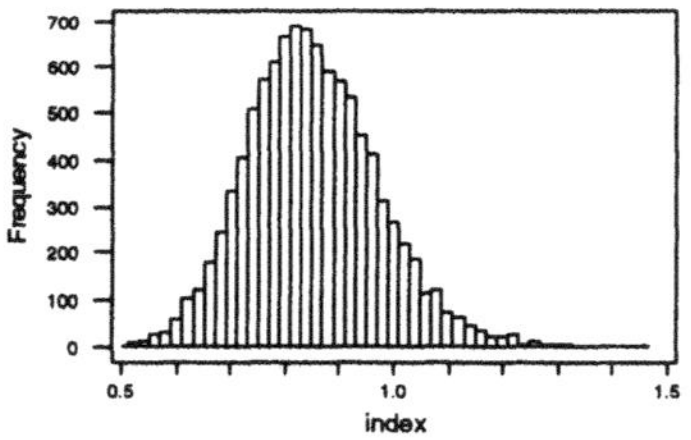

(b) μ =0, σ =1 and ρ =.5. $C_{pkr} \approx .87$

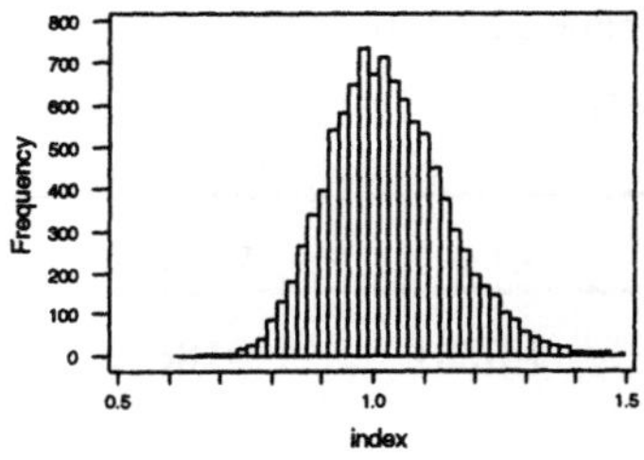

(c) μ =1.5, σ =.5 and ρ =0. $C_{pkr}=1$

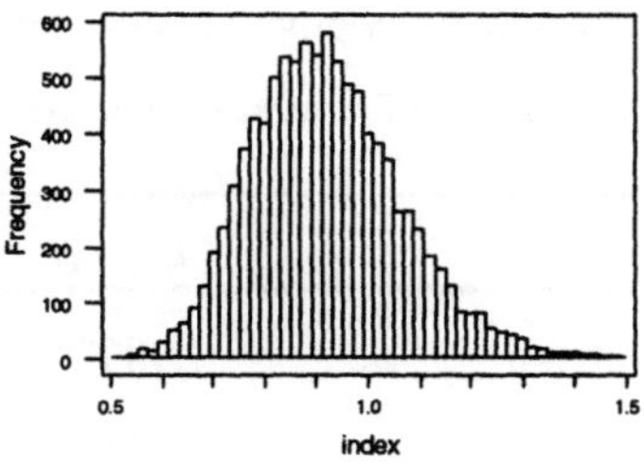

(d) μ =1.5, σ =.5 and ρ =.5. $C_{pkr} \approx .87$

Fig. 4.1 (a) - (d). Estimated sampling distributions for $\hat{C}_{pkr}$ for different sets of process parameters. Sample size n=50.

When comparing the two right hand panels to the two at the left, it is notable that the variance is higher but that the skewness and the kurtosis is lower, more 'normallike', when the autocorrelation is added to the process. Descriptive measures for the four estimated distributions are given in Table 4.2 and there it can be seen, that C_{pkr} appears to be underestimated when the process is on target, (a) and (b), and overestimated when the process is off target, (c) and (d).

In the Figures 4.1.2 (a) - (d) the sample size in increased to 200. For all cases it is reasonably to believe that the sampling distributions for $\hat{C}_{pkr}$ is closer to normal compared to the situation with n=50, which is also indicated by the decreasing skewness and kurtosis measures, see Table 4.2. The means in the table now indicate that the true C_{pk} is underestimated both when the process is on and off target.

From these simulations we conclude that $\hat{C}_{pkr}$ is approximately normally distributed with the moments discussed in sections 4.1.2 and 4.1.3.

Table 4.2. Descriptive measures for estimated sampling distributions for $\hat{C}_{pkr}$,for different sets of process parameters. Sample size n=50

	4.1 (a)	4.1 (b)	4.1 (c)	4.1 (d)
mean	.9892	.8500	1.0281	.9186
variance	.0110	.0145	.0134	.0212
skewness	.4322	.3776	.4308	.4517
kurtosis	.3415	.3717	.3291	.3264

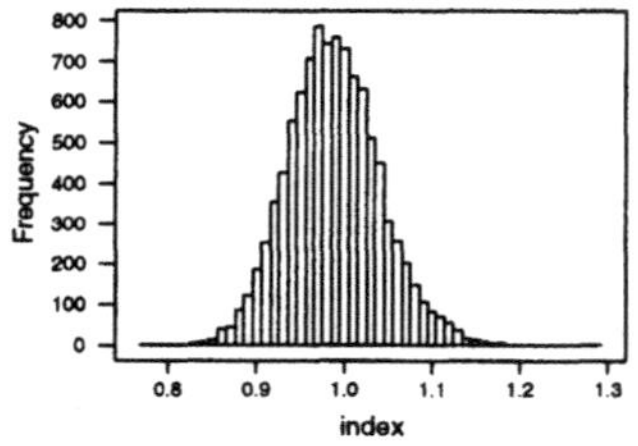

(a) μ =0, σ =1 and ρ =0. $C_{pkr}=1$

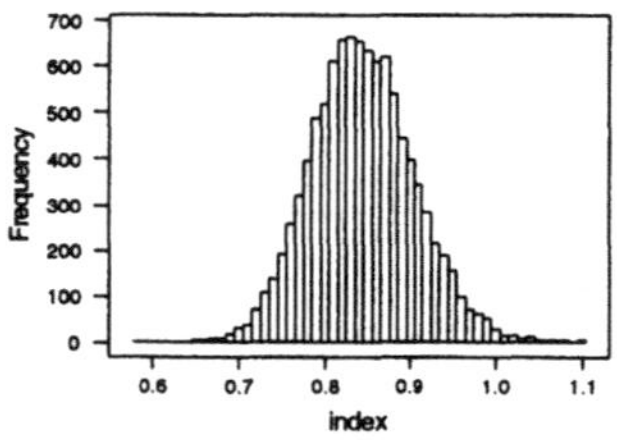

(b) μ =0, σ =1 and ρ =.5. $C_{pkr}\approx.87$

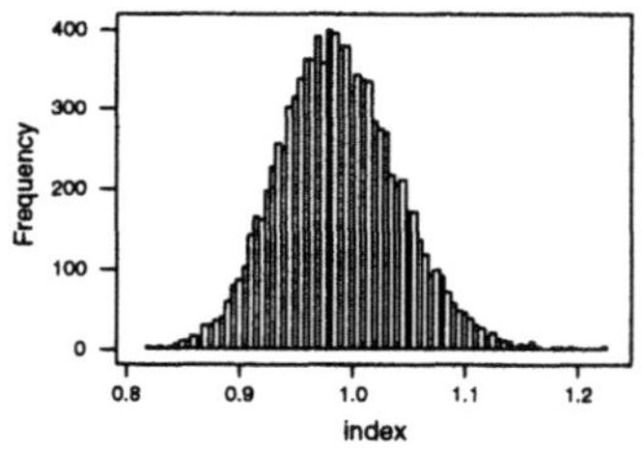

c) μ =1.5, σ =.5 and ρ =0. $C_{pkr}=1$

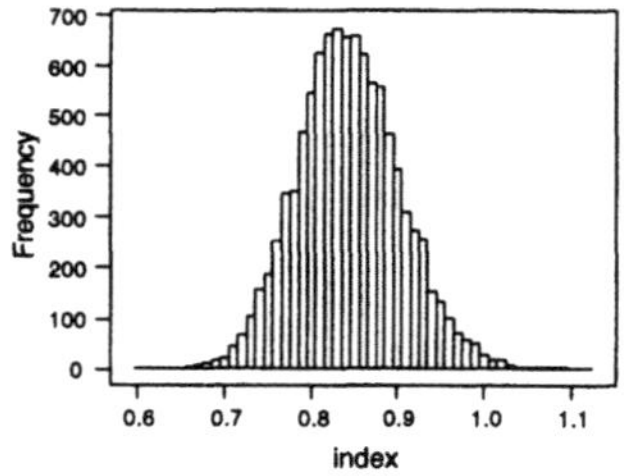

(d) μ =1.5, σ =.5 and ρ =.5. $C_{pkr}\approx.87$

Fig. 4.1.2 (a) - (d). Estimated sampling distributions for $\hat{C}_{pkr}$ for different sets of process parameters. Sample size n=200.

Table 4.2. Descriptive measures for estimated sampling distributions for $\hat{C}_{pkr}$ for different sets of process parameters. Sample size n=200

	4.2 (a)	4.2 (b)	4.2 (c)	4.2 (d)
mean	0.9877	0.8443	0.9878	0.8437
variance	0.0027	0.0036	0.0027	0.0035
skewness	0.2575	0.2041	0.2257	0.2080
kurtosis	0.1926	0.1294	0.0191	0.1279

4.4 Confidence Limits for C_{pkr}

Then approximate confidence limits for C_{pkr} can be calculated.

For the Bissel approach, using the variance (4.16), an approximate $(1-\alpha)\%$ lower confidence limit for C_{pkr} is

$$\hat{C}_{pkr} - z_{1-\alpha}\left\{\hat{C}_{pkr}^2 \cdot \left(1+\hat{\rho}^2\right)/\left[\left(1-\hat{\rho}^2\right)2(n-1)\right] + \left(1-\hat{\rho}^2\right)/9n\right\}^{1/2} \tag{4.25}$$

For the Taylor linearization approach, four different variance estimates are suggested, (4.21)-(4.24) and with the last estimate, an approximate $(1-\alpha)\%$ lower confidence limit for C_{pkr} is

$$\hat{C}_{pkr} - z_{1-\alpha} \cdot \left\{\hat{C}_{pkr}^2 \left(1+\hat{\rho}^2\right)/\left[\left(1-\hat{\rho}^2\right)2n\right] + (1+\hat{\rho})/\left[(1-\hat{\rho})9n\right]\right\}^{1/2} \tag{4.26}$$

5 Monte Carlo Simulation Studies

In this section the accuracy of the different estimations of an approximated lower confidence limits achieved in sections 4.1.2 and 4.1.3 is illustrated by means of a Monte Carlo simulation study.

For a given set of process parameters, a large number of samples, 10 000, from an AR(1)-process are simulated. For each sample the C_{pkr} -index and a $(1-\alpha)\%$ lower confidence limit for C_{pkr} are estimated. Then for each set of process parameters the empirical, percentage coverage rate, EPCR, is determined and compared to the nominal confidence level, NCL where

$$EPCR = \frac{\#\left(LL_i < C_{pkr}\right)}{10000} \cdot 100$$

and where LL_i is the lower confidence limit for C_{pkr} estimated from sample i, and NCL=$(1-\alpha)\%$.

The simulations are made for a nominal confidence of 95%, two different sample sizes, n=50 and n= 200 and autocorrelations in a range from -.9 to .9 in steps of .1.

Two parameter settings are used: case A with μ =0 and σ =1 and case B with: $\mu = 1.5$ and σ=.5. Together with the chosen upper and lower specification limits: 3 and -3, respectively, C_{pk} for both cases equals 1.

The result of the simulations are given in the following three sections:

5. Comparison of the Variance Estimates from the Taylor Linearization

As could be expected, the differences between the complete variance estimates, formula (4.21) and the successive reductions (4.22)- (4.24) diminish when the sample size increases. For n=200, the difference in EPCR when using (4.22) and the most simplified variance (4.24), see Figure 5.1.b, is never larger than 1%. For n=50 larger differences occur when the autocorrelation is high, $\rho > .5$, see Figure 5.1.a. For both sample sizes the largest differences occur when the EPCR is much lower than the NCL.

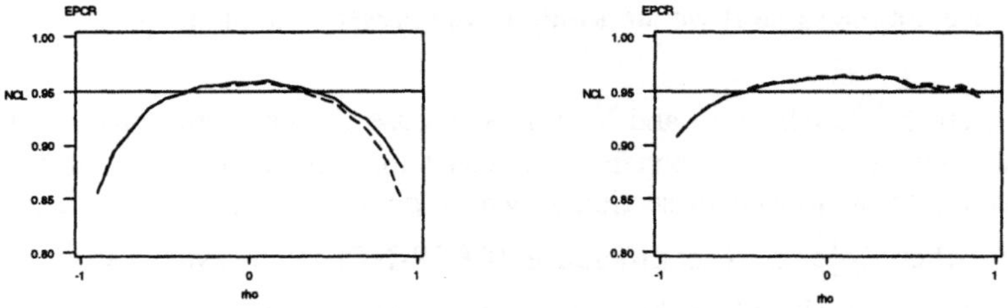

(a) Sample size 50 (b) Sample size 200

Fig. 5.1. EPCR for lower confidence limit (4.22) (—) and (4.24) (- - -) for autocorrelations (-.9:.9). Case A.

A summary of the first comparison is that whenever the Taylor series approach gives an approximate lower confidence limit for C_{pkr} with an EPCR reasonably close to the NCL, the unreduced variance (4.21) gives the best result but for practical reasons the most reduced variance (4.24) can be used.

5.2 EPCR for the Bissel and Taylor Linearization Approaches

A comparison between the variance estimates, Bissel, and Taylor series, is made by comparing the EPCR for lower confidence limits estimated by formulas (4.25) and (4.26). The difference between EPCR for the two estimates are neglectable, except for extremely high autocorrelations, $|\rho| > .8$, where (4.26) gives a better result.

Below the lower confidence limit (4.24) will be used in the following result section.

5.3 EPCR when Including or Ignoring the Presence of an Autocorrelation

In the following figures, the EPCR based on (4.26) are compared to the EPCR based on lower confidence limits (2.2), recommended by Bissel. It can be noted that the point estimate for C_{pkr} is the same for both limits since the sample mean and variance are the same if we are aware of the presence of an autocorrelation or not.

In the Figure 5.3.1, panels a) and b), refers to case A, where the process mean equals the target. For both panels it is clear that the confidence limits are conservative, as mentioned in section 2, when $\rho = 0$ but that the coverage rate decreases when $|\rho| > 0$. The difference (EPCR-NCL), however, varies if the autocorrelation is included in the estimations or not. Figure (b) shows that the EPCR is close to the NCL for all autocorrelations $\rho > -.5$ when autocorrelation is included but only for $|\rho| < .2$ when not. For *n*=50, Figure (a), the EPCR is close to the NCL for $|\rho| < .4$ when autocorrelation is included in the estimations.

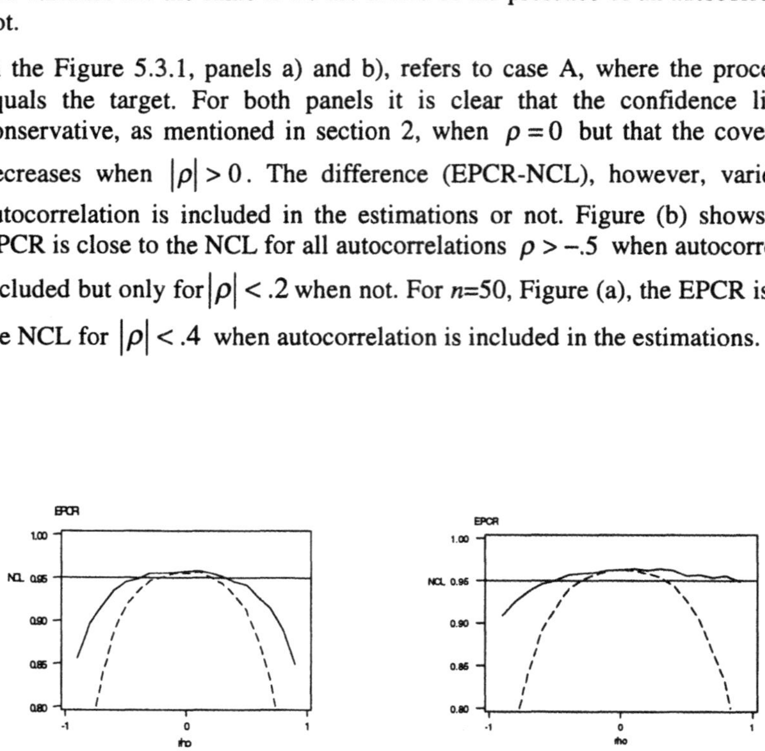

(a) (b)

Fig. 5.3.1. EPCR for lower confidence limit when autocorrelation is included (—) and not included (- - -) for autocorrelations (-.9:.9). Sample sizes: (a) 50 and (b) 200. Case A. NCL=95%

The Figures 5.3.2 panels (a) and (b), refers to case B, where the process mean is not equal to the target. Here the EPCR is always lower than the NCL, both if the autocorrelation is included or not. Again, the difference (EPCR-NCL) depends on the size of the autocorrelation. For large samples, panel (b), the confidence limits is reasonably accurate for autocorrelations $|\rho| < .6$ but only for $\rho < |.2|$ if the autocorrelation is ignored in the estimations. For n=50, an interval, panel (a), the difference (EPCR-NCL) is only acceptable small for autocorrelations closed to 0.

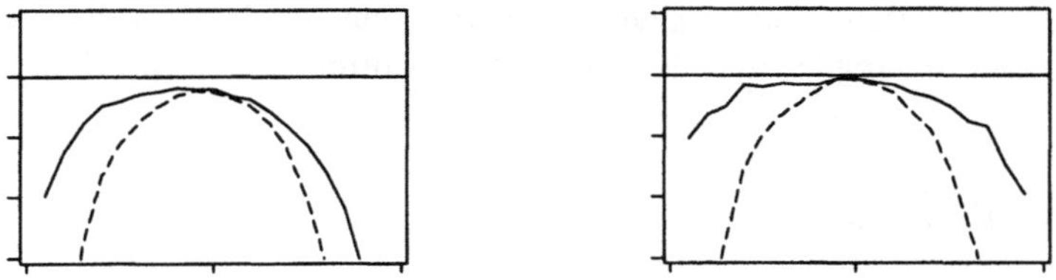

(a) (b)

Fig. 5.3.2. EPCR for lower confidence limit when autocorrelation is included (—) and not included (- - -) for autocorrelations (-.9:.9). Sample sizes: (a) 50 and (b) 200.

The simulations discussed in section 5 are also made for other confidence levels, 90% and 99%, and for intermediate sample sizes. The results from these are similar to the results given in sections 5.1 to 5.3.

6 Summary

If, in a production process, successive observations on a quality characteristic are autocorrelated, it is found that the traditional confidence limits for C_{pk} are not appropriate. An alternative index, C_{pkr}, was introduced. Approximate lower confidence limits for the index were developed based on two different approaches, Bissel and a Taylor linearization. For the Taylor approach three successive simplifications of the index variance were deduced. Large Monte Carlo simulations evaluated the confidence limits, where empirical coverage rates were compared to nominal confidence limits. The simulations showed that

large samples are necessary to achieve satisfactory appropriate coverage rates and sample sizes smaller than 50 can not be recommended.

the most simplified variance estimator of the Taylor approach gives a satisfactory coverage rate which corresponds with the coverage rate for the Bissel approach.

if the process mean equals the chosen process target, the empirical coverage rate will be closer to the nominal confidence level for most values of autocorrelation, especially for large samples, if an estimated autocorrelation coefficient is included in the estimation,

if the process mean does not equal the chosen process target, the empirical coverage rate will be closer to the nominal confidence level for a limited interval of the autocorrelation, if an estimated autocorrelation coefficient is included in the estimation. If the sample size is reduced, the autocorrelation range will decrease.

REFERENCES

1. Bissel, A.F. (1990). How reliable is Your Capability Index? Royal Statistical Society **39 (3)**, 331-340
2. Box, G.E.P., Jenkins, G.M. and Reinsel, G.C. (1994). Time Series Analysis. Prentice Hall, New Jersey
3. Carlsson, O. (1995). On the Use of the Cpk Index for Detecting Instability of a Production Process. Economic Quality Control **10**, 39-50.
4. Chou, Y.M. and Owen, D.B.,(1989). On the Distribution of the Estimated Process Capability Index. Communications in Statistics-Theory and Methods **18 (12)**, 4549-4560
5. Heavlin, W.D. (1988). Statistical properties of capability indices, Technical Report No. 320, Tech. Library, Advanced Micro Devices, Inc., Sunnyvale, California
6. Kane, V.E. (1986). Process Capability Indices. Journal of Quality Technology **18**, 41-52
7. Kendal, M.G. and Stuart, A. (1968). The advanced theory of statistics. Volume 3, Griffin &Co. Ltd. London, 450-451
8. Kotz, S. and Johnson, N.L. (1993). Process Capability Indices. Chapman & Hall, London
9. Kushler, R.H and Hurley, P. (1992). Confidence Bounds for Capability Indices. Journal of Quality Technology **24**, 188-195
10. Parzen, E. (1962). Stochastic Processes. San Francisco, CA, Holden Day
11. Wallgren, E. (1996). Properties of the Taguchi Capability Index for Markov Dependent Quality Characteristics. Lic.Thesis, Usala University
12. Zhang, N.F., Stenback, G.A. and Wardrop, D.M. (1990). Interval Estimation of Process Capability Index C_{pk}. Communications in Statistics-Theory and Methods **19**, 4455-4470
13. Zhang, N. F. (1998). Estimating process capability indexes for autocorrelated data. Journal of Alied Statistics **25 (4)**, 559-574

APPENDIX A

In this appendix we show that the covariance of $\overline{X}$ and $\text{cov}(X_t, X_{t-1})$ is zero, assuming that X_t is a stationary Gaussian process.

For a stationary time series, $\text{cov}(x_i, x_j)$ only depends on the difference *(j-i)* and this covariance is here denoted R_{j-i}. Then

$$\text{cov}\left(\overline{X}, \text{cov}(X_t X_{t-1})\right) = \text{cov}\left[\Sigma X_t, (\Sigma X_t X_{t-1} - \Sigma X_t \Sigma X_{t-1} / n)\right] / (n(n-1) =$$

$$\left[\sum_{i=1}^{n}\sum_{j=1}^{n} \text{cov}(X_i, X_j X_{j-1}) - \sum_{i=1}^{n}\sum_{j=1}^{n}\sum_{k=1}^{n} \text{cov}(X_i, X_j X_{k-1}) / n\right] / (n(n-1)$$

Let $X_i = \mu + Z_i$, where Z_i is normally distributed with zero mean. Then, since $E(Z_1 Z_2 Z_3) = 0$, Parzen (1962) and $\text{cov}(X_1, X_2 X_3) = \mu(R_{12} + R_{13})$, Zhang (1998) each covariance in the doublesum equals $\mu(R_{j-i} + R_{j-1-i})$ and each covariance in the triplesum is equal to $\mu(R_{j-i} + R_{k-1-i})$. The covariance becomes

$$\text{cov}\left(\overline{X}, \text{cov}(X_t X_{t-1}\right) =$$

$$\left[\sum_{i=1}^{n}\sum_{j=1}^{n} \mu(R_{j-i} + R_{j-1-i}) - \sum_{i=1}^{n}\sum_{j=1}^{n}\sum_{k=1}^{n} \mu(R_{j-i} + R_{k-1-i}) / n\right] / (n(n-1) =$$

$$\left[\sum_{i=1}^{n}\sum_{j=1}^{n} R_{j-i} + \sum_{i=1}^{n}\sum_{j=1}^{n} R_{j-1-i} - \sum_{i=1}^{n}\sum_{j=1}^{n}\sum_{k=1}^{n} R_{j-i} / n - \sum_{i=1}^{n}\sum_{j=1}^{n}\sum_{k=1}^{n} R_{k-1-ii} / n\right] \mu / (n(n-1) =$$

$$\left[\sum_{i=1}^{n}\sum_{j=1}^{n} R_{j-i} + \sum_{i=1}^{n}\sum_{j=1}^{n} R_{j-1-i} - \sum_{i=1}^{n}\sum_{j=1}^{n} R_{j-i} - \sum_{i=1}^{n}\sum_{k=1}^{n} R_{k-1-i}\right] \mu / (n(n-1) = 0$$

which completes the proof.

Part 4

Experimental Design

Split-Plot Experimentation for Process and Quality Improvement

Scott Kowalski[1] and G. Geoffrey Vining[2]

[1] University of Florida, Gainesville, Florida, USA
[2] Virginia Polytechnic Institute and State University, Blacksburg, VA, USA

Abstract. Industrial experiments often involve certain factors which are harder to change than others. For example, many process use furnaces, which take a significant amount of time to reach thermal equilibrium after a change in the temperature conditions. Thus, the thermal condition of the furnace is an example of a hard-to-change factor. A natural way to accommodate such a factor is through a split-plot experiment.

This paper reviews the current literature on the use of split-plot experiments in industry. It will outline some of the controversies present, the extension to robust parameter design, and the special problems presented by second-order experimentation.

1 Introduction

Many industrial experiments involve two types of factors: some with levels that are easy-to-change and one or more with levels that are difficult or costly to change. Consider a sintering process for battery anodes involving different densities of nickel powder and different furnace temperatures. The temperature profiles in a furnace require a great deal of time to reach equilibrium. Experimenters prefer to get as much information as they can for a single setting of the temperature conditions; consequently, furnace temperature is a hard-to-change factor. On the other hand, the operators can change nickel powders relatively easily. In this situation, a reasonable experiment fixes the level of furnace temperature and then runs all the levels of density for the nickel powder. The experiment then changes the setting of temperature and repeats the process. In the process, this experiment generates more information about the nickel powder, because it is replicated more often, than about the furnace temperature. Too often, experimenters analyze the data from such an experiment as if they were completely randomized, which can lead to incorrect conclusions as well as a loss of precision. Proper analysis of data obtained from such experiments must take the restricted randomization scheme into account.

Many text books (for example, Cochran and Cox (1957), Cox (1958), Hinkelmann and Kempthorne (1994), and Montgomery (1997)) outline "split-plot" experiments where the experimental unit for one factor (easy-to-change) is an observational or sub-unit for another factor (hard-to-change). Split-plot experiments originated in agriculture. Typically, we call the experimental unit for the hard-to-change factor the whole plot, and we call the sub-units the subplots. Every split-plot experiment requires two randomizations. Whole plot treatments are randomly assigned to whole plots based on the whole plot design. Within each whole plot, subplot treatments are randomly assigned to subplots with a separate randomization for each whole plot. The two separate randomizations cause two distinct error terms: one for the whole plot treatments and one for subplot treatments and the interaction between whole plot treatments and subplot treatments.

An example will help to illustrate the basic principles. Consider our sintering process in more detail. The engineers wish to study the furnace's belt speed (z_1), the furnace's peak temperature (z_2), the density of the nickel powder (x_1), and the amount of an additive necessary for bonding the material together during the sintering operation (x_2). Belt speed affects the furnace's equilibrium, so it along with peak furnace temperature are hard-to-change factors, denoted by the z's. Nickel density and the amount of the additive are easy-to-change factors, denoted by the x's. A reasonable design replicates a 2^2 factorial in the hard-to-change factors. Within each treatment combination of the hard-to-change factors, we run another 2^2 factorial in the easy-to-change factors. Table 1 illustrates the structure of this design. As shown, this design is blocked over time, which is typical in industrial experimentation.

Table 1. An Example of a 2^4 Factorial Design Run as a Split-Plot

$z_1 =$	−1	−1	1	1		−1	1	−1	1
$z_2 =$	1	−1	−1	1		−1	−1	1	1
$x_1 =$	−1	−1	−1	1		−1	1	−1	1
$x_2 =$	1	−1	1	1		−1	1	1	1
$x_1 =$	1	1	−1	−1		1	−1	1	−1
$x_2 =$	1	1	−1	−1		−1	1	−1	−1
$x_1 =$	1	1	1	−1		−1	−1	−1	−1
$x_2 =$	−1	−1	1	−1		1	−1	−1	1
$x_1 =$	−1	−1	1	−1		1	1	1	1
$x_2 =$	−1	1	−1	1		1	−1	1	−1

Classical split-plot analysis originated in agriculture where the treatments were typically categorical with a single whole plot factor and a single subplot factor. In our example, we can represent the four combinations of belt speed and peak temperature as four levels of a single whole plot treatment.

Similarly, we can represent the four combinations of density and amount of additive as four levels of a single subplot treatment. Let y_{ijk} denote the observation for subplot treatment k receiving whole plot treatment i in block j. Kempthorne (1952) uses as his model

$$y_{ijk} = \mu + \tau_i + \beta_j + \delta_{ij} + \gamma_k + (\tau\gamma)_{ik} + \epsilon_{ijk} \text{ for } i = 1, 2, \ldots, t \quad j = 1, 2, \ldots, b \quad k = 1, 2, \ldots, s$$

where

- t is the number of levels for the whole plot treatment (in our case, 4),
- b is the number of blocks or replicates of the basic whole plot experiment (in our case, 2),
- s is the number of levels for the subplot treatment (in our case, 4),
- μ is the overall mean,
- τ_i is the effect of the ith whole plot treatment,
- β_j is the effect of the jth block,
- δ_{ij} is the whole plot error term,
- γ_k is the effect of the kth subplot treatment,
- $(\tau\gamma)_{ik}$ is the whole plot treatment by subplot treatment interaction, and
- ϵ_{ijk} is the subplot error.

He uses randomization theory to derive the expected mean squares summarized in Table 2. In this table, σ_w^2 is the experimental error variance for the whole plot treatment, and σ_s^2 is the experimental error variance for the subplot effects.

Table 2. Expected Mean Squares Table under Randomization Theory

Source	df	Expected Mean Square
Whole Plot Treatment	$t-1$	$\sigma_w^2 + \frac{bs}{t-1}\sum_{i=1}^{t}\tau_i^2$
Whole Plot Error	$(t-1)(b-1)$	σ_w^2
Subplot Treatment	$s-1$	$\sigma_s^2 + \frac{bt}{s-1}\sum_{k=1}^{s}\gamma_k^2$
Whole Plot × Subplot	$(t-1)(s-1)$	$\sigma_s^2 + \frac{b}{(t-1)(s-1)}\sum_{i=1}^{t}\sum_{k=1}^{s}(\tau\beta)_{ik}^2$
Subplot Error	$t(b-1)(s-1)$	σ_s^2

Many analysts assume that the blocks are random effects and use an unrestricted mixed model to derive the appropriate mean squares. The most common model for this approach is

$$y_{ijk} = \mu + \tau_i + \beta_j + (\tau\beta)_{ij} + \gamma_k + (\tau\gamma)_{ik} + \epsilon_{ijk}$$

where $(\tau\beta)_{ij}$ is the random block by whole plot treatment interaction. Typically, we assume that $(\tau\beta)_{ij}$ follows a normal distribution with mean 0 and variance $\sigma^2_{\tau\beta}$. Table 3 summarizes the relevant expected mean squares. If we call the block by whole plot interaction the whole plot error, then Tables 2 and 3 suggest the same basic testing procedures.

Table 3. Expected Mean Squares Table under the Unrestricted Mixed Model

Source	df	Expected Mean Square
Whole Plot Treatment	$t-1$	$\sigma^2 + s\sigma^2_{\tau\beta} + \frac{bs}{t-1}\sum_{i=1}^{t}\tau_i^2$
Block × Whole Plot	$(t-1)(b-1)$	$\sigma^2 + s\sigma^2_{\tau\beta}$
Subplot Treatment	$s-1$	$\sigma^2 + \frac{bt}{s-1}\sum_{k=1}^{s}\gamma_k^2$
Whole Plot × Subplot	$(t-1)(s-1)$	$\sigma^2 + \frac{b}{(t-1)(s-1)}\sum_{i=1}^{t}\sum_{k=1}^{s}(\tau\gamma)_{ik}^2$
Subplot Error	$t(b-1)(s-1)$	σ^2

Montgomery (1997) uses a restricted mixed model as the basis for his analysis of the following form

$$y_{ijk} = \mu + \tau_i + \beta_j + (\tau\beta)_{ij} + \gamma_k + (\tau\gamma)_{ik} + (\beta\gamma)_{jk} + (\tau\beta\gamma)_{ijk} + \epsilon_{ijk}$$

where

- $(\tau\beta)_{ij}$ is the random block by whole plot treatment interaction,
- $(\beta\gamma)_{jk}$ is the random block by subplot treatment interaction,
- $(\tau\beta\gamma)_{ijk}$ is the random block by whole plot treatment by subplot treatment interaction.

Under this restricted mixed model, the random interactions involving a fixed effect are assumed subject to the constraint that the sum of that interaction's effects over the levels of the fixed effect is zero. We note that the structure of the design does not allow the estimation of the variance of ϵ_{ijk}. Table 4 gives the resulting expected mean squares, which suggests that we have three distinct error terms. We should use the block by whole plot by subplot interaction to test the whole plot by subplot interaction; we should use the block by subplot interaction to test the main effect of the subplot treatment; and we should use the block by whole plot interaction to test the main effect of the whole plot. This restricted analysis reduces to the other two analyses only if the block by subplot interaction is unimportant. In such a case, its contribution can be pooled with the block by whole plot by subplot interaction to form the same error term as the randomization and unrestricted mixed model analyses.

Table 4. Expected Mean Squares Table Under the Restricted Mixed Model

Source	df	Expected Mean Square
Whole Plot	$t-1$	$\sigma^2 + s\sigma^2_{\tau\beta} + \frac{bs}{t-1}\sum_{i=1}^{t}\tau_i^2$
Block × Whole	$(t-1)(b-1)$	$\sigma^2 + s\sigma^2_{\tau\beta}$
Subplot Treatment	$s-1$	$\sigma^2 + t\sigma^2_{\beta\gamma} + \frac{bt}{s-1}\sum_{k=1}^{s}\gamma_k^2$
Block × Sub	$(b-1)(s-1)$	$\sigma^2 + t\sigma^2_{\beta\gamma}$
Whole Plot × Subplot	$(t-1)(s-1)$	$\sigma^2 + \sigma^2_{\tau\beta\gamma} + \frac{b}{(t-1)(s-1)}\sum_{i=1}^{t}\sum_{k=1}^{s}(\tau\gamma)^2_{ik}$
Block × Whole × Sub	$(t-1)(b-1)(s-1)$	$\sigma^2 + \sigma^2_{\tau\beta\gamma}$

The analyses outlined so far ignore the basic factorial structure. One can use appropriate contrasts to partition the whole plot treatment into the effects of belt speed, peak temperature, and their interaction. Similarly, one can partition the subplot factor and the whole plot by subplot interaction into their corresponding factorial effects. The easiest way to perform this analysis involves two steps. First, analyze the data using the classical split-plot analysis as summarized by the above expected mean squares. Second, reanalyze the data as a typical 2^4 factorial, making sure to use the appropriate error terms from the first analysis. Table 5 illustrates the form of the ANOVA table for our example.

The concept of the split-plot design can be extended if further randomization restrictions exist. For example, suppose there are two levels of randomization restrictions within a block, then a split-split-plot design can be used. For a more detailed discussion of split-plot designs and their extensions see Yates (1937), Cox (1958) and Montgomery (1997).

Table 5. Analysis of Variance Table for the 2^4 Run as a Split-Plot Experiment Run

Source	df
Whole Plot Treatment	$t-1=3$
x_1 †	1
x_2 †	1
x_1x_2 †	1
Blocks	$b-1$
Block × Whole Plot	$(t-1)(b-1)$
Subplot Treatment	$s-1=3$
x_3 ††	1
x_4 ††	1
x_3x_4 ††	1
Whole × Subplot	$(t-1)(s-1)=9$
x_1x_3 ††	1
x_1x_4 ††	1
x_2x_3 ††	1
x_2x_4 ††	1
$x_1x_3x_4$ ††	1
$x_2x_3x_4$ ††	1
$x_1x_2x_3$ ††	1
$x_1x_2x_4$ ††	1
$x_1x_2x_3x_4$ ††	1
Subplot Error	$t(b-1)(s-1)$

† These terms are tested using the Block × Whole Plot Treatment interaction.

†† These terms are tested using the Subplot Error.

2 Split-Plots in Industry

Frankly, the split-plot error structure has been underutilized in industrial experimentation, and the literature reflects that point. Most industrial experiments assume a completely randomized error structure.

One area explored considers how to use confounding when the whole plot and/or subplot treatments are of a factorial nature, to reduce the number of whole plots and/or subplots needed since many industrial experiments face constraints which limit the size of the experiment. Bartlett (1935) suggested the possibility of confounding higher-order subplot interactions to reduce the number of subplots needed within each whole plot. Later, split-plot confounding was studied by Addelman (1964). He provided a table containing factorial and fractional-factorial arrangements that involve split-plot confounding. However, he did not consider confounding within the whole plots.

In some experiments, there are constraints on the number of subplots within each whole plot. When the whole plots are arranged in a CRD, Robinson (1967) discussed situations where the number of subplots per whole plot is less than the number of subplot treatments. The whole plots are treated as blocks and then a balanced incomplete block (BIB) design is used to allocate the subplot treatments to the whole plots. If the whole plots are arranged in an RCB design, the same procedure can be applied. If the number of whole plots per block is less than the number of whole plot treatments, then an incomplete block design can be used there as well. Robinson (1970) gave details on the case when both whole plot and subplot treatments are arranged in incomplete block designs. Essentially, the procedure amounts to arranging the whole plot treatments in blocks using a BIB design. Then, consider the whole plots as blocks and arrange the subplot treatments in another BIB design. Robinson (1970) provided formulas for the estimates of the main effects and interactions for three cases: within whole plot, between whole plot within block, and between blocks. Formulas are also given for the variance of the differences of the estimates for each case.

Lucas and Ju (1992) used a simulation study to investigate the use of split-plot designs in industrial experiments where one factor was difficult to change and served as the whole plot. They studied both a 2^4 full factorial and a four factor face-centered cube design with four center points. In each case, one factor was hard-to-change, and the others were easy-to-change. Their results confirm that split-plot designs will produce increased precision on the subplot factors while sacrificing precision on the whole plot factors.

Miller (1997) considers various fractional-factorial structures in strip-plot experiments. Strip-plot configurations can be applied when the process being investigated can be separated into two distinct stages and it is possible to apply the second stage simultaneously to groups of the first-stage product. Miller proposes a method for constructing strip-plot configurations for fractional-factorial designs which consists of three steps:

1. identify a suitable design for applying row treatments to rows ignoring columns;
2. identify a suitable design for applying column treatments to columns ignoring rows;
3. select a suitable fraction of the product of the row and column designs.

The method is applied for two-level designs and then extended to m-level and mixed-level designs.

Mee and Bates (1998) consider split-lot experiments involving silicon wafers. These experiments are performed in steps where a different factor is applied at each step. Thus, there are an equal number of steps and factors.

3 Bi-Randomization Designs

Letsinger, Myers, and Lentner (1996) introduce bi-randomization designs (BRD's). BRD's refer to designs with two randomizations similar to that of a split-plot design. The whole plot variables are denoted by $\boldsymbol{z} = (z_1, z_2, \ldots, z_a)'$, while the sub-plot variables are denoted by $\boldsymbol{x} = (x_1, x_2, \ldots, x_b)'$. Hence, the ith design run is $(\boldsymbol{z}_i, \boldsymbol{x}_i)$. BRD's are broken into two classes, crossed and non-crossed. Crossed BRD's are constructed as follows:

1. randomize the a unique combinations of $\boldsymbol{z}$ to the whole plot experimental units (EU's),
2. randomize the b levels of $\boldsymbol{x}$ to the smaller EU's within each whole plot (see Table 6).

Thus every level of $\boldsymbol{x}$ is "crossed" with every level of $\boldsymbol{z}$. These designs are the usual split-plot designs.

The non-crossed BRD's differ from the crossed BRD's in that not all levels of $\boldsymbol{x}$ are associated with z_i. The whole plots have different levels of the sub-plots and need not have the same number of levels. Non-crossed BRD's are constructed as follows:

1. randomize the a unique combinations of $\boldsymbol{z}$ to the whole plot EU's,
2. randomize the b_i levels of $\boldsymbol{x}$ to the smaller EU's within each whole plot (see Table 7).

Table 6. Crossed BRD From Letsinger, Myers, and Lentner (1996)

z_1	x_1	$\cdots$	x_b
z_2	x_1	$\cdots$	x_b
$\vdots$	$\vdots$	$\ddots$	$\vdots$
z_a	x_1	$\cdots$	x_b

Table 7. Non-Crossed BRD From Letsinger, Myers, and Lentner (1996)

z_1	x_{11}	$\cdots$	x_{1b_1}
z_2	x_{21}	$\cdots$	x_{2b_2}
$\vdots$	$\vdots$	$\ddots$	$\vdots$
z_a	x_{a1}	$\cdots$	x_{ab_a}

The distinction between these two can be thought of in terms of the sub-plot factors. The crossed BRD might be represented by a 2^k factorial in the sub-plot factors, while the non-crossed BRD might use a 2^{k-p} fractional-factorial in the sub-plot factors.

For both crossed and non-crossed BRD's, the two randomizations complicate the error structure. The first randomization leads to the whole plot error variance, σ_δ^2, while the second randomization leads to the sub-plot variance, σ_ϵ^2. It is assumed that the covariance between any two observations on the same whole plot is constant over all whole plots, and that observations on two sub-plots from different whole plots are uncorrelated. The response surface model is

$$\boldsymbol{y} = \boldsymbol{X}\boldsymbol{\beta} + \boldsymbol{\delta} + \boldsymbol{\epsilon}$$

where $\boldsymbol{\delta} + \boldsymbol{\epsilon} \sim N(0, \boldsymbol{V})$ with $\boldsymbol{V} = \sigma_\delta^2 \boldsymbol{J} + \sigma_\epsilon^2 \boldsymbol{I}$ where $\boldsymbol{J}$ is a block diagonal matrix of $\mathbf{1}_{b_i}\mathbf{1}'_{b_i}$ and b_i is the number of observations within the ith whole plot. Using generalized least squares (GLS), the maximum likelihood estimate (MLE) of the response surface model is

$$\widehat{\boldsymbol{\beta}} = \left(\boldsymbol{X}'\boldsymbol{V}^{-1}\boldsymbol{X}\right)^{-1} \boldsymbol{X}'\boldsymbol{V}^{-1}\boldsymbol{y} \tag{1}$$

with

$$\text{Var}\left(\widehat{\boldsymbol{\beta}}\right) = \left(\boldsymbol{X}'\boldsymbol{V}^{-1}\boldsymbol{X}\right)^{-1} . \tag{2}$$

From (1), it is seen that the model estimation depends on the matrix $\boldsymbol{V}$ and thus both σ_δ^2 and σ_ϵ^2.

Suppose that the response surface model is partitioned into the whole plot and sub-plot terms as

$$\boldsymbol{y} = \boldsymbol{Z}\boldsymbol{\gamma} + \boldsymbol{X}^*\boldsymbol{\beta}^* + \boldsymbol{Z}'\boldsymbol{\Delta}\boldsymbol{X}^*$$

where $\boldsymbol{\Delta}$ is a matrix of whole plot × sub-plot interaction parameters. The response surface design should be large enough to test for general lack of fit as well as lack of fit from the whole plots. Therefore, the number of whole plots available must exceed the number of parameters in $\boldsymbol{\gamma}$.

For the crossed BRD, there is an equivalence between ordinary least squares (OLS) and GLS. This equivalence means that (1) becomes

$$\widehat{\boldsymbol{\beta}} = \left(\boldsymbol{X}'\boldsymbol{X}\right)^{-1}\boldsymbol{X}'\boldsymbol{y},$$

and the model estimation no longer depends on the error variance. However, for testing purposes, the error variance must be estimated. Letsinger, Myers, and Lentner (1996) suggest augmenting the response surface model with insignificant whole plot terms, $\boldsymbol{Z}^*\boldsymbol{\rho}$, to saturate the a whole plot degrees of freedom. The whole plot saturated model can be used to calculate lack of fit sums of squares for both the whole plots and the sub-plots. Then, approximate t-tests can be formed by substituting the estimated error variances into Equation (2).

Non-crossed BRD's present a more complicated situation. The equivalency of OLS and GLS is only true in the case of a first order model. Although more complex, the above method can be adapted for the first order case. Letsinger, Myers, and Lentner (1996) compare three methods for the second order case. They are OLS, iterative reweighted least squares (IRLS), and restricted maximum likelihood (REML). Though IRLS and REML appear to be better methods, the "best" method depends on the design, model, and any prior information.

4 Split-Plots in Robust Parameter Designs

Genichi Taguchi proposed methods for designing experiments for product design that are robust to environmental variables. The goal of robust design is to design an experiment that identifies the settings of the design factors that make the product robust to the effect of the noise variables. The design factors, factors controlled during manufacturing, make up the inner array while the environmental factors, or noise factors, make up the outer array. Environmental factors are factors that are difficult to control but cause variation in products. The experiment consists of a cross-product of the experimental design of the inner array and the experimental design of the outer array. Unless these arrays are small, Taguchi's designs become large and expensive.

It should be noted that the "combined array" is an alternative to Taguchi's crossed array. The combined array utilizes a single experimental design in both the design and environmental factors. Therefore, the response is modeled directly as a function of the design factors and the environmental factors using a single linear model. More details on the combined array can be found in Welch et al. (1990), Shoemaker et el. (1991), and O'Donnell and Vining (1997). Box and Jones (1992) investigate the use of split-plot designs as an alternative to the crossed array.

Box and Jones (1992) consider three experimental arrangements for the robust parameter design with the use of split-plot designs,

1. arrangement (a), the whole plots contain the environmental factors and the subplots contain the design factors;
2. arrangement (b), the whole plots contain the design factors and the subplots contain the environmental factors;
3. arrangement (c), the subplot factors are assigned in 'strips' across the whole plot factors, this is commonly called a strip-block experiment.

These three arrangements are illustrated through an example seeking the best recipe for a cake mix. Three design factors have been identified as affecting taste. They are flour, shortening, and egg powder and are studied using a 2^3 factorial design. The consumer may have an oven in which the temperature is biased up or down. Also, the consumer may overcook or undercook the cake. Therefore, the recipe is to be robust to two environmental factors, oven temperature and baking time, in a 2^2 factorial design.

4.1 Arrangement (a)

Under this arrangement, the whole plots contain the environmental factors and the subplots contain the design factors. Suppose there are m levels of the environmental factors, $E_1, E_2, \ldots, E_j, \ldots, E_m$, applied to the whole plots, n levels of the design factors, $D_1, D_2, \ldots, D_i, \ldots, D_n$, applied to the subplots, and l replicates, $r_1, r_2, \ldots, r_u, \ldots, r_l$ with the whole plots in l randomized blocks. For the cake mix example, $m = 4$, $n = 8$, and $l = 1$. Arrangement (a) requires $m \times n \times l$ subplots and $m \times l$ whole plots. Thus for the cake mix example, $4 \times 8 \times 1 = 32$ cake mix batches are required but only $4 \times 1 = 4$ operations of the oven are necessary. By comparison, a completely randomized cross-product array would require 32 cake mix batches and 32 operations of the oven. Thus, the split-plot arrangement has saved time by reducing the number of operations of the oven.

The model for arrangement (a) is

$$y_{ijk} = \mu + \gamma_k + \alpha_j + \eta_{jk} + \delta_i + (\alpha\delta)_{ij} + \epsilon_{ijk} \,,$$

where y_{ijk} is the response of the kth replicate of the ith level of factor D, and the jth level of factor E, μ is the overall mean, γ_k is the random effect of the kth replicate, with $\gamma_k \sim N(0, \sigma_\gamma^2)$, α_j is the fixed effect of the jth level of factor E, δ_i is the fixed effect of the ith level of factor D, $(\alpha\delta)_{ij}$ is the interaction effect of the ith level of D and the jth level of E, $\eta_{jk} \sim N(0, \sigma_w^2)$ is the whole plot error, $\epsilon_{ijk} \sim N(0, \sigma_s^2)$ is the subplot error, and η_{jk} and ϵ_{ijk} are independent.

4.2 Arrangement (b)

With this arrangement, the whole plots contain the design factors while the subplots contain the environmental factors. Arrangement (b) requires only $8 \times 1 = 8$ cake mix batches but requires $4 \times 8 \times 1 = 32$ operations of the oven. Again, a completely randomized cross-product array would use 32 cake mix batches and 32 operations of the oven. Here, the savings of the split-plot design is not as great since only the number of cake mix batches is reduced. This is not an ideal situation for industrial experiments. First of all, the design factors are of greater interest. Therefore, applying the design factors to the whole plots results in a loss of precision for these factors. Hence, it is possible to have large differences between the levels of the design factors that are insignificant when tested. Also, from an economic point of view, arrangement (b) is costly. It requires an inefficient use of the environmental factors which in industrial experiments are typically the difficult- or expensive-to-change factors.

The model for arrangement (b) is

$$y_{ijk} = \mu + \gamma_k + \delta_i + \theta_{ik} + \alpha_j + (\alpha\delta)_{ij} + \epsilon_{ijk} ,$$

where y_{ijk} is the response of the kth replicate of the ith level of factor D, and the jth level of factor E, μ is the overall mean, γ_k is the random effect of the kth replicate, with $\gamma_k \sim N(0, \sigma_\gamma^2)$, α_j is the fixed effect of the jth level of factor E, δ_i is the fixed effect of the ith level of factor D, $(\alpha\delta)_{ij}$ is the interaction effect of the ith level of D and the jth level of E, $\theta_{ik} \sim N(0, \sigma_w^2)$ is the whole plot error, $\epsilon_{ijk} \sim N(0, \sigma_s^2)$ is the subplot error, and θ_{ik} and ϵ_{ijk} are independent.

4.3 Arrangement (c)

Now, consider the arrangement where the subplot treatments are randomly assigned in strips across each block of whole plot treatments (see Table 8). For the cake mix example, suppose the $n = 8$ batches of cake mix are subdivided into $m = 4$ subgroups. One subgroup from each batch is then selected and these eight are baked in the same oven at the appropriate temperature for the appropriate time. This arrangement requires only 8 cake mix batches and only 4 operations of the oven. Therefore, the strip-block experiment is easier to run than the completely randomized cross-product design, as well as both arrangements (a) and (b).

Table 8. Strip-Block Arrangement (Box and Jones (1992))

	Block 1				Block 2				Block 3		
	a_1	a_2	a_3		a_3	a_2	a_1		a_1	a_3	a_2
b_1				b_2				b_1			
b_2				b_1				b_2			

The model for the strip-block arrangement is

$$y_{ijk} = \mu + \gamma_k + \alpha_j + \eta_{jk} + \delta_i + \theta_{ik} + (\alpha\delta)_{ij} + \epsilon_{ijk}$$

where y_{ijk} is the response of the kth replicate of the ith level of factor D, and the jth level of factor E, μ is the overall mean, γ_k is the random effect of the kth replicate, with $\gamma_k \sim N(0, \sigma_\gamma^2)$, α_j is the fixed effect of the jth level of factor E, δ_i is the fixed effect of the ith level of factor D, $(\alpha\delta)_{ij}$ is the interaction effect of the ith level of D and the jth level of E. In arrangement (c), $\eta_{jk} \sim N(0, \sigma_E^2)$, $\theta_{ik} \sim N(0, \sigma_D^2)$, $\epsilon_{ijk} \sim N(0, \sigma^2)$ and η_{jk}, θ_{ik} and ϵ_{ijk} are independent.

Box and Jones discuss the ANOVA tables for all three arrangements. These tables indicate the appropriate denominators for tests involving the design factors, the environmental factors, and their interactions assuming replication. When there is no replication, they suggest using two separate normal probability plots, one for the whole plot effects and another for the subplot and subplot by whole plot interactions, to detect significant effects. Also, if the design and environmental factors are factorial combinations, it may be possible to pool negligible higher order interactions to get estimates of the whole plot and subplot errors.

It is of great interest to the researcher to learn how and which environmental factors influence the design variables. This information is contained in the subplot × whole plot interactions. However, Taguchi's analysis is commonly conducted in terms of a performance statistic, such as the signal to noise ratio (SNR). The SNR is calculated for each point in the inner array using data obtained from the outer array about that point. Therefore, Taguchi ignores any information contained in the interactions of the design and environmental factors. This is generally considered to be a serious drawback to the Taguchi designs.

Phadke (1989) presented an example, analyzed using Taguchi's SNR's, involving a polysilicon deposition process. Polysilicon film is typically deposited on top of the oxide layer of the wafers using a hot-wall, reduced pressure reactor. The reactant gases are introduced into one end of a three-zone furnace tube and are pumped into the other end. The wafers enter the low-pressure chemical vapor deposition furnace in two quartz boats, each with 25 wafers, and polysilicon is deposited simultaneously on all 50 wafers. The desired output of this process is a wafer which has a uniform layer of film of a specified thickness. Six design factors each at three levels were identified: temperature, pressure, nitrogen flow, silane flow, setting time, and cleaning method. Tube location and die location were considered noise factors. Three responses, film thickness, particle counts, and deposition rate, were of interest. The smaller the better SNR was used in the analysis for particles, the target is best SNR was used for thickness, and a 20 $\log_{10}$ transformation was used for deposition rate. The data were analyzed using ANOVA techniques to determine the effect of each design factor on the responses. A more detailed discussion of the selection of factors, design and analysis of the SNR's is contained in Chapter 4 of Phadke (1989).

The actual structure of this experiment was a split-split-plot design, because there are three sizes of experimental units with different sources of variation. The design factors are applied to the tube (run-to-run variability); the location in the tube affects the wafer (wafer-to-wafer variability); whereas location in the wafer affects the die (die-to-die variability). Therefore, using Taguchi's SNR's to analyze this experiment will result in a complete loss of information in the design × noise factor interactions. Cantell and Ramírez (1994) reanalyzed the data using a split-split-plot design. They pool higher order interactions to get the necessary error terms in order to perform hypothesis tests on the design factors and the design × noise factor interactions. Interaction plots were used to determine the level of the design factor that minimizes variation across the levels of the noise factors. Although the final recommendations on the design factor levels by Cantell and Ramírez (1994) differs from Phadke (1989) on only one of the six design factors, the use of the split-split-plot design has allowed the process engineer to have a better understanding of the sources of variation. This added information may lead to process improvement in the future.

Kempthorne (1952) and Box and Jones (1992) provide details on the relative efficiency of these split-plot designs compared to the CRD and RCB. A summary of their conclusions is provided here. Consider the split-plot experiment as a uniformity trial. If the uniformity trial was run as a CRD or a RCB experiment, then, for arrangements (a) and (b), the subplot factor effects and the subplot × whole plot interactions are estimated more precisely than the whole plot factor effects. Compared with arrangements (a) and (b), the strip-block design estimates the subplot × whole plot interactions more precisely but the subplot factor effects with less precision. However, the whole plot factor effects are estimated with equal precision. Based on these results, arrangement (a) with the environmental factors applied to the whole plots is generally preferred over arrangement (b). Both the strip-block design and the split-plot design with the design factors applied to the subplots can be extremely useful in robust parameter design.

5 Conclusion

Industrial experimentation often involves the use of both hard-to-change and easy-to-change factors. Engineers routinely run split-plot experiments without realizing the implications for their analysis. In many cases, they believe they are running true factorial experiments. As a result, they often reach inappropriate conclusions from their experiments. More work needs to be done extending basic split-plot analysis to industrial experiments, especially for second-order models where the crossed BRD is no longer appropriate due to the size of the experiment in the subplots.

References

1. Addelman, S. (1964) Some Two-Level Factorial Plans With Split-Plot Confounding. Technometrics **30**, 253–258
2. Bartlett, M. S. (1935) Discussion of Complex Experiments by F. Yates. Journal of the Royal Statistical Society, Ser. B, **2**, 224–226
3. Box, G. E. P. and Jones, S. (1992) Split-Plot Designs for Robust Product Experimentation. Journal of Applied Statistics **19**, 3–26
4. Cochran, W. G. and Cox, G. M. (1957) Experimental Designs. 2nd. ed. John Wiley & Sons, New York, NY
5. Cox, D. R. (1958) Planning Experiments. John Wiley & Sons, New York, NY
6. Kempthorne, O. (1952) The Design and Analysis of Experiments. John Wiley & Sons, New York, NY
7. Letsinger, J. D. ; Myers, R. H. ; and Lentner, M. (1996) Response Surface Methods for Bi-Randomization Structures. Journal of Quality Technology **28**, 381-397
8. Lucas, J. M. and Ju, H. L. (1992) Split Plotting and Randomization in Industrial Experiments. ASQC Quality Congress Transactions. American Society for Quality, Milwaukee, WI, 374-382
9. Mee, R. W. and Bates, R. L. (1998) Split-Lot Designs: Experiments for Multistage Batch Processes. Technometrics **40**, 127–140
10. Miller, A. (1997) Strip-Plot Configurations of Fractional Factorials. Technometrics **39**, 153–161
11. Montgomery, D. C. (1997) Design and Analysis of Experiments. 4th ed. John Wiley & Sons, New York, NY
12. O'Donnell, E. M. and Vining, G. G. (1997) Mean Squared Error of Prediction Approach to the Analysis of the Combined Array. Journal of Applied Statistics **24**, 733–746
13. Robinson, J. (1967). Incomplete Split Plot Designs. Biometrics **23**, 793–802
14. Robinson, J. (1970) Blocking in Incomplete Split Plot Designs. Biometrika **57**, 347–350
15. Shoemaker, A. C. ; Tsui, K. ; and Wu, C. F. J. (1991) Economical Experimentation Methods for Robust Designs. Technometrics **33**, 415–427

16. Welch, W. J. ; Kang, T. ; and Sacks, J. (1990) Computer Experiments for Quality Control by Parameter Design. Journal of Quality Technology **22**, 15–22
17. Yates, F. (1937) The Design and Analysis of Factorial Experiments. Imperial Bureau of Soil Sciences, London, England

An Alternative Analysis Method to the ANOVA for p^k Unreplicated Fractional Factorial Experiments

Yoshikazu Ojima[1], Akira Kouda[2] and Masao Okuhara[3]

[1] Department of Industrial Administration, Science University of Tokyo, 278-8510, Japan
[2] Faculty of Economics, Higashi Nippon International University, 970-8023, Japan
[3] Suwa Junior College, Science University of Tokyo, 391-0292, Japan

Abstract. An alternative analysis method to the conventional analysis of variance is proposed for the problem of identifying the active factors in p^k unreplicated fractional factorial experiments (orthogonal arrays). The sums of squares calculated are arranged in ascending order, as $S(1) \leq S(2) \leq S(3) \leq ... \leq S(n)$. Assuming the least sum of squares, S(1) is inactive, i.e. having zero effect, the next candidate of inactive sums of squares, S(2), is tested. The following sums of squares, S(i)'s, i = 3, ... , n, are tested by the pooled inactive sums of squares, on the condition that S(2), ... , S(i−1) are not significant at the preceding tests, successively. The significance level of each test, α_S , is set the same value and the overall risk of the first kind of the whole procedure, α_T , is controlled to be 1%, 5% or 10%. The critical values at each step is numerically obtained for 3^3 orthogonal designs, and empirically obtained for 2^4 orthogonal designs by the Monte Carlo method. The proposed method can be applied not only to 2^k, 3^k orthogonal designs but also to Plackett-Burman designs.

1 Introduction

Unreplicated fractional factorial designs are widely used in industrial experiments, because there are many factors to be investigated in the screening stage for effective quality improvements. Such experiments are sometimes referred saturated experiments, and there is no estimate of pure error variance. Box and Meyer (1986) introduced an assumption of "effect sparsity", i.e. there are a few active factors and the others are having no effects. Some of the least sums of squares are usually used to estimate the error variance based on this assumption. The assumption is sometimes pointed out to be untestable. However, it is reasonable to assume only the least sum of squares is inactive, because the experiments will be useless in practical applications, if the least sum of squares is also active.

For the case of 2^k designs, we have $n = 2^k - 1$ sums of squares, S_1 , S_2 , ... , S_n calculated. Each S_i , $i = 1, \ldots, n$ corresponds to the estimated effect T_i , where T_i is $N(\theta_i, \sigma^2)$-distributed, with θ_i (the actual effect) and σ^2 all unknown. We call θ_i active if it is not 0, and inactive otherwise. The sums of squares, S_i 's, $i = 1, \ldots, n$ are arranged in ascending order, and symbolized as $S(1) \leq S(2) \leq S(3) \leq \ldots \leq S(n)$.

Many authors addressed the problem after Daniel (1959) proposing a half-normal plot as a graphical tool. Zahn (1975a, b) discussed modifications of the plot. Voss (1988) and Berk and Picard (1991) proposed several further procedures of this type. From the estimation point of view, Wilk, Gnanadesikan and Freeny (1963) presented a maximum likelihood estimate of σ^2. Lenth (1989), Dong (1993), Schneider, Kasperski, and Weissfeld (1993), Haaland and O'Connell (1995) and Kunert (1997) all proposed robust estimation methods of σ^2. Box and Meyer (1986) who applied a Baysian approach made another important contribution. Benski (1989) presented a method on a combination of testing for normality and outliers. Loughin and Noble (1997) proposed a permutation test. Holms and Berrettoni (1969), and Okuhara and Tsubaki (1987) proposed chain-pooling procedures for the analysis of variance. We propose a revised method of Okuhara and Tsubaki (1987) in this article. Venter and Steel (1996) proposed another hypothesis-testing approach. The procedures of Holms and Berrettoni (1969), Okuhara and Tsubaki (1987), and Venter and Steel (1996) are briefly reviewed in Section 2.1, because there are some similarities between their test statistics.

The method, which we propose in this article, is an overall test procedure for identifying the active factors. Assuming only the least sum of squares, $S(1)$ is inactive, the next candidate of inactive sums of squares, $S(2)$, is tested. The following sums of squares, $S(i)$'s, $i = 3, \ldots, n$, are tested by the pooled inactive sums of squares, on the condition that $S(2), \ldots, S(i-1)$ are not significant at the preceding tests, successively. The significance level of each test, α_S , is set the same value and the overall risk of the first kind of the whole procedure, α_T , is controlled to be 1%, 5% or 10%. The critical values at each step for 2^4 are given in Section 2.2. The proposed method can be applied to three or more level orthogonal designs. For the case of 3^k designs, we have $n = (3^k - 1)/2$ sums of squares. The critical values are given in Section 2.3 for the 3^3 design, which is referred as L_{27} by Taguchi (1987). The method is illustrated with examples from Box and Meyer (1986) in Section 3.

2 An overall test

2.1 A sequential testing procedure

In the notation introduced previously, the null hypothesis H_0 is that all θ_i 's are inactive. In other words, H_0 is that each S_i/σ^2, $i = 1, \ldots, n$ is independently and identically central χ^2-distributed with $p-1$ degrees of freedom for p^k unreplicated fractional factorial designs.

To describe precisely, the proposed method is a set of sequential tests for identifying the inactive factors. We assume only the least sum of squares, $S(1)$, is inactive at first. The next candidate of inactive sums of squares, $S(2)$, is tested by $S(1)$. The test statistic is

$$RS(2) = S(2)/S(1). \tag{1}$$

The critical value $RS(2; \alpha_S)$ at the significance level, α_S , is obtained from the distribution under H_0.

If $S(i)$ is not significant at level α_S , $S(i)$ is also assumed inactive and be pooled to other inactive S 's. Let $C(i)$ be an average of the i-th smallest sums of squares, i.e.

$$C(i) = \frac{1}{i}\sum_{j=1}^{i} S(j), \tag{2}$$

the next test statistic is

$$RS(i+1) = S(i+1)/C(i). \tag{3}$$

The critical value $RS(i+1; \alpha_S)$ is obtained from the distribution of $RS(i+1)$ under H_0 and the condition that all $S(1), \ldots, S(i)$ are not significant at the level α_S. The testing process will be successively performed. If $S(i)$ is significant at level α_S, all of $S(i)$, $S(i+1)$, ... , $S(n)$ are indicated to be active.

The risk of the first kind exists in each step. For testing $S(2)$, the risk is α_S as described. And for testing $S(3)$, the risk becomes $\alpha_S(1-\alpha_S)$, because it is only tested in the case $S(2)$ is not significant. Similarly, for testing $S(i)$, the risk becomes $\alpha_S(1-\alpha_S)^{i-2}$. Consequently, the overall risk of the first kind of the whole procedure α_T has a relation to α_S as,

$$1-\alpha_T = (1-\alpha_S)^{n-1}. \tag{4}$$

Then we have

$$\alpha_S = 1-(1-\alpha_T)^{1/(n-1)}. \tag{5}$$

The critical values are given in Sections 2.2 and 2.3.

The procedures of Holms and Berrettoni (1969), Okuhara and Tsubaki (1987), and Venter and Steel (1996) should be reviewed, because there are some similarities between their test statistics. The procedure of Okuhara and Tsubaki (1987), which is an original version of ours, used the same test statistics. But the tests are performed unconditionally, and the critical values are marginally obtained on the null hypothesis.

The procedure of Holms and Berrettoni (1969) is preliminary for the usual analysis of variance. Their test statistics are $S(i)/C(i)$, and the critical values at 0.001%, ... , 0.75%, are cited from Holms' unpublished Ph.D. thesis. These critical values are thought to be obtained from unconditional distributions.

Venter and Steel (1996) discussed the test statistics generally and found out

$$V_i = \sqrt{S(i+1)/C(i)} \tag{6}$$

best. The procedure requires marginal distribution functions, F_i's of all V_i's under H_0. Moreover, let $P_i = 1 - F_i(V_i)$ so that P_i is a p value associated with V_i, the overall test statistic is expressed as $min\{\ P_i\!: \ l \le i \le n-1\}$, where l is the least number of inactive θ_i's that may safely be assumed to be present. Only the critical values for the overall test statistic is given, but no way is described to obtain P_i's in the article.

2.2 Critical values for the 2^4 saturated design

It is impossible to obtain the critical values analytically for the 2^k, so the Monte Carlo method must be used to approximate the values of $RS(i;\ \alpha_S)$, i=2, ... , n. We generate n_{emp} observations for each conditional distribution for 2^4. The critical values are approximately obtained as the corresponding quantiles of the empirical distribution of the n_{emp} observations. If the critical value has not enough precision, the generation of empirical distributions is iterated n_{iter} times.

For the case of $S(i)$, a set of 15 random numbers of $N(0, 1)$, is generated, squared, sorted in ascending order, and then labeled $S(1)$, ... , $S(15)$. If all of $S(1)$, ... , $S(i-1)$ are not significant at the level α_S, we can add one observation $S(i)/C(i-1)$ to the empirical distribution of $RS(i)$. This process is iterated until we get n_{emp} observations, and the $RS(i;\ \alpha_S)$ is approximately obtained as the upper α_S quantile of the distribution. After that we can proceed to the case of $S(i+1)$. The values of n_{emp} and n_{iter} are 2,000,000 and 30, respectively.

The values of α_S are obtained from (5) as 7.176×10^{-4}, 3.657×10^{-3} and 7.498×10^{-3}, for α_T=0.01, α_T=0.05 and α_T=0.10, respectively. The critical values are given in Table 1. The observed α_S values of each step and the observed α_T values are successfully tested by the χ^2 goodness-of-fit tests and the binomial tests with 10,000,000 sets of sums of squares.

Table 1. Critical values for 2^4

i	α_T=0.01 (α_S=7.176×10^{-4})	α_T=0.05 (α_S=3.657×10^{-3})	α_T=0.10 (α_S=7.498×10^{-3})
2	2.00×10^6	7.63×10^4	1.811×10^4
3	2.23×10^3	4.41×10^2	2.15×10^2
4	2.51×10^2	85.2	52.8
5	87.6	38.9	26.9
6	48.1	24.8	18.41
7	32.6	18.70	14.52
8	25.2	15.53	12.45
9	21.3	13.77	11.28
10	19.10	12.81	10.63
11	17.90	12.42	10.44
12	17.70	12.54	10.65
13	18.80	13.46	11.45
14	22.1	15.81	13.47
15	33.3	23.7	20.0

2.3 Critical values for the 3^3 saturated design

For the case of 3^k designs, we have $n = (3^k - 1)/2$ sums of squares. The 3^3 design is referred as L_{27} by Taguchi (1987), and widely used in Japan. The null distribution of each sum of squares is central χ^2 with 2 degrees of freedom, i.e. the exponential distribution. The critical values for 3^k designs can be mathematically obtained, because the differences between successive ordered exponential variates are also independently and exponentially distributed. The values of α_S are obtained from (5) as 8.371×10^{-4}, 4.265×10^{-3} and 8.742×10^{-3}, for α_T=0.01, α_T=0.05 and α_T=0.10, respectively. The critical values are given in Table 2 for the 3^3 design.

Table 2. Critical values for 3^3

i	α_T=0.01 (α_S=8.371×10^{-4})	α_T=0.05 (α_S=4.265×10^{-3})	α_T=0.10 (α_S=8.742×10^{-3})
2	1.294×10^3	2.54×10^2	1.238×10^2
3	55.3	24.4	16.93
4	20.7	11.83	9.20
5	13.04	8.47	6.95
6	10.13	7.07	5.99
7	8.76	6.40	5.52
8	8.08	6.08	5.31
9	7.84	6.02	5.30
10	7.96	6.20	5.47
11	8.53	6.67	5.91
12	10.02	7.81	6.91
13	14.79	11.33	9.92

3 Examples

To illustrate our method, we now apply it to the four examples labeled I, II, III and IV from Box and Meyer (1986). The results of the procedure at the significance level α_T=0.05 are shown in Table 3.

In Example I, 15 sums of squares are calculated and arranged in ascending order. The least sum of squares $S(1)$ is S_7, i.e. the sum of squares for the column 7 in original Table 3 of Box and Meyer (1986), and the second least $S(2)$ is S_5. The column "c. n." in Example I in Table 3 indicates that $S_7 \leq S_5 \leq S_{10} \leq \ldots \leq S_8 \leq S_2 \leq S_4$. For this case, both $S(1)$ and $S(2)$ are 0.0001 (more precisely, 0.00005625), then $RS(2)=S(2)/C(1)=1$. The result obviously indicates $S(2)$ inactive. The next sum of squares $S(3)=S_{10}=0.0002$ (more precisely 0.00015625) then $RS(3)=2.78$ and $S(3)$ is found also inactive. The process is continued successively. The result of testing $RS(13)=S(13)/C(12)$ indicates the $S(13)=S_8$, $S(14)=S_2$, $S(15)=S_4$ are active.

The result of our procedure indicates the $S(14)=S_{14}$, $S(15)=S_{15}$ are active in Example II, and $S(13)=S_{13}$, $S(14)=S_4$, $S(15)=S_{12}$ are active in Example III. But no active factors have been found in Example IV.

We have applied the procedure at different significance levels, i.e. α_T=0.01 and α_T=0.10, but the same results have been obtained. The results agree with those of Box and Meyer (1986).

Table 3. Results for Examples I - IV (at α_T=0.05)

	Example I			Example II			Example III			Example IV		
order	c. n. (*)	sum of squares	test statistic	c. n. (*)	sum of squares	test statistic	c. n. (*)	sum of squares	test statistic	c. n. (*)	sum of squares	test statistic
1	7	0.0001		6	0.0025		10	0.04		3	0.0000	
2	5	0.0001	1.00	9	0.0100	4.00	14	0.04	1.00	13	0.0002	25.00
3	10	0.0002	2.78	1	0.0625	10.00	6	0.16	4.00	15	0.0014	17.31
4	14	0.0008	8.44	11	0.0625	2.50	11	0.36	4.50	2	0.0018	3.45
5	3	0.0008	2.95	12	0.0625	1.82	7	0.36	2.40	12	0.0028	3.27
6	15	0.0011	2.96	4	0.0900	2.25	2	0.64	3.33	5	0.0046	3.72
7	6	0.0018	3.82	2	0.0900	1.86	3	1.44	5.40	6	0.0176	9.86
8	13	0.0018	2.72	3	0.3600	6.63	1	1.44	3.32	4	0.0233	5.76
9	11	0.0023	2.80	7	0.5625	6.08	15	1.44	2.57	11	0.0410	6.37
10	9	0.0033	3.42	13	0.5625	3.89	9	1.96	2.98	14	0.0613	5.96
11	12	0.0068	5.67	8	0.6400	3.43	5	3.24	4.11	7	0.0885	5.76
12	1	0.0127	7.40	5	0.6400	2.81	8	5.76	5.70	9	0.1040	4.72
13	8	0.0770	<u>29.36</u>	10	0.7225	2.76	13	57.76	<u>41.06</u>	1	0.1463	5.07
14	2	0.2525	<u>30.26</u>	14	18.4900	<u>62.15</u>	4	84.64	14.74	10	0.2525	6.66
15	4	0.9950	<u>38.59</u>	15	38.4400	<u>24.07</u>	12	121.00	10.64	8	0.2998	5.63

* c. n. : Original column number in Table 3 of Box and Meyer (1986)

4 Conclusion

We proposed an overall testing method for the problem of identifying the active factors in saturated fractional factorial experiments. The proposed method is an alternative analysis method to the conventional ANOVA. The testing method consists of sequential tests which are applied to the ordered sums of squares. The significance level of each test, α_S , is set the same value and the overall risk of the first kind of the whole procedure, α_T , is controlled to be 1%, 5% or 10%. The critical values at each step are numerically obtained for the 3^3 orthogonal designs, and empirically obtained for the 2^4 orthogonal designs. The proposed test can be applied not only to 2^k, 3^k orthogonal designs but also to Plackett-Burman designs.

Acknowledgment

The authors are grateful to Professor Y. Iizuka of the University of Tokyo, Professor H. Tsubaki of the University of Tsukuba, and Mr. S. Nishikawa of Kumika Industry Co., Ltd. for a number of valuable comments and suggestions.

References

1. Benski, H. C. (1989) Use of a Normality Test to Identify Significant Effects in Factorial Designs. Journal of Quality Technology **21**, 174-178
2. Berk, K. N., and Picard, R. R. (1991) Significance Tests for Saturated Orthogonal Arrays. Journal of Quality Technology **23**, 79-89
3. Box, G. E. P., and Meyer, R. D. (1986) An Analysis for Unreplicated Fractional Factorials. Technometrics **28**, 11-18
4. Daniel, C. (1959) Use of Half-Normal Plots in Interpreting Factorial Two-Level Experiments. Technometrics **1**, 311-341
5. Dong, F. (1993) On the Identification of Active Contrasts in Unreplicated Fractional Factorials. Statistica Sinica **3**, 209-217
6. Haaland, P. D., and O'Connell, M. A. (1995) Inference for Effect-Saturated Fractional Factorials. Technometrics **37**, 82-93
7. Holms, A. G., and Berrettoni, J. N. (1969) Chain-Pooling ANOVA for Two-Level Factorial Replication-Free Experiments. Technometrics **11**, 725-746
8. Kunert, J. (1997) On the Use of the Factor-Sparsity Assumption to Get an Estimate of the Variance in Saturated Designs. Technometrics **39**, 81-90

9. Lenth, R. V. (1989) Quick and Easy Analysis of Unreplicated Factorials. Technometrics **31**, 469-473

10. Loughin, T. M., and Noble, W. (1997) A Permutation Test for Effects in an Unreplicated Factorial Design. Technometrics **39**, 180-190

11. Okuhara, M., and Tsubaki, H. (1987) New Sequential Pooling Procedure of ANOVA in L16 Orthogonal Array. Proceedings of International Conference on Quality Control 1987 - Tokyo, 513-518

12. Schneider, H., Kasperski, W. J., and Weissfeld, L. (1993) Finding Significant Effects for Unreplicated Fractional Factorials Using the *n* Smallest Contrasts. Journal of Quality Technology **25**, 18-27

13. Taguchi, G. (1987) System of Experimental Design, UNIPUB/ Kraus International Publications, New York

14. Venter, J. H., and Steel, S. J. (1996) A Hypothesis-Testing Approach Toward Identifying Active contrasts. Technometrics 38, 161-169

15. Voss, D. T. (1988) Generalized Modulas-Ratio Tests for Analysis of Factorial Designs with Zero Degrees of Freedom for Error. Communications in Statistics - Theory and Methods **17**, 3345-3359

16. Wilk, M. B., Gnanadesikan, R., and Freeny, A. E. (1963) Estimation of Error Variance from Smallest Ordered Contrasts. Journal of the American Statistical Association **58**, 152-160

17. Zahn, D. A. (1975a) Modifications of and Revised Critical Values for the Half-Normal Plot. Technometrics **17**, 189-200

18. Zahn, D. A. (1975b) An Empirical Study of the Half-Normal Plot. Technometrics **17**, 201-211

Modeling and Analysis of Dynamic Robust Design Experiments

Kwok-Leung Tsui

Georgia Institute of Technology, School of Industrial & Systems Engineering, Atlanta, USA

Abstract. This paper investigates the modeling issue of the dynamic robust design problem. Taguchi (1986) proposes a loss model approach which models the loss measures as functions of the control factor effects. Miller and Wu (1996) propose a response function model approach which models the loss measures as functions of the effects of both the control and noise factors. In this paper we propose a response model approach which directly models the response as a function of the effects of the control, noise, and signal factors. It is found that the effect estimates of the intercept and slope functions from the three approaches are identical. The significant effects identified from the loss model and the response function model approaches may be spurious. The loss model approach may lead to non-optimal solutions and information loss. The three modeling approaches are compared in a real example.

1 Introduction

Robust Design is an important method for improving product quality, manufacturability, and reliability at low cost. The method has been applied to problems with static and dynamic systems. Static systems are defined as those whose desired output of the system has a fixed target value. Dynamic systems are those whose target value depends on the input signal set by the system user or operator. Most research in robust design has been focused on static systems. Until recently researchers started working on robust design problems with dynamic systems. (See Miller and Wu,1996; Lunani, Nair, and Wasserman, 1995; and McCaskey and Tsui, 1997.)

The objective of robust design method is to identify control factor combinations which make product and process designs insensitive (i.e., robust) to hard-to-control variations (noise). Robustness is quantified by some loss measures such as signal-to-noise ratios. For static problems, it has been shown in Tsui (1994) and Steinberg and Bursztyn (1994) that directly modeling the loss measures, as proposed by Taguchi, creates unnecessary biases for the factorial effect estimates and may lead to non-optimal solutions. The response model approach, an alternative proposed in Box and Jones (1990), Lucas (1989), Shoemaker, Tsui and Wu (1991), and Welch, Yu, Kang, and Sacks (1990), does not have this shortcoming.

For dynamic problems, Taguchi proposes the loss model (LM) approach to model the loss measures as functions of the effects of the control factors only.

Miller and Wu (1996) propose the response function model (RFM) approach to model the loss measures as functions of the effects of both the control and noise factors. We propose the response model (RM) approach to directly model the response as a function of the effects of the control, noise, and signal factors. Below we compare the three approaches in detail. In Section 2 we discuss the similarities and differences of the three modeling approaches. We show that the effect estimates of the intercept and slope functions from different approaches are identical. The significant effects identified from the LM and RFM approaches may be spurious. The LM approach may create unnecessary biases in variance effect estimates and may lead to non-optimal solutions and information loss. Section 3 illustrates these results with a real example. Section 4 concludes with a discussion.

2 Comparison of LM, RFM, and RM Approaches

2.1 Analysis Procedures

In this section we investigate the relationships of the effect estimates in the LM, RFM, and RM approaches under an additive model. Assume that the true model of the response (Y) is linear in p control factors $(C_1, ..., C_p)$, linear in q noise factors $(N_1, ..., N_q)$, linear in k known functions of the signal factor $(f_1(M), ..., f_k(M))$, and all these factors and functions interact with each other, i.e.,

$$
\begin{aligned}
Y = & (\alpha_0 + \boldsymbol{\alpha}^T \mathbf{C}) + (\beta_{10} + \boldsymbol{\beta_1}^T \mathbf{C}) f_1(M) + ... + (\beta_{k0} + \boldsymbol{\beta_k}^T \mathbf{C}) f_k(M) \\
& + \sum_{h=1}^{q} [(\gamma_{h0} + \boldsymbol{\gamma_h}^T \mathbf{C}) + (\rho_{1h0} + \boldsymbol{\rho_{1h}}^T \mathbf{C}) f_1(M) + ... \\
& + (\rho_{kh0} + \boldsymbol{\rho_{kh}}^T \mathbf{C}) f_k(M)] N_h + \epsilon \ , \qquad (1)
\end{aligned}
$$

where $\mathbf{C} = (C_1, ..., C_p)^T$, $\mathbf{f}(M) = (f_1(M), ..., f_k(M))^T$, $\boldsymbol{\alpha} = (\alpha_1, ..., \alpha_p)^T$, $\boldsymbol{\beta_1} = (\beta_{11}, ..., \beta_{1p})^T$, $\boldsymbol{\beta_k} = (\beta_{k1}, ..., \beta_{kp})^T$, $\boldsymbol{\gamma_h} = (\gamma_{h1}, ..., \gamma_{hp})^T$, $\boldsymbol{\rho_{1h}} = (\rho_{1h1}, ..., \rho_{1hp})^T$, $\boldsymbol{\rho_{kh}} = (\rho_{kh1}, ..., \rho_{khp})^T$, N_h's are indepentdently distributed with mean zero and variance ${\sigma_{N_h}}^2$, ϵ distributed with mean zero and variance ${\sigma_\epsilon}^2$, and N_h and ϵ are independent. When the ideal signal-response function is linear, $\mathbf{f}(M) = M$. When it is a polynomial function, $f_1, ..., f_k$ are chosen as the polynomial contrasts of the discrete signal levels being tested in the experiment. This will be illustrated in Section 3.

In the LM approach, at any fixed values of $\mathbf{C}$, the response model in (1) is expressed as a linear regression model in $f_1(M), ..., f_k(M)$, i.e.,

$$
Y|\mathbf{C} = \alpha'(\mathbf{C}) + \beta_1'(\mathbf{C}) f_1(M) + ... + \beta_k'(\mathbf{C}) f_k(M) + e(\mathbf{C}, M, \mathbf{N}, \epsilon) \ , \qquad (2)
$$

where

$$\alpha'(\mathbf{C}) = (\alpha_0 + \boldsymbol{\alpha}^T\mathbf{C}), \tag{3}$$

$$\beta'_i(\mathbf{C}) = (\beta_{i0} + \boldsymbol{\beta_i}^T\mathbf{C}), \tag{4}$$

$$e(\mathbf{C}, M, \mathbf{N}, \epsilon) = \sum_{h=1}^{q}[(\gamma_{h0} + \boldsymbol{\gamma_h}^T\mathbf{C}) + (\rho_{1h0} + \boldsymbol{\rho_{1h}}^T\mathbf{C})f_1(M) + ...$$
$$+(\rho_{kh0} + \boldsymbol{\rho_{kh}}^T\mathbf{C})f_k(M)]N_h + \epsilon,$$

with $\mathbf{N} = (N_1, ..., N_q)^T$. It follows that $e(\mathbf{C}, M, \mathbf{N}, \epsilon)$ is independently distributed with mean zero and variance

$$\sigma_e^2(\mathbf{C}, M) = \sum_{h=1}^{q}[(\gamma_{h0} + \boldsymbol{\gamma_h}^T\mathbf{C}) + (\rho_{1h0} + \boldsymbol{\rho_{1h}}^T\mathbf{C})f_1(M) + ...$$
$$+(\rho_{kh0} + \boldsymbol{\rho_{kh}}^T\mathbf{C})f_k(M)]^2{\sigma_{N_h}}^2 + \sigma_\epsilon^2.$$

As this variance depends on the signal factor M, we consider integration of $\sigma_e^2(\mathbf{C}, M)$ over M for determination of optimal control factor settings, i.e.,

$$\sigma_e^2(\mathbf{C}) = \int_M \sigma_e^2(\mathbf{C}, M)dM \tag{5}$$

The analysis procedure of the LM approach is to first estimate the parameters in model (2), $\alpha'(\mathbf{C})$, $\beta'_1(\mathbf{C})$, ..., $\beta'_k(\mathbf{C})$, and $\sigma_e^2(\mathbf{C})$, at each fixed value of $\mathbf{C}$. As these parameters are functions of the control factors, the second step is to treat each of these parameter estimates as a separated response, then estimate the control factor effects and determine the factor settings that optimize the dynamic system.

In the RFM approach, a different regression model is considered. At any fixed values of $\mathbf{C}$ and $\mathbf{N}$, the response model in (1) can be expressed as a linear regression model in $f_1(M), ..., f_k(M)$, i.e.,

$$Y|\{\mathbf{C}, \mathbf{N}\} = \alpha''(\mathbf{C}, \mathbf{N}) + \beta''_1(\mathbf{C}, \mathbf{N})f_1(M) + ... + \beta''_k(\mathbf{C}, \mathbf{N})f_k(M) + \epsilon\ , \tag{6}$$

where

$$\alpha''(\mathbf{C}, \mathbf{N}) = (\alpha_0 + \boldsymbol{\alpha}^T\mathbf{C}) + \sum_{h=1}^{q}(\gamma_{h0} + \boldsymbol{\gamma_h}^T\mathbf{C})N_h,\ \textit{and} \tag{7}$$

$$\beta''_1(\mathbf{C}, \mathbf{N}) = (\beta_{i0} + \boldsymbol{\beta_i}^T\mathbf{C}) + \sum_{h=1}^{q}(\rho_{ih0} + \boldsymbol{\rho_{ih}}^T\mathbf{C})N_h. \tag{8}$$

It follows from (1) that the error term ϵ is identically and independently distributed with mean zero and variance σ_ϵ^2. The analysis procedure of the RFM approach is to first estimate the parameters in model (6), $\alpha''(\mathbf{C}, \mathbf{N})$, $\beta_1''(\mathbf{C}, \mathbf{N})$, ..., $\beta_k''(\mathbf{C}, \mathbf{N})$, and σ_ϵ^2, at each fixed value of $\mathbf{C}$ and $\mathbf{N}$. As these parameters are functions of the control and noise factors, the second step is to treat each of these parameter estimates as a separated response, then estimate the control and noise factor effects and determine the control factor settings that optimize the dynamic system.

In the RM approach, the procedure is to first estimate all the parameters in model (1) by using the combined array design and the overall raw data. The second step is to use these parameter estimates to approximate the intercept, slope, and variance functions, $\alpha'(\mathbf{C})$, $\beta_1'(\mathbf{C})$,..., $\beta_k'(\mathbf{C})$, and $\sigma_e^2(\mathbf{C})$ in equations (3) to (5), then determine the control factor settings that optimize the dynamic system.

Although the estimation procedures of the three approaches are different, it can be shown that the effect estimates of the intercept and slope functions from different approaches are identical. However, even though the estimates are identical, the identified significant effects in the LM and RFM approaches may be spurious. As for the effect estimates of the variance function, similar to the result in the static case (see Tsui, 1994), the LM approach may create unnecessarily biased estimates. Both the spurious effects and unnecessary biases may lead to non-optimal solutions. Below we will discuss each of these issues in detail.

2.2 Identical Effect Estimates in Intercept and Slope Functions

To show that the intercept and slope effect estimates from different approaches are identical, we need the following Lemma.

Assume that the expected response $E(y)$ at $\mathbf{x}$ and $\mathbf{z}$ can be described by the linear model:

$$E(y) = \sum_{j=1}^{k} \theta_{0j} f_j(\mathbf{z}) + \sum_{i=1}^{h} \sum_{j=1}^{k} \theta_{ij} g_i(\mathbf{x}) f_j(\mathbf{z}), \tag{9}$$

where $f_1(\mathbf{z}) \equiv 1$. Suppose the design is a product array (see Shoemaker et al., 1991, for definition) of an inner design with n runs at ($\mathbf{x}_1$,..., $\mathbf{x}_n$) and an outer design with m runs at ($\mathbf{z}_1$,..., $\mathbf{z}_m$). We denote the corresponding response of the design as $\mathbf{y}^T = (\mathbf{y_1}^T, ..., \mathbf{y_n}^T)$, with $\mathbf{y_i}$ being the $m \times 1$ response vector corresponding to the m outer design runs at the ith run of the inner design. It follows that the regression matrix of the linear model (9), X, can be expressed as the Kronecker product of the corresponding regression matrices, i.e.,

$$X = (X_x' \otimes X_z) = \left(\mathbf{1}_n \otimes X_z \; X_x \otimes X_z \right) \;, \tag{10}$$

where

$$X'_x = \begin{pmatrix} 1 & g_1(\mathbf{x}_1) & \dots & g_h(\mathbf{x}_1) \\ \cdot & \cdot & \dots & \cdot \\ \cdot & \cdot & \dots & \cdot \\ 1 & g_1(\mathbf{x}_n) & \dots & g_h(\mathbf{x}_n) \end{pmatrix} = (\mathbf{1}_n, X_x),$$

$$X_z = \begin{pmatrix} f_1(\mathbf{z}_1) & \dots & f_k(\mathbf{z}_1) \\ \cdot & \dots & \cdot \\ \cdot & \dots & \cdot \\ f_1(\mathbf{z}_m) & \dots & f_k(\mathbf{z}_m) \end{pmatrix},$$

with $\mathbf{1}_n$ being a n-dimensional vector of 1's and $A \otimes B$ being the Kronecker product of matrices A and B.

Lemma. Assume that the column vectors in matrix X'_x are orthogonal to each other. Then the least squares estimates of the parameters in model (9) are identical to the estimates obtained by the following sequential steps of least squares estimations. (i) For each row of matrix X_x, compute the least squares estimates using matrix X_z and corresponding response vector y_i, i.e., $\mathbf{b}_i = (b_{i1}, \dots, b_{ik})^T = (X_z^T X_z)^{-1}(X_z^T \mathbf{y}_i)$, i= 1 ,..., n. (ii) For j=1, ..., k, compute the least squares estimates of $\theta_j = (\theta_{0j}, \theta_{1j}, \dots, \theta_{hj})$ using matrix X'_x and response vector $\mathbf{b'}_i = (b_{1j}, \dots, b_{nj})^T$.

Proof. Without loss of generality, we assume that each column vector in X_x is normalized to constant n. From equation (10), we have

$$X^T X = \begin{pmatrix} (\mathbf{1}_n^T \otimes X_z^T)(\mathbf{1}_n \otimes X_z) & (\mathbf{1}_n^T \otimes X_z^T)(X_x \otimes X_z) \\ (X_x^T \otimes X_z^T)(\mathbf{1}_n \otimes X_z) & (X_x^T \otimes X_z^T)(X_x \otimes X_z) \end{pmatrix}.$$

As the columns in X'_x are orthogonal, the four blocked matrices of $X^T X$ can be simplified as follows.

$$\begin{aligned} (X^T X)_{11} &= (\mathbf{1}_n^T \mathbf{1}_n) \otimes (X_z^T X_z) = n(X_z^T X_z), \\ (X^T X)_{12} &= (\mathbf{1}_n^T X_x) \otimes (X_z^T X_z) = \mathbf{0}, \\ (X^T X)_{21} &= (X_x^T \mathbf{1}_n) \otimes (X_z^T X_z) = \mathbf{0}, \\ (X^T X)_{22} &= (X_x^T X_x) \otimes (X_z^T X_z) = n I_k \otimes (X_z^T X_z), \end{aligned}$$

where I_k is a k dimensional identity matrix. Thus $X^T X$ can be simplified to the following blocked diagonal matrix.

$$X^TX = n \begin{pmatrix} X_z^T X_z & 0 \ldots & 0 \\ \cdot & \cdot \ldots & \cdot \\ \cdot & \cdot \ldots & \cdot \\ 0 & \cdot \ldots & X_z^T X_z \end{pmatrix}.$$

Similarly, $X^T y$ can be simplified as follows.

$$X^T\mathbf{y} = \begin{pmatrix} X_z^T \sum_{i=1}^n \mathbf{y_i} \\ X_z^T \sum_{i=1}^n g_1(\mathbf{x_i})\mathbf{y_i} \\ \cdot \\ \cdot \\ X_z^T \sum_{i=1}^n g_h(\mathbf{x_i})\mathbf{y_i} \end{pmatrix}.$$

Therefore, the least squares estimates in model (1), $\hat{\theta}_{LS} = ((\hat{\theta}_{01}, \ldots, \hat{\theta}_{0k}), (\hat{\theta}_{11}, \ldots, \hat{\theta}_{1k}), , \ldots, (\hat{\theta}_{h1}, \ldots, \hat{\theta}_{hk}))^T$, can be calculated as follows:

$$\hat{\theta}_{LS} = (X^TX)^{-1}(X^T\mathbf{y}) = n^{-1} \begin{pmatrix} \sum_i (X_z^T X_z)^{-1}(X_z^T \mathbf{y_i}) \\ \sum_i g_1(\mathbf{x_i})(X_z^T X_z)^{-1}(X_z^T \mathbf{y_i}) \\ \cdot \\ \cdot \\ \sum_i g_h(\mathbf{x_i})(X_z^T X_z)^{-1}(X_z^T \mathbf{y_i}) \end{pmatrix}.$$

By re-arranging the least squares estimates in the equation above, they are clearly equivalent to the least squares estimates obtained from the sequential steps stated in the Lemma. This finishes the proof.

First we compare the effect estimates of the intercept and slope functions for the LM and RM approaches. As described in Section 2.1, the second step of the LM approach is to estimate the control factor effects that affect the intercept and slope functions. As shown in (3) and (4), these effect parameters are corresponding to the control factor main effect and control-by-signal interaction parameters in model (2.1). According to the analysis procedure, the effect estimates are obtained through two sequential least squares estimations: (i) At each row of the control array design, consider the noise runs as replications and compute the least squares estimates of the regression model for the signal functions, $f_1(M), \ldots, f_k(M)$. (ii) For each of the coefficients in the fitted regression models, compute the least squares estimates according to the control array.

Following the Lemma above, matrix X_x is corresponding to the control array and matrix X_z is corresponding to the regression matrix of the fitted regression model. As the column vectors in the control array are orthogonal, it follows from the Lemma that the control effect estimates of the intercept and slope functions from the LM approach are identical to the least squares estimates of the RM approach.

Next we compare the effect estimates of the intercept and slope functions for the RFM and RM approaches. As described above, the second step of the RFM approach is to estimate the control and noise factor effects that affect the intercept and slope functions. As shown in (7) and (8), the effect parameters for the intercept function are corresponding to the control and noise main effects and control-by-noise interactions in (1). The effect parameters for the slope functions are corresponding to the control-by-signal, noise-by-signal, and control-by-noise-by-signal interactions in (1). According to the analysis procedure, the effect estimates are obtained through two sequential least squares estimations: (i) At each row of the control-by-noise product array design, compute the least squares estimates of the regression model for the signal functions, $f_1(M), ..., f_k(M)$. (ii) For each of the coefficients in the fitted regression models, compute the least squares estimates according to the control-by-noise product array.

Following the Lemma above, matrix X_x is corresponding to the control-by-noise product array regression matrix and matrix X_z is corresponding to the regression matrix of the fitted regression model. As the column vectors in the product array regression matrix are orthogonal, it follows from the Lemma that the control and noise effect estimates of the intercept and slope functions from the RFM approach are identical to the least squares estimates of the RM approach.

2.3 Spurious Significant Effects

Although the effect estimates of the intercept and slope functions from different approaches are identical, the significant effects identified from common statistical techniques can be different. This is because the groupings of the effect estimates are different in different approaches. As an error estimate is often unavailable, normal probability plots or other non-replicated experiments analysis techniques (e.g., Lenth, 1989) are commonly used for identifying significant effects. In the RFM approach, the analysis technique is applied to each group of control and noise effects separately from the intercept function and the slope functions. Sometimes the identified significant effects may be spurious as the number of effects in each group is small and thus the error estimate may be unreliable. In the RM approach, the analysis technique is applied to the control and noise effects from all the intercept function and slope functions together so that the error estimate will be more reliable. In addition, it allows us to compare the effect estimates across the intercept and slope functions and thus provides more reliable results.

For the LM approach, it suffers from an even more serious problem than the RFM approach as each group contains only the control factor effects. The number of effects in each group is even smaller and thus the error estimate may be more unreliable.

2.4 Biased Effect Estimates in Variance Function

As for the control factor effect estimates of the variance function, there is no identical relationship between the three approaches. In the LM approach, similar to the static case discussed in Tsui (1994), there exists a potential bias problem in estimating the control factor effects that affect the variance function, σ_e^2, and thus may lead to non-optimal solutions. Suppose the experiment for model (1) is a saturated or a resolution III design for the control factors (control array) crossed with a similar design for the noise factors. This implies that the control main effects are confounded with the control-by-control interactions and the noise main effects are confounded with the noise-by-noise interactions. Since there are no control-by-control and noise-by-noise interactions in model (1), if the experimenter chooses the RFM or the RM approaches, he/she can obtain unbiased estimates of the control and noise main effects and the control by noise interactions in (1). In the LM approach, as shown in (5), the variance $\sigma_e^2(\mathbf{C})$ is quadratic in the control factors ($\mathbf{C}$) even though the original response is linear in C. When the experimenter models $\sigma_e^2(\mathbf{C})$, the control main effects will be confounded with the control-by-control interactions. Obviously, if these interactions are large, the main effect estimates of the variance model will be seriously biased.

In general, under model (1), if the experimental design allows confounding between main effects and 2-factor interactions or between 2-factor interactions themselves in the control array (i.e., resolution III or IV), there will be a bias problem in the LM approach except for the extremely special cases discussed in Tsui (1994). Therefore, the effect estimates calculated from the LM approach under a highly fractionated control array are likely to be biased. As illustrated in the example in next section, the biased effect estimates may lead to non-optimal factor settings and misleading results.

2.5 Information Loss

In addition to the potential bias problem, the LM approach also suffers from the information loss problem similar to that described in Shoemaker et al. (1991) when comparing to the RFM and RM approaches. As the LM approach directly models the variance function in (5), it will not provide the information on how the control factors dampen the variation caused by individual noise factors. On the other hand, the RFM and RM approaches model the control-by-noise interaction effects and thus provide this information.

Moreover, the RFM and RM approaches also provide the information on variance reduction which is not caused by the explicit noise factors. In models (1) and (6), the variance estimates are corresponding to the variance of ϵ, σ^2. This variance represents the part of process variation which is not caused by the explicit noise factors, $N_1, .., N_q$. In the cases where all major noise sources are identified, this variance should be a constant independent of the control and noise factors. However, it is possible that some important noise sources are missed so that σ^2 depends on the control and noise factors. In this case, it is important to model the residual variance with respect to the control and noise factors in both the RFM and RM approaches. Note that although the two approaches are estimating the same variance parameter, the variance estimates are generally different. In particular, the variance estimate in the RFM approach measures the deviation of the data from the fitted local regression model (6) for each group of raw data. The variance estimate in the RM approach measures the deviation of the data from the fitted global model (1) for all raw data together. In either case, the modeling of this residual variance reveals the potential for variance reduction which is not caused by the explicit noise factors.

3 An Injection Molding Example

In this Section, we compare the LM, RFM, and RM approaches with a real example. DeMates (1990) applied the robust design method to optimize an injection molding process with a dynamic characteristic. Miller and Wu (1996) analyzed the experiment with the RFM approach. Below we first describe the process and the experiment, then analyze the experiment and compare the results of the LM, RFM, and RM approaches.

In the injection molding process the system was required to inject different amounts of material for different applications. The part weight was adopted as the response and high injection pressure was chosen as the signal factor to change the amount of material injected. An experiment was run to study the relationship and to optimize the design. Seven control factors were identified in the experiment and listed in Table 1. Each factor was tested at two levels using a standard orthogonal array OA8.

Table 1. Control Factors for Injection Molding Experiment

Factor	+1	-1
A: injection speed	2.0	0.0
B: clamp time	44 s	49 s
C: high injection time	6.3 s	6.8 s
D: low injection time	17 s	20 s
E: clamp pressure	1900 psi	1700 psi
F: water cooling	70 F	80 F
G: low injection pressure	650 psi	550 psi

For each of the eight rows in the array, data for four replications at each of eight signal levels were collected for each of two noise levels. The experiment, therefore, required 512 (8x4x8x2) observations. Summaries of the control, signal, and noise factors used and the raw data from the experiment can be found in Miller and Wu (1996).

For the LM approach, a linear regression model was first fitted for each row of the control array using least squares estimation. From these fitted models, the estimates of the intercept, the slopes, and the error variance were used as performance measures to identify the control factor settings that optimize the system.

Following McCaskey and Tsui (1997), for an additive model, an appropriate two-step procedure is to first minimize the error variance, then shift the intercept and slope functions to their targets. As suggested in Miller and Wu (1996), the best response-to-signal model for this experiment is a quadratic model with linear and quadratic contrasts, $f_1(M) = (-7, -5, -3, -1, 1, 3, 5, 7)$ and $f_2(M) = (7, 1, -3, -5, -5, -3, 1, 7)$. Thus a linear regression model with regressors $f_1(M)$ and $f_2(M)$ was fitted to the data for each row of the control array and the error variance was computed. Tables 2 and 3 show the summary and effect estimates for the intercept, slope, and variance functions. The following significant effects were identified based on half normal plots and subjective judgement: factors A, C, E, F, G are significant for the intercept, factor C is significant for the linear slope function $f_1(M)$, factors B, D, E are significant for the quadratic slope function $f_2(M)$, and factors A, B, D, F are significant for the variance. To minimize the variance, factors A, B, D, F are chosen as $(-1, -1, 1, 1)$ respectively. As factor E is significant for $f_2(M)$ but not significant for the variance, it can be used to adjust the quadratic slope function to its target. Similarly, as factor C is significant for $f_1(M)$ but not significant for $f_2(M)$ and the variance, it can be used to adjust the linear slope function to its target. Finally, as factor G is significant for the intercept but not significant for $f_1(M)$, $f_2(M)$, and the variance, it can be used to adjust the intercept to its target.

Table 2. Intercept, Slopes, and Error Variance of Loss Model

Run	Intercept	$f_1(M)$	$f_2(M)$	Error Variance
1	665.82	5.01	1.25	8.39
2	662.15	4.91	1.46	19.70
3	666.76	4.93	1.25	9.06
4	666.35	4.66	1.40	10.40
5	665.29	4.56	1.38	4.34
6	674.31	4.33	1.35	9.75
7	666.40	4.92	1.31	1.55
8	664.30	4.97	1.27	3.34

Table 3. Loss Model Effect Estimates

Effect	Intercept Estimate	$f_1(M)$ Estimate	$f_2(M)$ Estimate	Variance Estimate
A	-1.151	0.090	0.008	3.572
B	0.471	-0.082	0.027	2.230
C	-1.754	0.166	-0.009	-0.070
D	-0.356	0.070	-0.037	-2.479
E	1.375	0.023	-0.054	-0.681
F	-0.982	0.014	-0.008	-1.697
G	1.799	-0.055	-0.005	-0.793

For the RFM approach, we refer to the analysis done in Miller and Wu (1996). Based on their analysis results, the fitted models for the intercept and slope functions in (7) and (8) are as follows.

$$\hat{\alpha''} = 666.4 - 1.2A - 1.8C + 1.4E - 1.0F + 1.8G + 1.1N, \qquad (11)$$

$$\hat{\beta''_1} = 4.79 + 0.16C, \qquad (12)$$

$$\hat{\beta''_2} = 1.33 + 0.03B - 0.04D - 0.05E - 0.04N + 0.01AN$$
$$-0.03FN - 0.02GN. \qquad (13)$$

As described in Miller and Wu (1996), the dynamic system is optimized by the interpretation of these fitted models. These models will be compared with the fitted models in the RM approach later.

In the RM approach, we first fit all the raw data to a single model on the control factors, noise factors, and linear and quadratic contrasts of the signal factor. Model (1) is fitted using a t-test by estimating the pure error with the remaining degrees of freedom. The parameter estimates and their t-ratio are summarized in Table 4. All terms at 10% significant level are included in the following selected model:

$$\hat{y} = 666.4 - 1.16A + 0.47B - 1.75C - 0.36D + 1.37E - 0.98F + 1.80G$$
$$+(4.79 + 0.09A - 0.08B + 0.17C + 0.07D - 0.05G)f_1(M)$$
$$+(1.33 - 0.05E)f_2(M) + (1.07 + 0.54A - 0.29E - 0.27G)N$$
$$+(0.05 + 0.05A - 0.05E)f_1(M)N. \qquad (14)$$

Table 4. Response Model Parameter Estimates and t-Ratios

Effect	Estimate	t-ratio	Effect	Estimate	t-ratio
			N	1.070	9.407
A	-1.152	-10.128	AN	0.541	4.757
B	0.471	4.139	BN	-0.058	-0.512
C	-1.754	-15.424	CN	-0.134	-1.178
D	-0.356	-3.132	DN	-0.177	-1.559
E	1.374	12.083	EN	-0.291	-2.562
F	-0.982	-8.634	FN	0.074	0.653
G	1.798	15.813	GN	-0.273	-2.397
f_1	4.786	192.833	f_1N	0.053	2.135
Af_1	0.089	3.600	Af_1N	0.045	1.819
Bf_1	-0.083	-3.339	Bf_1N	0.028	1.128
Cf_1	0.166	6.680	Cf_1N	-0.011	-0.438
Df_1	0.069	2.791	Df_1N	-0.009	-0.357
Ef_1	0.022	0.895	Ef_1N	-0.052	-2.081
Ff_1	0.014	0.544	Ff_1N	-0.015	-0.597
Gf_1	-0.055	-2.224	Gf_1N	-0.022	-0.875
f_2	1.333	53.715	f_2N	-0.040	-1.605
Af_2	0.007	0.298	Af_2N	-0.027	-1.082
Bf_2	0.027	1.085	Bf_2N	0.003	0.139
Cf_2	-0.009	-0.378	Cf_2N	0.011	0.426
Df_2	-0.037	-1.501	Df_2N	0.011	0.425
Ef_2	-0.054	-2.192	Ef_2N	0.004	0.152
Ff_2	-0.008	-0.325	Ff_2N	-0.026	-1.029
Gf_2	-0.006	-0.239	Gf_2N	-0.022	-0.870

By comparing the effect estimates of the RM approach with those of the LM approach, it is found that the control main effect estimates in the intercept and slope functions in Table 3 are identical to the control main effect and control by signal interaction estimates in Table 4. By comparing the effect estimates of the RM approach with those of the RFM approach, it is found that the control and noise main effect and interaction estimates in (11) to (13) are identical to the corresponding estimates in Table 4. These verify the identical relationships explained in Section 2.2.

Below we will apply the RM approach to determine the optimal settings of the control factors. To minimize the process variance, we follow the approach in Box and Jones (1990) to estimate the process variance. By assuming that the noise variable N in (1) is random with unit variance, the process variance caused by the noise in (5) can be estimated by

$$\begin{aligned}\hat{\sigma_e} &= (1.07 + 0.54A - 0.29E - 0.27G)^2 \\ &\quad + (0.05 + 0.05A - 0.05E)^2 \Big(\int f_1^2(M) dM \Big) \\ &= (1.07 + 0.54A - 0.29E - 0.27G)^2 + 21(0.05 + 0.05A - 0.05E)^2 \quad (15)\end{aligned}$$

which is a quadratic function of the control factors A, E, and G. By considering the choices of control factor values, the combination of $A = -1, E = 1$, and $G = 1$ minimizes the predicted process variance $\hat{\sigma_e}$.

To determine the optimal settings of the other control factors, we will approximate the intercept and slope functions in (2) as follows.

$$\begin{aligned}\hat{\alpha'} &= 666.4 - 1.16A + 0.47B - 1.75C - 0.36D + 1.37E - 0.98F + 1.80G, \\ \hat{\beta_1'} &= 4.79 + 0.09A - 0.08B + 0.17C + 0.07D - 0.05G, \\ \hat{\beta_2'} &= 1.33 - 0.05E.\end{aligned}$$

As factors B, C, and D are significant for the linear slope function but not significant for process variance, any of the three factors can be used to adjust the linear slope function to its target value. As factor F is significant for the intercept function but not significant for both the linear slope and process variance, it can be used to adjust the intercept to its target. For the quadratic slope function, no adjustment can be made as there does not exist any adjustment factor. In case a specific target value is needed in practice, factors A, E, and G will be used to minimize the process variance under the on-target constraint.

By comparing this result with the result of the LM approach, they are obviously different. The difference is caused by the different significant effects in the intercept, slope, and variance functions. For the intercept and slope functions, although the effect estimates are the same, the methods to identify significant effects are different. In the LM approach, the significant effects are identified for each of the intercept and slope functions separately. However, the number of effects is so small that identification methods such as half probability plots can be misleading. In the RM approach, the effects are identified all together and the result is thus more reliable. For the variance function, as pointed out in Section 2.4, the effects are potentially biased as the design in the experiment is a resolution III design. According to equation (5), factors A, E, G are significant but the effects are quadratic in the variance, i.e., the second order terms $A^2, E^2, G^2, AE, AG, EG$ are also significant for the variance. For a resolution III design, it is known that squared terms are confounded with the overall mean and AE, AG, EG are confounded with D, F, B, respectively. These confounding relationships explain that the significant effects D, F, B in the variance function of the LM approach are actually spurious interaction effects. This illustrates the potential bias problem of the LM approach.

By comparing the RM result to the RFM result, the significant effects in the intercept and slope functions are obviously different. By re-grouping the significant effects in (5) into the intercept and slope functions in (7) and (8), we have the following models:

$$\begin{aligned}
\hat{\alpha}'' &= 666.4 - 1.16A + 0.47B - 1.75C - 0.36D + 1.37E - 0.98F + 1.80G \\
&\quad +1.07N + 0.54AN - 0.29EN - 0.27GN, \\
\hat{\beta}_1'' &= 4.79 + 0.09A - 0.08B + 0.17C + 0.07D - 0.05G + 0.05N + 0.05AN \\
&\quad -0.05EN, \\
\hat{\beta}_2'' &= 1.33 - 0.05E.
\end{aligned}$$

Clearly these models are very different from those in (11) to (13) resulting from the RFM approach. As mentioned before, although the effect estimates in the two approaches are the same, the identified significant effects are different, which results into different fitted models. Again, as in the RFM approach the significant effects are identified for each of the intercept and slope functions separately, the number of effects is so small that identification methods such as half probability plots can be misleading. In the RM approach, the effects are identified all together and the result is thus more reliable. Nevertheless, the RFM approach does not suffer from the potential bias problem as the control-by-noise interactions have been modeled in the intercept and slope functions.

4 Discussion

For the static robust design problem, the response model approach has been shown to be a useful alternative to Taguchi's loss model approach for analyzing experiments. This paper compares two different response model (RFM and RM) approaches to the LM approach.

As the effect estimates in the intercept and slope functions are obtained by linear procedures, the corresponding effect estimates are identical in all three approaches. However, although the effect estimates are identical, the identified significant effects from the three approaches are different as the groupings of the effect estimates are different. As shown in the example, both the LM and RFM approaches suffer from the problem of spurious significant effects as the effects were identified in small groups separately. On the other hand, the RM approach pools all the estimates together for identifying significant effects and thus provides more reliable results.

In addition, the LM approach suffers from the potential bias problem as in the static case. As illustrated in the example, the problem led to misleading optimal factor settings. Both the RFM and RM approaches do not suffer from this problem as the control-by-noise interactions have been modeled explicitly. The LM approach also suffers from the information loss problem as in the static case. As illustrated in Miller and Wu (1996), the residual variance of the fitted models were analyzed to further understand how the control factor affect the process variance.

Acknowledgements
This research was supported by the National Science Foundation under Grants DDM-9257918 and DDM-9312698 and the Croucher Foundations under Grant CF96/97.EG02.

References

1. Box, G.E.P. and Jones, S. (1990) Designs for Minimizing the Effects of Environmental Variables. Technical Report, University of Wisconsin
2. DeMates, J. (1990) Dynamic Analysis of Injection Molding Using Taguchi Methods. Eight Symposium on Taguchi Methods, 313-331, American Supplier Institute, Dearborn, MI
3. Lenth, R.V. (1989) Quick and Easy Analysis of Unreplicated Experiments. Technometrics **31**, 469-473
4. Lucas, J.M. (1989) Achieving a Robust Process Using Response Surface Methodology. Proceedings of the American Statistical Association: Sesquicentennial Invited Paper Sessions, 579-593
5. Lunani, M., Nair, V., Wasserman, G.S. (1995) Robust Design with Dynamic Characteristics: A Graphical Approach to Identifying Suitable Measures of Dispersion. Technical Report **#253**, Department of Statistics, University of Michigan, Ann Arbor
6. McCaskey, S.D. and Tsui, K.-L. (1997) Analysis of Dynamic Robust Design Experiments. International Journal of Production Research **35**, 1561-1574
7. Miller, A. and C. F. J. Wu (1996) Parameter Design for Signal-Response Systems: A Different Look at Taguchi's Dynamic Parameter Design. Statistical Science **11**, 122-136
8. Shoemaker, A.C., Tsui, K.-L., and Wu, C.F.J. (1991) Economical Experimentation Methods for Robust Parameter Design. Technometrics **33**, 415-427
9. Steinberg, D.M. and Bursztyn, D. (1994) Dispersion Effects in Robust Design with Noise Factors. Journal of Quality Technology **26**, 12-20
10. Taguchi, G. (1986) Introduction to Quality Engineering: Designing Quality into Products and Processes. Asian Productivity Organization, Tokyo, Japan

11. Tsui, K.-L. (1994) Avoiding Unnecessary Bias in Robust Design Analysis. Computational Statistics and Data Analysis **18**, 535-546
12. Welch, W.J., Yu, T.K., Kang, S.M. and Sacks, J. (1990) Computer Experiments for Quality Control by Parameter Design. The Journal of Quality Technology **22**, 1-12

GPSR Compliance

The European Union's (EU) General Product Safety Regulation (GPSR) is a set of rules that requires consumer products to be safe and our obligations to ensure this.

If you have any concerns about our products, you can contact us on ProductSafety@springernature.com

In case Publisher is established outside the EU, the EU authorized representative is:

Springer Nature Customer Service Center GmbH
Europaplatz 3
69115 Heidelberg, Germany

Batch number: 10725973

Printed by Printforce, the Netherlands